Logistic Core Operations with SAP

Jens Kappauf • Bernd Lauterbach
Matthias Koch

Logistic Core Operations with SAP

Inventory Management, Warehousing, Transportation, and Compliance

Jens Kappauf
SAP AG, Walldorf
Germany
info@books.kappauf.eu

Bernd Lauterbach
SAP AG, Walldorf
Germany
b.lauterbach.logistics@googlemail.com

Matthias Koch
SAP AG, Walldorf
Germany
m.koch.logistics@googlemail.com

ISBN 978-3-642-43593-5 ISBN 978-3-642-18202-0 (eBook)
DOI 10.1007/978-3-642-18202-0
Springer Heidelberg Dordrecht London New York

Softcover reprint of the hardcover 1st edition 2012

Printed on acid-free paper

Springer is part of Springer Science+Business Media (www.springer.com)

Contents

Chapter 1
Introduction

The most imperative challenge facing a number of managers in recent years has been to regain lost market share and secure new competitive advantages. The impetus for this trend continues to be the ubiquitous tendency toward globalization and the ensuing intensification of international competition. *Customer orientation, lean management* and *re-engineering* are the buzzwords that characterize these efforts.

In fact, several companies need to reorganize their value-added processes, whereby special attention should be paid to the interfaces between the sales and procurement markets, which are increasing in importance. Within this context, there is hardly a corporate function that has grown in significance in recent years as much as *logistics*.

Treated until just a few years ago as an operational aid and an object of isolated rationalization efforts, logistics – especially in the age of *supply chain management* – now is considered an essential element of strategic corporate leadership. More and more, logistics is being functionally mapped in standard business software. Accordingly, there is great demand for logistical expertise in connection with the know-how surrounding the implementation of logistics in complex information systems.

Definition of supply chain management

Supply chain management (SCM) is the observation and administration of logistical processes along the entire value creation chain, which includes suppliers, customers and end consumers.

1.1 Purpose of This Project

We have divided our presentation of logistics operations with SAP into two volumes. The purpose of them is to provide you with an introduction to the world of logistics with SAP software and assist you in understanding the terminology, concepts and technological components as well as their integration. Because the described processes are complex and include a number of functional details, we

J. Kappauf et al., *Logistic Core Operations with SAP*,
DOI 10.1007/978-3-642-18202-0_1, © Springer-Verlag Berlin Heidelberg 2012

have attempted to make the examples as representative as possible in terms of the presentation and functional explanation of the SAP system components, SAP ERP (Enterprise Resource Planning) and SAP SCM (Supply Chain Management). This means that the two volumes cover all components of SAP Business Suite and core functions within the context of logistics. A few components, especially technical ones, and functional areas (e.g. disposal, maintenance and service management) will not be covered.

We have taken care to explain business-related questions and their SAP-specific solutions as well as technical terms, and illustrate their relationships. The books are designed to provide an easily understandable yet well-substantiated look at the respective process chains, and be a useful source of information for everyone – from IT experts with only basic knowledge of the business-related issues, to employees in logistics departments who are not yet familiar with SAP terms and applications.

1.2 Whom Do These Books Address?

Logistic Core Operations with SAP cannot answer every query, but we hope to give you the tools with which to ask the right questions and understand the essential issues involved. The contents of this book are thus aimed at the following target groups:

1.2.1 SAP-Beginners

The books are dedicated to everyone looking for a lucid, informed introduction to logistics with SAP. Thus, each chapter describes in detail a specific logistics field and provides an overview of the functionality and applications of the respective components in practical business use. In this regard, we address SAP beginners and employees in departments where SAP is to be implemented, as well as students wishing to obtain an impression of the logistic core processes and their mapping in SAP software.

1.2.2 Ambitious Users

We also speak to ambitious users of SAP Business Suite who, in addition to relevant logistical processes, want a look at process integration and the functions up- or downstream, as well as their mapping in SAP Business Suite.

1.2.3 Managers and IT-Decision-Makers

Last but not least, we turn to management staff and IT decision-makers who are considering the implementation of SAP Business Suite or its individual components and wish to obtain an overview of logistical processes with SAP systems.

1.3 Operational Significance of Logistics

The operational significance of logistics for many companies still lies in its rationalization potential. In general, a reduction of logistics costs should improve corporate success by achieving a competitive advantage. Surveys of businesses have demonstrated that, for the coming years, companies are still counting on a considerable cost-reduction potential of 5–10% of total costs (see 3PL Study 2009, The State of Logistics Outsourcing 2009, Third-Party Logistics). This statement does not contradict the fact that the share of logistics costs of many companies was more likely to increase in the past because, for instance, it is directly related to which operational processes are included in the logistics process.

Thus, the scope of logistics in recent years has continually expanded, for example to include production planning and control (PPS systems) or quality control. In addition, significant investments are being made in IT technology, in areas such as supply chain management planning. In the near future, this will lead to a decrease in administrative logistics costs (e.g. through shipment tracking, transport organization or Internet-based ordering).

Further savings are expected in commercial and industrial firms by subcontracting logistics services (logistics outsourcing). In particular, operative logistics tasks, such as transport, storage, commissioning and packaging, have already been outsourced to a high degree to external logistics service providers. However, since a lack of quality in logistics services is generally not blamed on the service providers involved, but rather on the supplier, outsourcing logistics functions can be problematic.

1.3.1 Supplementary Logistics Services

When the quality of competitor products continues to become more comparable and there is hardly room to lower prices any further, competition takes place on the level of service performance. Within these services, logistics ranks highly: Delivery dependability, rapid returns processing and a high degree of customer service quality are characteristics with which a company can set itself apart from its competitors.

1.3.2 Customer-Oriented Logistics

Several logistics processes either include interfaces with customers or have effects on the customer. That is why logistics processes must be oriented toward customer needs and performed in a service-friendly manner. In an era when logistics demands are becoming ever more exacting and, by the time the consumer is reached, ever more customized, companies that master these processes to the advantage of their customers will experience a competitive edge that, at least in the short term, cannot be bridged by the competition. Firms boasting excellent logistics management can hardly be replaced by other suppliers. Thus, there are cases in which logistics in commercial and industrial businesses is one of the core competences for which outsourcing should *not* be considered. This is not to say that the fulfillment of basic logistics functions (e.g. transport or storage) cannot be outsourced, as there are plenty of providers on the market that are capable of taking over the logistics tasks of a previous supplier on a short-term basis without detriment to quality (make-or-buy decision).

1.3.3 Definition of "Logistics" for This Book

There are a growing number of ever-changing definitions and classification options offered in print as well as on the Internet for the term *logistics*. Of these, we would like to use the functional, flow-oriented definition of the American logistics society "Council of Supply Chain Management Professionals" as the basis for this book and its journey into the logistics options of the SAP Company:

> Logistic management is that part of supply chain management that plans, implements, and controls the efficient, effective forward and reverse flow and storage of goods, services, and related information – between the point of origin and the point of consumption in order to meet customers' requirements.(Source: Council of Supply Chain Management Professionals)

According to this definition, logistics serves to move goods within the entire value chain and requires coordination and integration between companies. It focuses primarily on real goods, tangible assets and services that provide benefit to the customer, and it integrates them into the core logistics functions of transport, transfer and storage.

Logistics therefore comprises the planning, control and execution of goods and information flow – between a company and its suppliers, within a company, and between a company and its customers.

1.3.4 Differentiation from Materials Management

Materials management, on the other hand, includes all activities involved in supplying a company and its production processes with all necessary materials at optimal cost. Logistics takes into account spatial and temporal gaps involved in supply processes, not only with regard to the materials, but also to the information to be exchanged between business partners. That is why we consider materials management to be not only a part of logistics, but its center, whereby the functions of logistics are more comprehensive than those of materials management.

1.3.5 Functional Classification of Logistics

A further possibility of logistics classification is the differentiation of logistics phenomena according to functional aspects. As a cross-sectional function, logistics maintains interfaces with the primary functional business areas of acquisition, production and sales.

1.3.6 Classic Logistics Core Areas

We thus traditionally differentiate between the following logistics areas in the order in which goods flow through a company, from the acquisition to the sales market:

- Procurement logistics
- Production logistics
- Distribution logistics

1.3.7 Expansion of the Traditional Core Areas

Current logistics definitions augment these traditional core areas to include further aspects. These include disposal logistics and operational maintenance, or service management. Spare parts logistics ensures the materials-management-related supply and availability of spare parts.

1.3.8 Objective: As Comprehensive a Portrayal As Possible

With this project, we have attempted to provide as comprehensive a portrayal of logistics processes and issues as possible, which include not only theoretical principles but also problems of practical use, as well as their implementation in SAP Business Suite. Therefore, we have expanded the traditional cross-sectional business functions to include the following logistics areas, to which chapters in Volume 2 are dedicated:

- Transport logistics
- Warehouse logistics and inventory management

1.3.9 Related Literature in the Appendix

Due to our goal of providing basic knowledge of logistics core processes and their mapping in SAP Business Suite within the framework of conceptual possibilities, disposal logistics and service management and maintenance (as well as compliance) are not within the scope of this book. For more information on these topics, refer to the bibliography. There you will find information concerning all of the books or sources we have quoted or referenced.

1.3.10 Logistics Functional Areas

Figure 1.1 shows the classic and expanded functional areas of logistics that will be discussed in more detail in *Logistic Core Operations with SAP.*

1.3.11 Procurement Markets

On the side of the *procurement markets,* it is the task of *procurement logistics* to acquire the articles as well as raw materials and supplies necessary for the operational processes of manufacturing and distribution. Procurement is carried out with reference to a certain procurement and stock situation, especially based on materials management planning as part of *production logistics.* The result of such planning may be a purchase requisition. The purchase requisition is cleared for procurement, converted into an order and transferred to the determined source for internal or external procurement. The conclusion of a procurement transaction may involve receiving and paying a supplier invoice, in addition to goods receipt into the

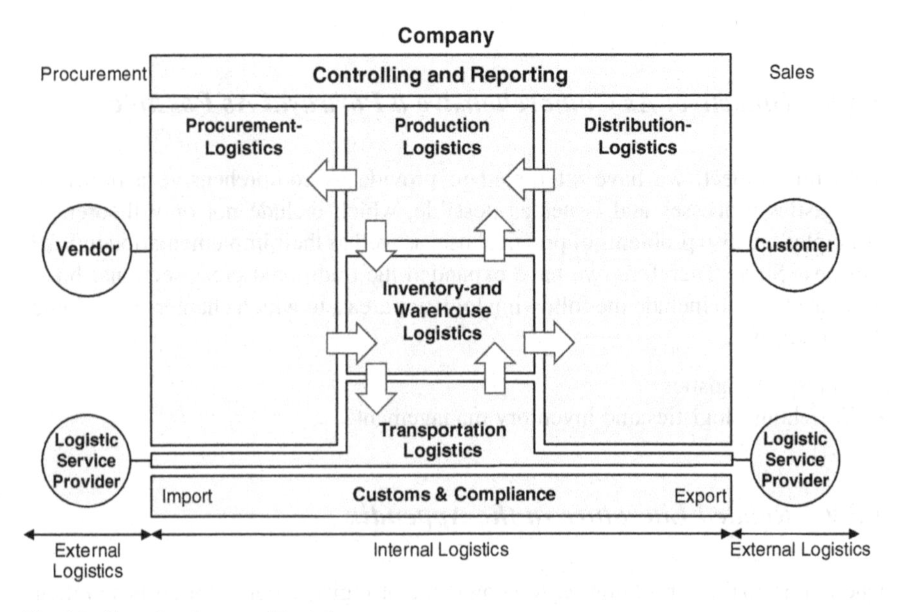

Fig. 1.1 Functional areas of logistics

warehouse. The Goods Receipt Department not only records stock but also its valuation for the Accounting Department. Transfer of the materials to stock, their quality inspection and inventory management are part of *warehouse logistics* and *inventory management.*

1.3.12 Sales Markets

Distribution logistics primarily concerns sales processes that generally begin with a customer ordering materials and indicating a desired delivery date. Using this information, a sales order is generated. Depending on the delivery date, shipping activities are initiated such that the materials reach the customer in a timely manner. *Warehouse logistics* takes over the task of commissioning and material provision. As soon as the materials have left the warehouse, a goods issue is booked to update stock and inventory management values.

A carrier can be commissioned to deliver the materials. *Transport logistics*, in a cross-sectional function, takes on the task of booking the transport planning and the transport itself. At the end of the sales procedure, an invoice is produced for the customer. As soon as the customer has paid for the materials, payment receipt is recorded in Accounting.

1.4 Structure of This Book

Logistic Core Operations with SAP is divided into two volumes. *Inventory Management, Warehousing, Transportation and Compliance* represents the second volume.

1.4.1 Transport Logistics

Due to its logistic significance and related SAP applications, we have dedicated the entire second chapter following this introduction to "Transport Logistics" (Chap. 2), which illustrates the various SAP solutions regarding the topic of transport. We not only shed light on the perspective of shipping agents from the realms of manufacture and trade, but also that of transport service providers. In addition to the basics of transport logistics and its business significance and transport from the standpoint of shippers and logistics service providers, we explain in detail the individual systems and applications, and their integration into the procurement and distribution logistics systems as well as the necessary master data.

1.4.2 Warehouse Logistics and Inventory Management

Chapter 3, "Warehouse Logistics and Inventory Management", describes warehouse logistics as a link between the internal and external logistics systems. Thus, we present those SAP processes in the realm of inventory management, goods movement and warehouse management. In doing so, we make a clear, systematic distinction between stock and warehouse management. This chapter also discusses warehouse management using the Warehouse Management solution in SAP ERP (WM) as well as in SAP SCM, SAP Extended Warehouse Management (SAP EWM).

Besides the application-specific description of the basic warehouse processes in the areas of goods receipt and goods issue, we also have a look at the fundamentals of inventory management, its evaluation and the integration of system components. We illustrate special stock and special procurement forms, consignment and sub-contract order processing based on their significance to central logistics, as well as the technical differences between WM (ERP) and EWM (SCM) processes.

1.4.3 Trade Formalities

Chapter 4 focuses on "Trade Formalities: Governance, Risk, Compliance". This chapter offers an overview of the functions of foreign trade and customs processing with SAP ERP and SAP Business Objects Global Trade Services.

1.4.4 Controlling and Reporting

Logistic control and the related reporting process are the focus of Chap. 5, "Controlling and Reporting", which also covers integration into the SAP logistics processes. We primarily examine SAP Event Management as a *Track & Trace* system for the tracking of shipments and events, the classic SAP ERP-based functions in the realm of distribution and logistics information systems, as well as SAP NetWeaver Business Warehouse (SAP NetWeaver BW). The classic reporting functions are complemented by SAP BusinessObjects. SAP thus offers the necessary tools to support users in the generation, formatting and distribution of conclusive reports, or so-called *dashboards*. Dashboards enable more than simple data evaluation, focusing on the integration and generation of intuitive visualizations that immediately display where there is a need for action.

1.4.5 Overview of Volume 1

The first volume of this series *procurement, production and distribution logistics,* provides a detailed look at the components and functions of SAP Business Suite and SAP NetWeaver as they relate to logistics. It begins with an explanation of master data and the organizational structures used in the logistics components of SAP Business Suite. This is followed by an examination of the integration of these components and functions in specific logistics applications. Internal and external procurement are featured, including requirements determination, order processing, invoicing and purchasing optimization. It also highlights production logistics, with an emphasis on the deployment side. Sales and procurement planning using SAP ERP(SAP Enterprise Resource Planning) and SAP APO (SAP Advanced Planning & Optimization) are also discussed. Finally, the book illustrates how distribution logistics can be managed with SAP, including order processing in SAP ERP and SAP CRM (SAP Customer Relationship Management), the processing of inquiries, quotations, orders, and deliveries, and special distribution situations, such as returns processing, returnable packaging and consignment processing.

1.4.6 Appendix

At the end of both volumes, you will find a glossary (Appendix A), a bibliography (Appendix B) and a detailed index that can help you find important terms and their definitions quickly.

1.5 Thanks

This book was created with the aid and the direct and indirect expertise of several SAP colleagues, whom we wish to sincerely thank.

We are very grateful to Dorothea Glaunsinger and Hermann Engesser at Springer Verlag for their guidance and support. Thanks also to Frank Paschen and Patricia Kremer at SAP-Press for their first-rate assistance during the making of the original German publication of this book. We also wish to thank translator Andrea Adelung and copy editor George Hutti.

We would like to express our special thanks to our wives and families:

- Susanne Kappauf with Leni and Anni
- Yumi Kawahara with Kai and Yuki
- Susanne Koch with David and Leah

You are the ones who, through your patience and willingness to do without a great deal of things, have made this book possible.

Chapter 2
Transport Logistics

Transport logistics refers to the transport of goods of all kinds using a variety of means, such as trains, trucks, airplanes, ships or parcel services.

Transport logistics is a major component of business process networks. Its significance has increased in recent years due to increasing globalization. Whereas companies in the 1980s and 1990s frequently focused on reducing internal costs by introducing ERP systems, among other measures, now rising energy costs are shifting that focus to logistics outside the company. In recent years, we have seen similar cost optimization tendencies in the realm of transport.

2.1 The Fundamentals of Transport Logistics

The topic of transport can be considered from the perspective of various business models, such as that of logistics service providers and carriers, or from the view of a producing or trading company (called the *shipper view* in this context). Both business models exhibit their own special features in the business process. The cooperation of business partners in the network has its own character and set of rules. In addition, goals can be different. Figure 2.1 illustrates the transportation relationships between various business partners. The transport logistics involved in external logistics can be organized by the shippers themselves as well as by logistics service providers.

From the perspective of transport logistics, we generally differentiate between local and long-haul transport. Local transport involves a vehicle executing pick-up or delivery and returning to the starting point on the same day. This category generally includes the delivery of cargo that has been fed from a long-haul into a local transport network (*on-carriage*), and pick-ups that have been transferred from the local to the long-haul transport network (*pre-carriage*). The truck is the most commonly used vehicle in local transport.

Long-haul transport is either carried out as direct long-haul transport (*direct leg*) or via line haul. For direct long-haul transportation, a means of transport containing the goods to be conveyed is sent directly from the shipper to the recipient over a

J. Kappauf et al., *Logistic Core Operations with SAP*,

DOI 10.1007/978-3-642-18202-0_2,

long distance. For line-haul processing, goods picked up on local transport routes are transferred to another means of transport (airplane, ship, train or truck) in a logistics center and transported along with goods from other shippers. Several transfer processes are also possible along the entire transport route. Figure 2.2 illustrates the transportation lanes and the transportation network in local and long-haul transport.

For long-haul transport, organizational processing efforts are generally much higher than for local transport. Depending on the type of transportation (air, sea, etc.), the type of goods (dangerous goods, foodstuffs, etc.) and geographic circumstances of the origin, points of transit and destination, long-haul transport can require the following additional tasks:

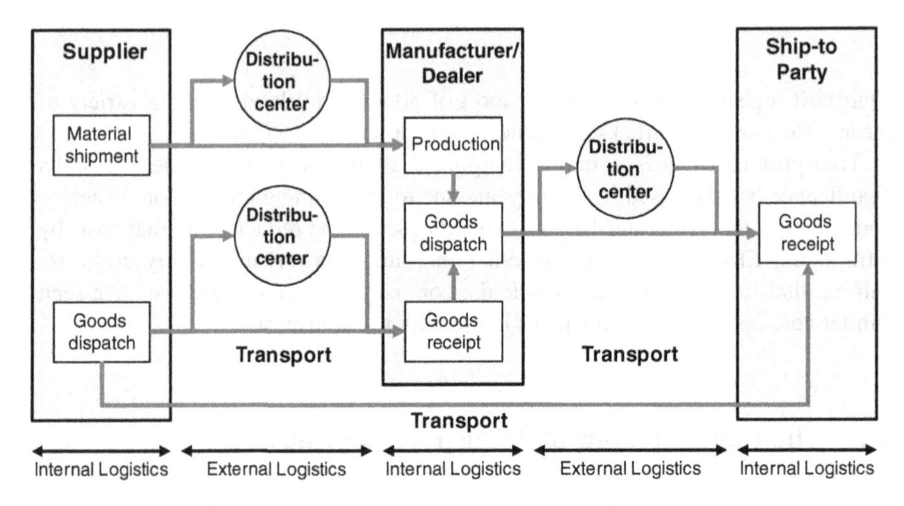

Fig. 2.1 Transportation relationships between various business partners

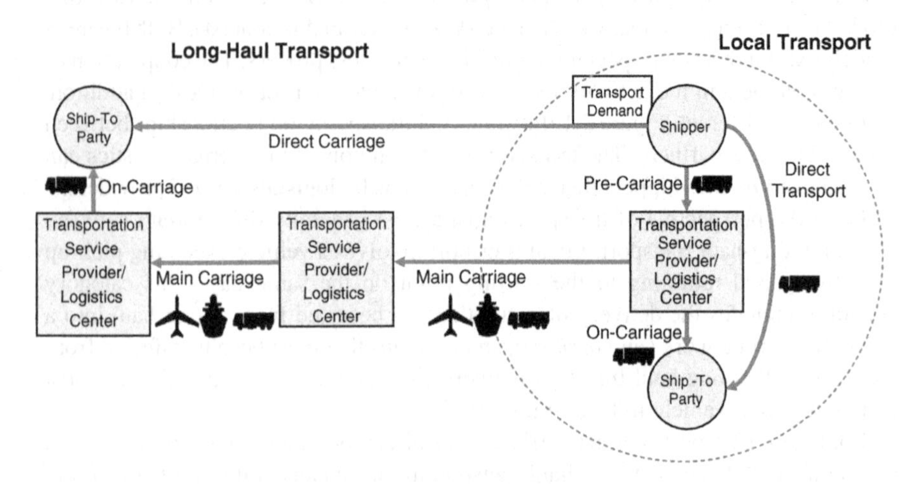

Fig. 2.2 Transportation network in local and long-haul transport

- Shipping space reservation on ships or airplanes
- Foreign trade processing with export and import permits, customs fees and embargo checks
- Dangerous goods processing with various national or transport mode-specific regulations
- Coordination and seamless planning of goods movement at the various load transfer points and on various modes of transport
- Cost calculation, processing and risk responsibility according to various Incoterms

International air and sea transport can become very complex as a result.

2.1.1 Business Significance

Transport logistics does not occur in an isolated fashion. It is always connected to other business processes, whether they be in one's own company or at a business partner's location. It organizes the exchange of goods between business partners. Bad organization can harm or impede the subsequent business processes. In times of very advanced internal process optimization, a well-functioning, optimized transport logistics is becoming ever more important. Here, even greater savings potential, total process optimization and service advantages can be achieved.

The optimization goal of transport logistics is to process all planned goods transports in such a way that:

- Existing transport means are used as optimally as possible
- As few empty runs as possible occur
- Available service providers can be contracted economically and in close observation of agreements
- All goods are transported according to laws and regulations (dangerous goods regulations, trade regulations, etc.)
- Operating supply, human resources and service provider costs are minimized
- Service times and stipulated service grade and levels (such as 24-h delivery) are observed

Many of these optimization opportunities can be utilized to their fullest with the aid of a suitable software system.

2.1.2 Transportation from the Shipper View

From the view of a shipper, there are three major process types that must be supported in transport logistics:

- *Incoming shipments*, in which ordered goods are picked up or material replenishment is acquired for production
- *Outgoing shipments*, in which produced or deliverable materials or goods are transported from one plant or warehouse to a goods recipient
- *Third-party transactions*, in which the shipper has the goods transported directly from a supplier to the recipient, without the shipper physically receiving the goods.

The desired direction of the transport demand, that is, where the goods to be shipped are staged and where they are ultimately delivered, is virtually insignificant for actual processing. However, you need to observe the stipulated tariffs and any related Incoterms.

A forwarding agent can conduct transport logistics in three different ways:

- **Completely autonomous transport logistics**
  The shipper maintains his own fleet and drivers, and, with them, attempts to achieve optimal capacity utilization with the goods to be transported. Cost minimization is a primary goal. The objective is to execute all goods shipments using as few vehicles as possible. This type of organization can be observed more often in smaller producing enterprises or retail companies. Transport logistics is primarily in the form of local transports to and from factories and distribution centers.
- **Internal transportation planning with external logistics service providers**
  The shipper can master the planning of the shipments in an optimal manner. However, he does not own his own fleet, and thus commissions a logistics service provider or carrier to execute the shipments according to precise instructions. This type of organization is often seen in companies in which a number of independent suborganizations transmit their transport demands to one central planning office.
- **Complete outsourcing of transportation tasks and services**
  The shipper transfers the individual transport demand tasks to a logistics service provider and has that agent decide on their processing. In a more extreme form, the service provider is given further tasks for the external logistics chain (warehouse management, order processing, and inventory control) and takes on increased responsibility.

2.1.3 Transportation from the View of the Logistics Service Provider

Logistics service provider is the collective term for carriers as well as freight forwarding agents. In both types of companies, the core process and value creation concern the processing of shipments. For the purposes of this book, freight forwarding agents refers to companies that organize the transportation of goods, while carriers execute the physical transport of goods.

Both kinds of companies work closely together. Freight forwarders who do not own their own fleets are dependent upon carriers who act as the executing business partner. Larger logistics enterprises often comprise both types of companies; the logistics service provider organization accepts, plans and processes aggregated orders, and then passes them on to internal carriers and other external carriers.

Carriers have responsibilities in the following realms:

- Provision of mode-specific transportation capacities (on rail, air, sea, and road)
- Optimized use of an internal fleet and thus the opportunity to offer attractive prices for transportation services and the provision of transport means (such as containers)

The responsibilities of a logistics service provider include the following:

- The consolidation of goods from various customers to achieve maximum profitability
- Complete processing of goods transport for a customer, including the performance of all legally required services (customs clearance, dangerous goods treatment, paper printouts, import/export processing, goods movement) and the professional subcontracting of all involved carriers

Consolidation and profitability. The consolidation of goods from various customers gives the logistics service provider the opportunity to optimize profits. For instance, he can commission a carrier to execute a container shipment (full container) for $1,000 and subsequently resell the available 24 pallet spaces to a customer for $100 each. The service provider turns a profit from the 11th sold pallet space. Of course, he also carries risk of incomplete capacity utilization.

2.1.4 Shipper and Service Provider Hybrids

From the viewpoint of the shipper, transportation processing is generally not a core competence upon which he wishes to concentrate, but rather a necessary task in the completion of the process chain. Instead of completely outsourcing all transportation services, the shipper can outsource his shipping department (with or without a fleet) in order to optimize his own competences and also provide them to other business partners for their transportation processing needs. With such efforts, outsourced logistics departments of larger enterprises increasingly represent direct competition to logistics service providers.

2.2 SAP Systems and Applications

Especially in the realm of transportation, several transpiration solutions have been developed in the history of SAP systems, each of which had a certain user group and focus (see Fig. 2.3):

1. In 1987, the first transportation solution in the mainframe system SAP R/2 was introduced to the market (*Realtime Vertrieb*, RV, with *Realtime Transport*, RT), whose functions were strongly influenced by shippers in the chemical industry.
2. In 1993, with SAP R/3, the transportation processing solution SD-TRA entered the market, which represents a generic solution from the view of the shipper. Release SAP R/3 4.6 saw the categorization of the solution in the *Logistic Execution System* (LE-TRA).
3. In 2000, as a supplement to SAP ERP Transport, SAP introduced transportation planning and optimization for shippers in SAP APO (APO-TP/VS, *Transportation Planning/Vehicle Scheduling*).
4. *SAP Event Management* (SCM-EM) was introduced to the market in 2001 as a tracking and tracing solution for shippers as well as logistics service providers and integrated into ERP Transport for purposes of tracking shipments.
5. With *SAP Transportation Management* (SAP TM) in 2007, SAP supplied a comprehensive, independent transportation solution that served the needs of logistics service providers as well as shippers.

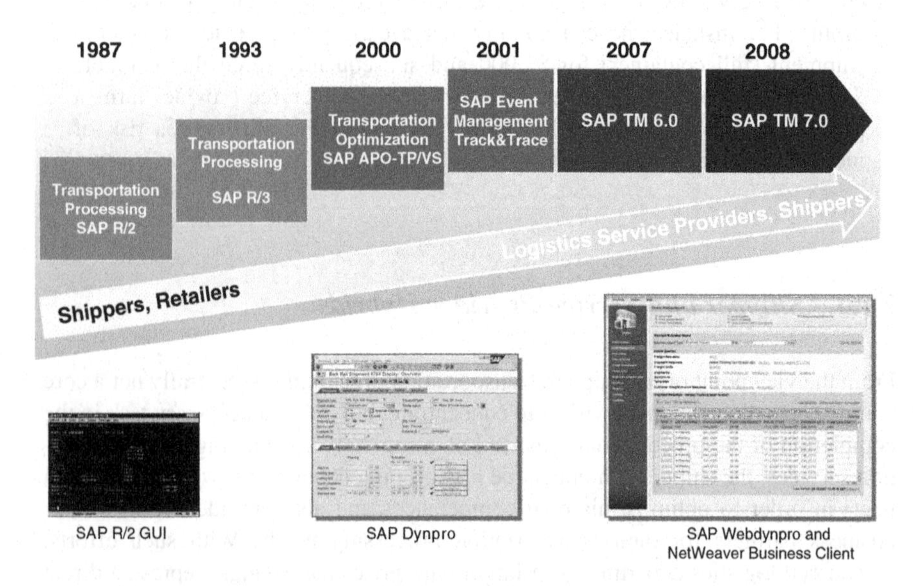

Fig. 2.3 History of transportation solutions in SAP

2.2.1 *Subprocesses and Components of SAP Transportation Solutions*

An overview of the transportation solution components is illustrated in Fig. 2.4. It shows how the components and individual subprocesses are integrated to enable transportation processing.

The main components of this transportation solution include:

- **SAP ERP: Sales and Distribution (SD) and Logistics Execution System (LES) for sales orders and delivery**
  The sales order (see Volume 1, Chap. 6, "Distribution Logistics") represents the starting point for the outgoing transport demands of a shipper. The goods purchased by a customer stemming from one or more points of departure generate the transport or individual shipment requirements. These individual requirements are defined in the deliveries that are put together based on the sales order.
- **SAP ERP: Materials Management (MM) and LES for purchase orders, stock transport orders and incoming deliveries**
  The purchase order (see Volume 1,, Chap. 4, "Procurement Logistics") is the source document for the transport demand of a shipper, in which the goods to be procured are defined along with their procurement locations. A purchase order can lead to deliveries, which then represent the individual shipment

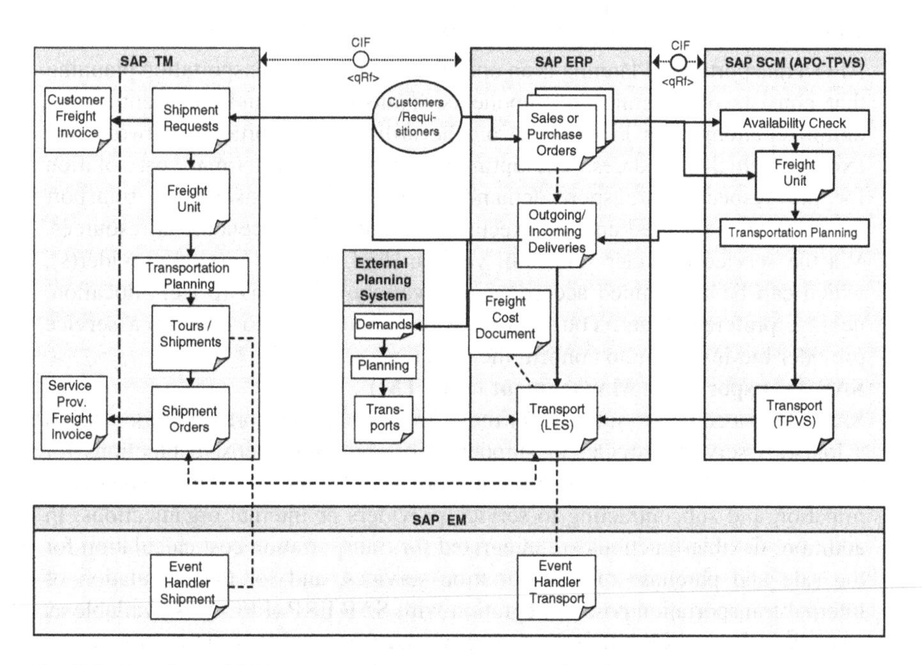

Fig. 2.4 Overview of SAP transportation components and their integration

requirements. Stock transfer orders (see Volume 1,, Chap. 4, "Procurement Logistics") between plants are a special type of order. In SAP ERP, they are treated similarly to a normal order, with the single exception that, in addition to an incoming delivery, an outgoing delivery is also generated, which represents the outgoing side from the stock view.

- **SAP ERP: Logistics Execution System (LES) for shipment and freight cost documents**
  The ERP shipment document and corresponding freight cost document are planning, execution and billing documents for transportation processing. With the component *Logistics Execution System*, you can create shipments and transportation chains, consolidate deliveries either manually or in a rule-based way, and document the process. Shipping costs can be calculated here as service provider costs from the shipper view and can be invoiced. In the customer order, you have the opportunity to access transport costs and pass them on to the customer in the normal invoice of that particular order.
- **SAP APO: global Available-to-Promise (gATP)**
  The global Available-to-Promise (see Volume 1, Chap. 6, "Distribution Logistics") in the APO system supports sales order processing by determining the best source for the materials ordered by a customer (sourcing). A customer's requested delivery dates, as well as shipping times, available and reserved material quantities, and available material alternatives, are also taken into consideration. SAM SCM 5.0 and later releases include this global availability check in APO Transportation Planning, where you can consult a detailed transportation plan to schedule shipments.
- **SAP APO: Transportation Planning/Vehicle Scheduling (TP/VS)**
  APO Transportation Planning is an optimization tool for transportation planning that consists of several subcomponents. Transport demands are sent to the optimizer along with information on the utilized transportation network and existing vehicle resources. The optimizer calculates an optimal cost solution for the respective transport demands: Routes with consolidated transport demands are generated and are executed using the most economical resources. Via the service provider selection, you can find the best service provider(s), which can be determined according to a variety of criteria (price, allocation, quality, preference, etc.)You then have the opportunity to conduct a service provider bid invitation to confirm the selection.
- **SAP Transportation Management (SAP TM)**
  SAP TM is a complete solution for the processing of transportation processes as a logistics service provider or shipper. It offers comprehensive functions for quotation and order management, transportation planning, posting, route determination and subcontracting to service providers or internal organizations. In addition, flexible functions are integrated for transportation cost calculation for the sale and purchase of transportation services, and for the calculation of internal transportation costs. Integration with SAP ERP (FI/CO) is available as a standard feature for billing customer and service provider freight costs.

- **SAP Event Management**
  SAP Event Management (see Chap. 5, "Controlling and Reporting") is a universal and very flexible tool that supports all types of visibility and status tracking processes (Tracking & Tracing: Shipment Tracking). It enables you to record performance data on your own and your partners' processes and thus generate a performance evaluation in connection with SAP NetWeaver BW.
  SAP Event Management is integrated with ERP transportation processing as well as SAP TM, and a variety of standard tracking scenarios are configured.
- **Special components for particular industries**
  Within the framework of the SAP portfolio, you can use further components for special industrial requirements that are not depicted in Fig. 2.4. These components include:
  - *SAP Oil & Gas Traders and Schedulers Workbench* (TSW) to plan and execute tanker transports while especially considering the raw material sale of in-transit stock.
  - *SAP Oil & Gas Transportation and Distribution* (TD) for the processing of bulk commodity transports in the downstream realm (such as for the supply of gas stations). Meter readings, temperature-dependent volume changes of bulk commodities and the compatibility of previous and subsequent tank loads are among the elements taken into account.
  - *SAP Rail Car Management* (RCM) for the processing of rail transports with a company's or a railway's freight cars. RCM, which is used by several companies in the chemical industry, is based on SAP Event Management, which it uses for freight car tracking. In addition, you can plan and execute the individual activities of the cars and manage your own loading railway stations and marshaling yards with *Onsite Event Management* (OSEM).

The number of resulting solutions is a direct reflection of the diversity of the transportation industry.

2.2.2 Transportation Processing Scenarios and Their Integration in Procurement and Distribution Logistics

Using the subcomponents and processes mentioned above, you can select various approaches for transportation processing with SAP. Each of these approaches offers a basic transport functionality that can be tailored through add-ons and integration mechanisms and is thus especially suited to support the demands of its respective user groups. The following rough guidelines can aid in the selection of the transportation solutions, described subsequently in more detail:

- **Traditional transportation processing for shippers (SAP ERP, Logistics Execution System)**
  Production or commercial enterprises with general transport demands that do not need complex strategies for source determination or availability check processes involved in transportation planning.
- **Traditional transportation processing with add-ons (SAP APO Transportation Planning and Service Provider Selection)**
  Production and commercial enterprises that have increased demands on transportation planning and optimization or service provider selection and bid invitation processes, but do not require the integration of an availability check.
- **Shipper solution with global availability check and transportation optimization (SAP APO-TP/VS)**
  Production and commercial enterprises for whom optimal transportation processing and minimized transport costs play a great role and for whom transportation is strongly dependent upon source determination and the availability of goods. This is especially true where issues such as material substitution or decisions regarding international source determination and supply sources are of great importance.
- **Traditional shipper solution with support by SAP TM**
  Production or commercial enterprises that already use the traditional SAP shipper solution for transportation processing can use this variant when transitioning to a new SAP TM system. The sales order integration with freight cost billing has been preserved, but transportation planning is transferred to the much more powerful TM system. Processing of transports can either be done in SAP ERP or directly in SAP TM.
- **Shipper solution with service provider reference (SAP TM in combination with SAP ERP Distribution Logistics)**
  Production or commercial enterprises where transportation processing is a multi-departmental or outsourced function. Such companies often have their own transportation departments that receive transport demands from several company divisions (or, under certain circumstances, from various ERP systems). However, the processing of these requirements should be consolidated to keep costs down. The transport departments often serve as a transportation service provider within the company.
- **Transportation service provider solution (SAP TM in combination with SAP ERP Financials)**
  Transportation service providers who sell transportation as a service to other companies and purchase transportation services from other companies (carriers).

Now we will take a closer look at these transportation solutions.

2.2.2.1 Traditional Transportation Processing for Shippers (SAP ERP, Logistics Execution System)

The traditional SAP transportation solution for shippers, used by more than 2,000 SAP customers around the world, is transportation processing with the SAP ERP component *Logistics Execution System.* It supports the shipments of outgoing goods, goods to be picked up and to be transferred. Figure 2.5 provides an overview of this transportation solution. The standard process for sold goods begins with an order initiated by the customer (see ❶ in Fig. 2.5). The customer order documents the goods sold that are to be transported and have to be delivered from one or more plants. Based on the customer order, one or more deliveries are generated (Distribution/Shipping) ❷. Through manual or rule-based planning, you can then put together shipments that contain one or more deliveries. You can also consolidate deliveries from various plants. To map long-haul transports, you have the opportunity to create individual shipment documents for pre-carriage, main and on-carriage legs, each of which reflects different legs of the same delivery. For each shipment document, you can create an *event handler* in SAP Event Management that enables tracking of the shipment ❸. With reference to the data cited in the shipment document and the indicated delivery dates, you can create a freight cost document and calculate the freight charges to be paid to the service provider ❹.

The sales price calculation based on the sales order for a material and the subsequently generated invoice can include the conditions used for the freight charges. This allows you to pass the charges paid to the transportation service

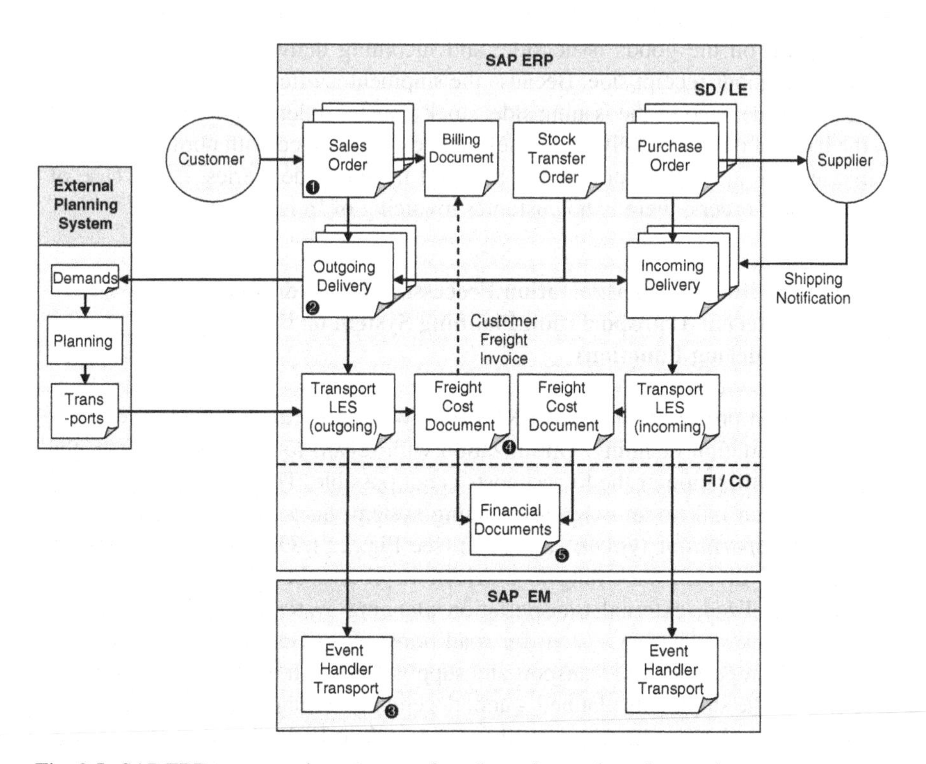

Fig. 2.5 SAP ERP transportation processes for sales orders and purchase orders

provider on to the customer. You can then use the freight cost document to trigger the transfer of the service provider costs to Financial Accounting, including the accrual of reserves ❺.

The process described above is the same for incoming deliveries whose transport demands result from purchase orders; in this case, charging the customer with freight costs is not possible. For an order initiated from within your company that is sent to a supplier, one or more deliveries are generated. Each incoming delivery can be organized into shipments in the same way as outgoing deliveries.

Special case when forming shipments with ERP Shipment Processing. Please note that for transportation processing using the component *Logistics Execution System* it is not possible to consolidate incoming and outgoing deliveries into a single shipment. A shipment is thus always *only in one direction*. If both incoming and outgoing deliveries must be planned, you need to create separate shipments for them.

Stock transfer orders between plants are created as a special type of order. From these stock transfer orders, outgoing deliveries are generated from the issuing plant on the goods issue side, and incoming deliveries to the receiving plant on the goods receipt side. Because the shipment is often necessarily based on the transport demand of the issuing side, stock transfer order shipments are created on the basis of outgoing deliveries. They can be consolidated with normal outgoing deliveries into a single shipment, but not with incoming deliveries. In the case of stock transfer orders, there is no customer invoicing of freight costs.

2.2.2.1 Traditional Transportation Processing with Add-Ons (External Transportation Planning System or Bid Tendering Function)

Transportation processing in SAP ERP offers you the options of manual or rule-based transportation planning. Optimization with regard to the shortest route, the best vehicle utilization or the lowest costs is not possible. To achieve such optimization, you can link to an external planning system via a *standard interface for external transportation systems* (SD-TPS) (see Fig. 2.5). Outgoing and incoming deliveries are divided according to a selection process based on preset rules or several specialized, external transportation planning systems. For instance, it is possible to link a planning system for road transport in Germany and a planning system for Europe-wide rail transport and supply them with the respective delivery documents. The shipments planned – and, depending on functionality, optimized – in the external planning systems are then sent back to ERP Shipment Processing, where they trigger the respective shipment documents. You can determine whether the external planning system should maintain the planning authority over the shipments from that point on or whether they are allowed to be edited in ERP

Shipment Processing. A resynchronization of changes made in the ERP system does not take place.

2.2.2.2 Traditional Shipper Solution with Extended Tendering Function

Another add-on to the traditional shipper solution for transportation processing with SAP ERP is the linkage of SAP TM with the use of the new service provider selection and bid tendering functions. Shipments that have been planned in the ERP system are tendered to a "pseudo" service provider, supplied by SAP TM, via the bid tendering interface. Shipment and freight orders as well as routes and freight units are created in Transportation Management based on the ERP shipments (see Fig. 2.6).

You can now use the service provider selection in SAP TM in order to determine the best service providers. You can then use the new tendering functions in SAP TM to conduct either sequential, simultaneous or open tendering. SAP Event Management controls the tendering process and, if necessary, the required reactions when a tendering deadline has been reached. In the case of a positive response to a tender, the respective shipment and freight orders are synchronized back into the ERP shipping documents.

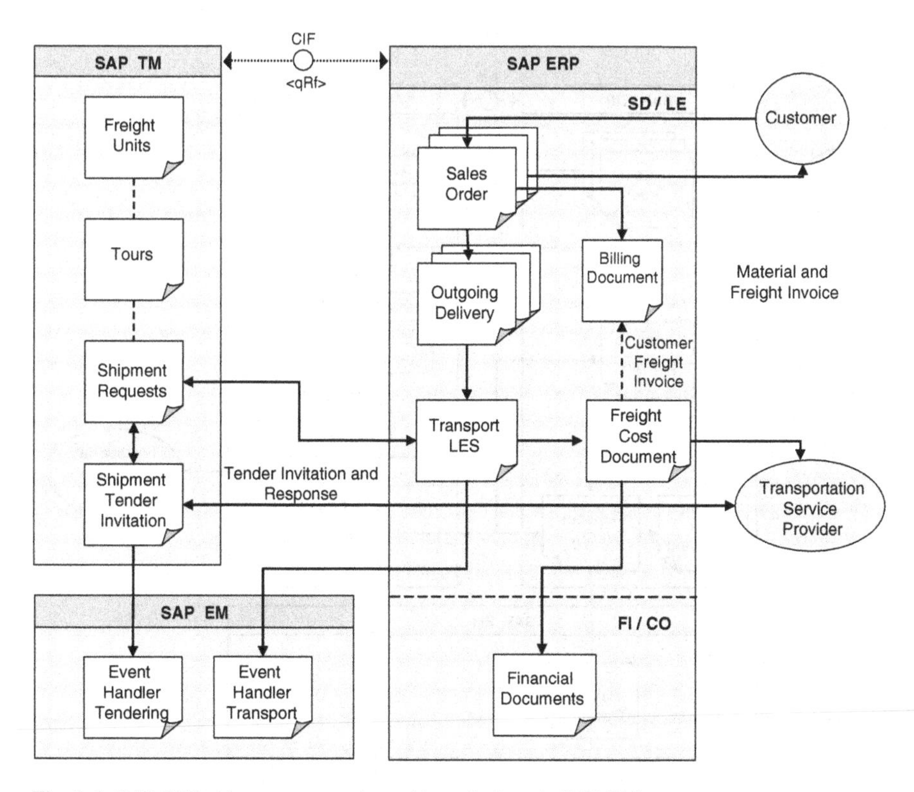

Fig. 2.6 SAP ERP shipment processing with tendering via SAP TM

2.2.2.3 Shipper Solution with the Global Available-to-Promise (ATP) Availability Check and Transportation Optimization (SAP APO-TP/VS)

If a shipping company has high demands with regard to transportation optimization and close integration with source determination and global availability, a transportation solution can be employed from SAP ERP Logistics and SAP APO. This solution can be used for purchasing- as well as sales-based processes. Figure 2.7 shows an overview of the process flow for the sales and distribution process.

Based on the *sales order*, a global Available-to-Promise (ATP), or availability check, is performed in the APO system. Within the context of this availability check, freight units can be generated with the *Routing Guide*, which can then either be scheduled forward or backward. The schedule determined by the availability check is then transferred back into the sales order. The planning remains in the APO system as a temporary plan until the sales order is saved. When it is saved, the temporary transportation plan is saved along with it. Based on this plan, a service provider selection and tendering process can be performed. Subsequently, outgoing deliveries and shipment documents are generated in the ERP system. Actual processing is performed on the basis of the ERP shipment documents.

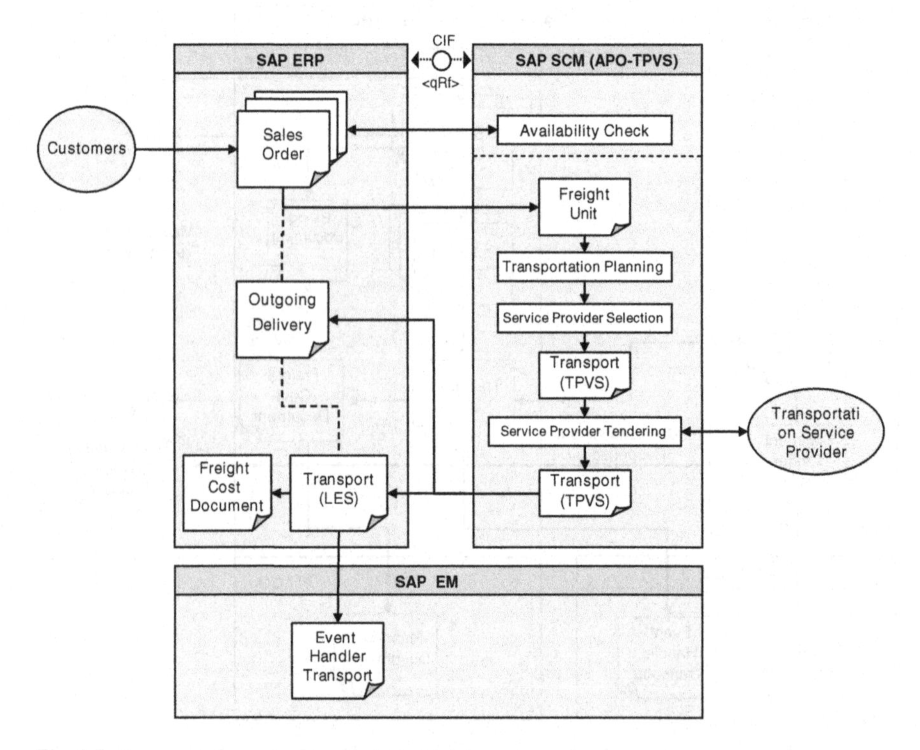

Fig. 2.7 Transportation process with SAP APO for sales order processing

Just as in the traditional shipper solution, you can also calculate and pay service provider costs and pass them on to the customer.

2.2.2.4 Traditional Shipper Solution with SAP TM Support

If a company is already employing the traditional SAP ERP transportation solution and is considering the transition to the new SAP TM, there is the possibility of doing so incrementally by transferring functions such as shipment planning to the SAP TM systems. Figure 2.8 shows the process.

Sales and purchase order processing and the creation of deliveries are conducted in the component *Sales and Distribution* (SD). The generated incoming and outgoing deliveries are then communicated as transport demands via the service interface to the TM system. In SAP TM, transportation planning and optimization, route determination, delivery creation and finally the creation of delivery and freight orders are performed. The freight orders are then sent back to the ERP system via the service interface, where they generate shipment documents. Transportation processing ultimately takes place in the *Logistics Execution System*, where freight cost calculation and customer freight billing are generated.

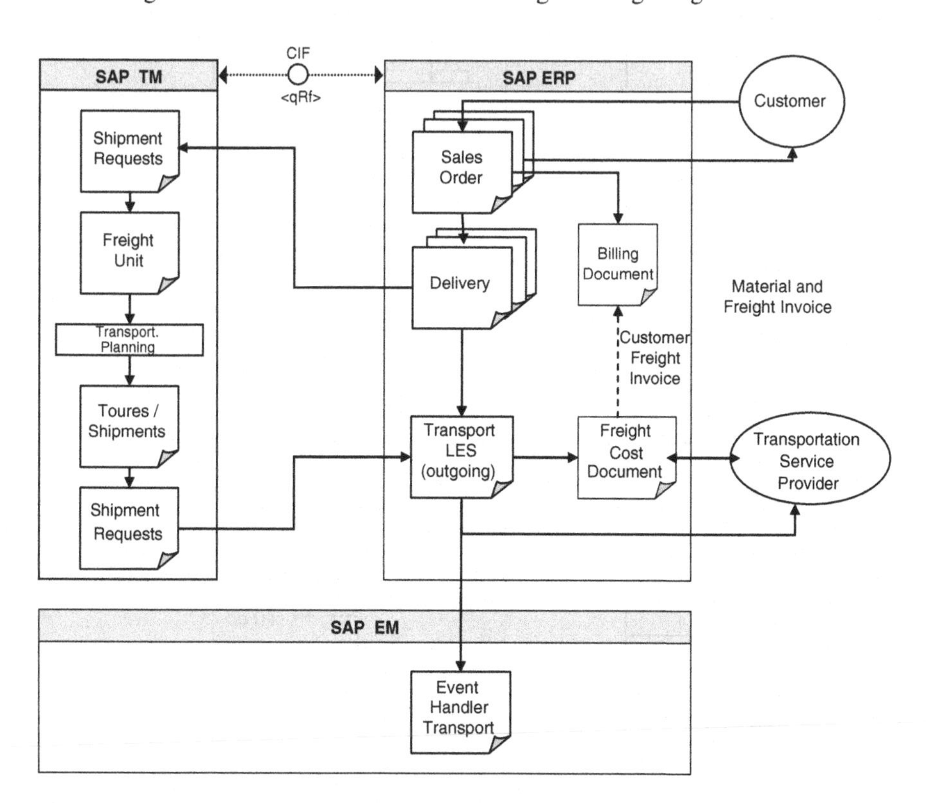

Fig. 2.8 SAP ERP transportation processing with SAP TM as a planning tool

2.2.2.5 Shipper Solution with Service Provider Reference (SAP TM in Combination with SAP ERP Logistics)

A shipping company with an outsourced transportation planning department generally requires transportation functionality that enables cross-ERP system transportation planning or has a strong relationship with a service provider. The transport demands can be generated in various systems, depending on the business area, such as in several ERP systems in which distribution logistics is processed separately. If a reduction in costs is targeted through a consolidation of transport demands from various systems, it is not possible with the traditional shipper solution, because the shipments created there each require references to delivery documents that are distributed among several systems.

Here, *Transportation Allocation* is generally used when processing is done through SAP TM. Sales orders, purchase orders and delivery documents are generated in several systems and sent to SAP TM via the service interface (see Fig. 2.9). In contrast to the previously described solution, the shipment and freight orders created no longer need to be sent back to the ERP system. This would

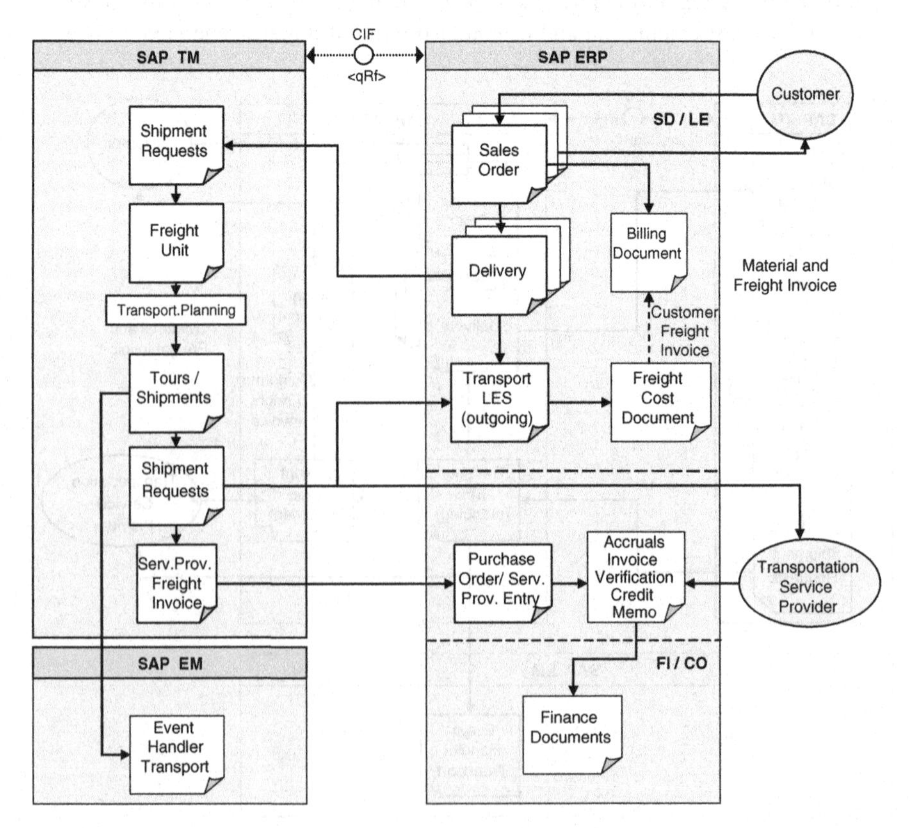

Fig. 2.9 Transportation processing for shippers with SAP TM and SAP ERP as the order processing system

only be done for purposes of customer freight calculation. However, freight cost calculation in SAP TM and SAP ERP must be done separately.

All steps in transportation planning, allocation and processing are then executed in SAP TM, whose flexible planning and freight cost tools offer a higher performance level than those of SAP ERP. Shipment tracking in this case is also accomplished via integration of the SAP TM business object "Shipment" with the respective event handler in SAP Event Management.

2.2.2.6 Transportation Service Provider Solution (SAP TM in Combination with SAP ERP Financials)

Transportation service providers who execute logistics for other companies can utilize a solution based on SAP TM that offers a flexible basis upon which to map their processes. Unlike transportation processing with SAP ERP, SAP TM does not require a reference to delivery documents or material master records, but rather can be employed for the entire transportation processing independent of master records and sales and distribution processing. The basic procedure is illustrated in Fig. 2.10.

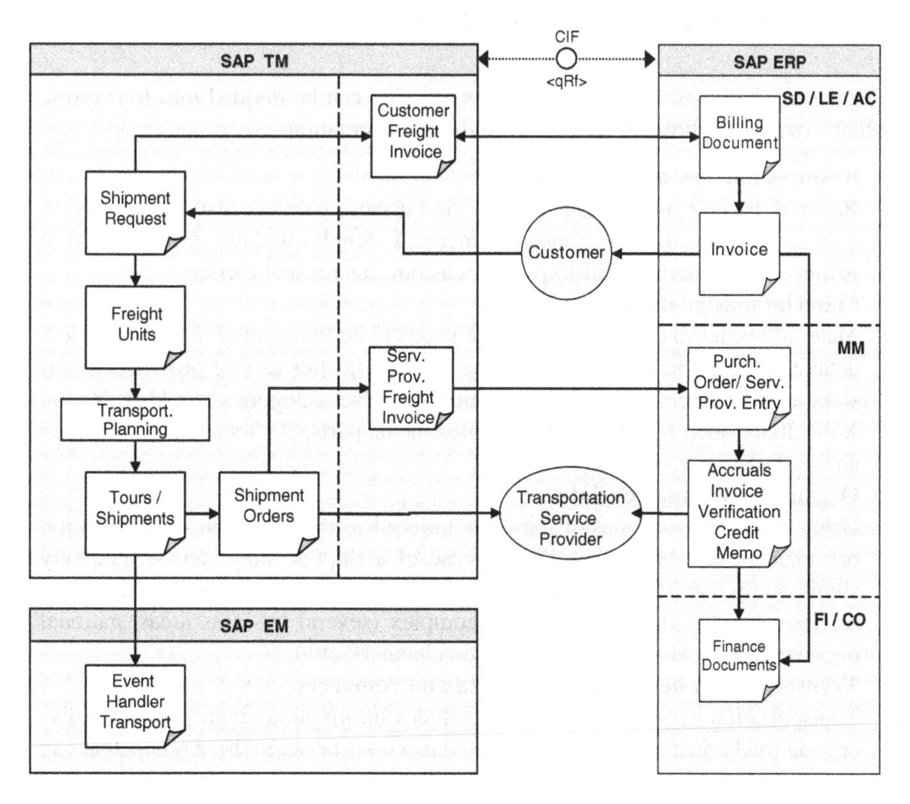

Fig. 2.10 Transportation management process with SAP TM

The transport demand is transferred directly by the customer as a transport request and made into a *shipment request*. Based on this, freight units are assembled that are consolidated, routed and scheduled in Transportation Planning. After optimized or manual planning, tours and shipments are formed that ultimately culminate in shipment orders, which are used to subcontract jobs to additional service providers and freight carriers.

Based on shipment requests, which also contain cost segments for the invoicing of services, customer freight invoice requests can also be created. They can be printed as pro-forma invoices as well as transferred to Billing in the ERP system, where customer invoices are generated and data is transferred to Financial Accounting.

The *shipment orders* form the basis from which you can create service provider freight invoice requests, which, once transferred to the ERP system, generate service provider orders and data sheets. Via accounting integration, you can then check incoming service provider invoices or, using the credit procedure, pay the invoiced amounts.

2.3 Master Data in Transport Logistics

The master data used in transportation processing can be divided into four types, which cover the following aspects of logistics information:

- **Business partner master data**
  Business partner master data defines the business partners that are directly or indirectly involved in the transport process. Examples include shippers, ship-to parties, sold-to parties, bill-to parties, customs agents and carriers.
- **Material master data**
  Material master data either defines the goods to be transported in a more or less detailed or categorized form, or, in the case of logistics service providers, maps shipping containers or services. Examples for data categories would be Yowai XVR-2030 video recorders, automobile chassis parts, 20-ft standard containers and 24-h delivery service.
- **Organizational master data**
  Using organizational master data, the units of a transportation or distribution organization can be defined. In the case of a shipper, they can be relatively simple (only one transport allocation organization), but in the case of a logistics service provider, they can get very complex (several business areas, national organizations, sales offices, distribution channels, etc.).
- **Transportation network master data and resources**
  Transportation network master data includes information about pick-up, delivery and load transfer points, about the connections between these points that can

be used as transport routes, and about transport means (resources) that travel between these points and can move goods.

One could consider freight cost data (such as installment tables) as a fifth type of master data. However, they are only treated as application data here. Table 2.1 lists the master data and their system applications, of which there are three types:

- **Obligatory**
  The *obligatory* use of master data means that transportation processing cannot be carried out without the respective master data.
- **Recommended**
  A *recommended* use indicates that the transportation process is significantly easier and more consistent with the respective master data. One example would be the use of customer master data in connection with ERP invoicing from the SAP TM system.
- **Optional**
  An *optional* usage means that this master data is integrated into the process in a reasonable manner, yet must only be used as needed.

The SCM master data listed in Table 2.1 is a technical component of SAP SCM and serves as the basis for APO Transportation Planning (TP/VS) as well as for SAP TM. Although the data is used by both systems, SAP TM exhibits more comprehensive functionality. Master data integration between ERP and SCM is accomplished via the CIF (Core Interface; for more information, see Volume 1, Chap. 3, "Organizational Structures and Master Data").

Table 2.1 Master data in transportation and its use in transportation solutions (use: ● obligatory, **○** recommended, ○ optional)

Master data entity	Type	System	Transportation solution		
			ERP	TM	TP/VS
Customer, vendor	Partner	ERP	●	**○**	●
Vendor	Partner	ERP	●	**○**	●
Material	Material	ERP	●	○	●
Packaging material	Material	ERP	○		
Route, leg	Network	ERP	●		
Plant, warehouse location	Network	ERP	●		●
Shipping point	Network	ERP	●		●
Transport allocation point	Organization	ERP	●		●
Freight conditions	Costs	ERP	●		
Business partner	Partner	SCM		**○**	
Location	Network	SCM		●	●
Product	Material	SCM		○	●
Resource	Network	SCM		○	●
Timetable	Network	SCM		○	

2.3.1 Customers and Vendors in SAP ERP

The significance of *customers* and *vendors* in SAP ERP was discussed in detail in Volume 1, Chap. 3, "Organizational Structures and Master Data"; Chap. 4, "Procurement Logistics"; and Chap. 6, "Distribution Logistics". For this reason, we will only refer to their transport-specific aspects here.

In transportation processing, customers can take on several roles. They are predefined in the ERP customer master as *partner roles*. You can expand these partner roles via Customizing settings. A customer, for instance, used in a transport process can serve as a sold-to party (who places an order that triggers a transport demand), a ship-to party, bill-to party or payer. Customer master maintenance enables you to maintain the necessary data for every role a customer has. A ship-to party does not require bank transfer data, and you need not keep shipping information for a payer. Important information that you can define in the customer master includes:

- Address information including international versions
- Payment transaction and bank transfer information
- Customer unloading points
- Export data
- Contact person

The address information and unloading points are significant to actual transportation allocation, the contact person and export data are used in the realm of logistics processing, and payment information is used for invoicing.

By defining partner roles in the customer master, you can establish a relationship between various customer master records. For instance, Sold-To Party A (the branch office of a car dealer in Chicago) can submit an order to be delivered to Ship-To Party B (a repair shop in St. Louis), yet have the invoice sent to the head office in Detroit (Bill-To Party C), which is ultimately paid by Payer D.

In the ERP system, you can allocate customers to various sales areas. This function is used for shipper processing in which a specific customer is served by a certain part of the sales organization. However, this data is not transferred to the SCM system. You can also save other shipping-specific attributes of a customer in the sales-related data, such as delivery priority and delivering plant.

Vendors can primarily take on two functions in transportation processing. To categorize vendors in one or both functions, you have to allocate them to an account group:

- **Transportation service provider**
  You can not only allocate freight forwarders and shipping companies to transportation service providers, but also customs agents, packing services, cleaning companies or other service providers in the transportation industry.
- **Supplier of goods**
  From a logistics standpoint, suppliers of goods mainly serve to determine the pick-up address for purchase orders and incoming deliveries.

The vendor master data functions similarly to the customer master data. Here, too, several views are available in which you can enter transport-related data.

2.3.2 Plants, Storage Locations, Shipping Points and Loading Points in SAP ERP

Plants, storage locations, shipping points and loading points form the logistical structure of a company in SAP ERP (see also Volume 1, Chap. 3, "Organizational Structures and Master Data"):

- **Plant, storage location**
  Organizational unit of logistics that categorizes a company in the views Production, Procurement, Maintenance and Planning. In a plant, materials are produced and/or goods and services are provided.
- **Shipping point, loading point**
  Organizational unit of logistics that executes shipping processing. Plants, shipping points and their subunits are technically not master data, but defined by SAP Customizing as basic organizational structures. Nevertheless, they are still considered master data from a logistics standpoint, and are matched in the same way as customers and vendors with the locations to be created in the SCM master data.

The primary characteristic of the organizational units mentioned here is their location, which is defined as an address and, in the case of a factory, is used for incoming and outgoing deliveries as a delivery or outflow address. In the case of a shipping point that is only used for outgoing deliveries, it is defined as the outflow location of a shipment. A further characteristic is the allocation to a plant calendar that defines the "active" work days. In the case of a shipping point, the respective loading and pick/pack time is also defined as a general, that is, material- and quantity-specific value.

2.3.3 Business Partners in SAP SCM

Business partners are organizations, companies and people that have a permanent or independent work or order relationship with a shipper or logistics service provider. The business partners defined in SAP SCM are used exclusively in SAP TM. For the default process, these business partners are automatically generated through master data transfer of customers and vendors from SAP ERP. Manual entry into SAP TM is thus only necessary in special cases.

In Table 2.1, you can see that SAP TM offers the opportunity to perform logistics processes to a large extent without the presence of business partner master data. However, a business partner master is practically essential for efficient billing and invoice management.

You can create business partner master records for business partners of the category *Customer* for a sold-to party, shipper and ship-to party of goods or for payers. Supplier-type business partners can be defined as forwarding agents, carriers, customs agents or operators of load transfer sites. The respective role is determined in the business partner master via the role definition function. You also have the option of assigning more than one role to a partner (for instance, "General Business Partner", "Financial Services Business Partner" and "Payer"). Figure 2.11 shows the respective entries for such a business partner.

For every general business partner, you can enter the following information:

- The main address of the business partner and additional addresses with notes (such as that it is a postal address or delivery address)
- Additional identification numbers to identify a business partner (such as the IATA agent code of an air freight service provider or Standard Carrier Alpha Code)
- Business hours and tax classification
- Information regarding payment transactions with bank transfer data and payment card information
- Status information and lock flags

Business partner for new customers. If you have to accept an order by telephone from a new customer for whom you do not yet have a business partner master record, you can create a special business partner as a new customer. You can then use it in a TM shipment order (shipment request) and

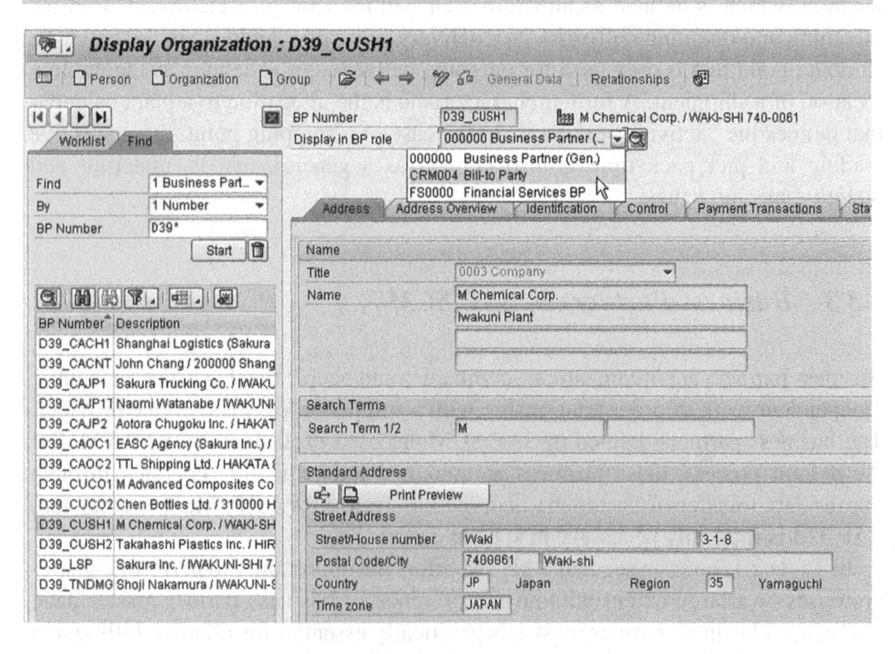

Fig. 2.11 Definition of a business partner SAP TM with several roles

provide it with the individual data from the order as a one-time address. After the new business partner is centrally generated and distributed, the partner *New Customer* can be easily replaced in the order.

2.3.3.1 Business Partner and Transportation Service Provider Profile for Vendors

Business partners that are defined as transportation service providers require additional, logistics-relevant attributes for efficient execution of transportation planning, allocation, tendering and subcontracting in SAP TM. These attributes define the authorization and service level of the service providers. You can maintain a *transportation service provider profile* with the following attributes for your business partners:

- Routes served in the transportation network
- Types of goods handled, product freight and transport groups
- Utilized/available transport equipment
- Fixed and dimension-based transport costs for transportation optimization

You can define an employee of a business partner as the type *Person*, as subordinate to a business partner. These employees are users in SAP TM for Internet collaboration in the transport bid tendering process.

If you maintain organizational units in SAP TM (see Sect. 2.3.8, "Organizational Data in SAP ERP and in SAP SCM"), a business partner of the type *Organizational Unit* is automatically generated for each unit. You can use these business partners directly in SAP TM to perform such tasks as recording the subcontract of a transportation job to a local company.

2.3.4 Materials in SAP ERP

From the shipper view, the material master data defined in SAP ERP includes the deliverable, producible and sellable goods that a transport demand can produce in a logistics process. The materials can be maintained here with their attributes and various quantities, and allocated to organizations.

In addition, you can define various types of transport materials and equipment in the material master (such as pallets, pallet cages and cardboard boxes), which can also represent transport demand through their use in packaging one or more other materials.

Packing hierarchy and transport demand. The actual transport demand can be created on various levels. If 9,600 bags of flour are to be sold and transported, the transport demand might look like this:

9,600 bags of flour, 960 cardboard boxes, each containing 10 bags of flour, 20 pallets of 48 cardboard boxes each or a 20-ft container with 20 pallets.

The bags and boxes of flour each represent their own sales quantity unit, and the pallets and containers are defined as packaging materials.

The basic characteristics of the material master were described in Volume 1, Chap. 4, "Procurement Logistics". In this chapter, we refer to their transport-specific attributes.

In addition to the obligatory definition of material number and description, you also have to define the *base unit of measure* (such as count, box or kilogram). Via the base unit, you can also define further quantity units with the conversion factors. The

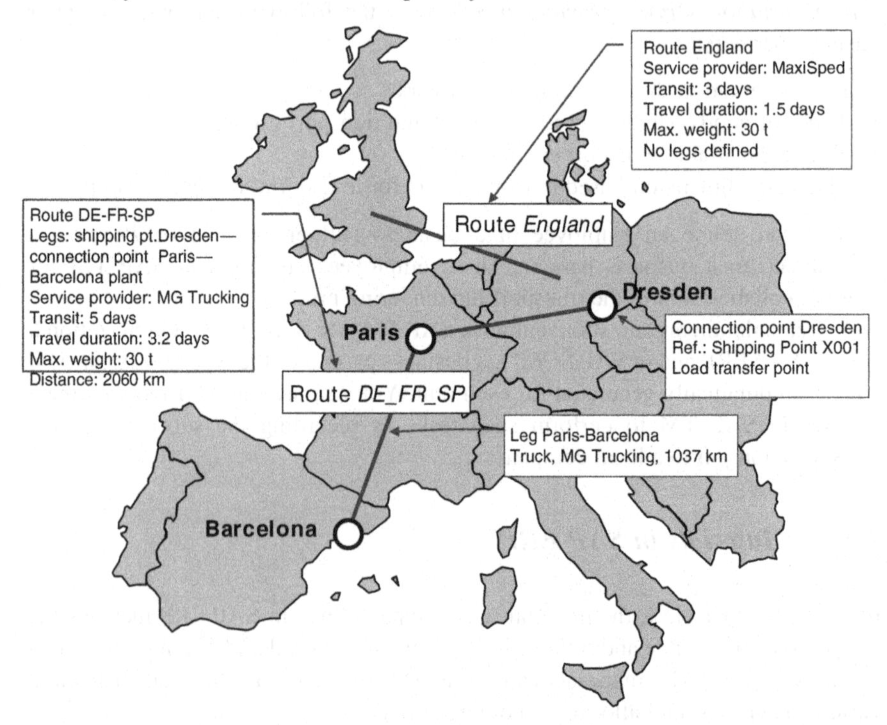

Fig. 2.12 Geographic elements of the transportation network in SAP ERP

Change View "Route Stages": Overview

Display BC Set Copy BC Set Double-Column Value Proposal

Dialog Structure
Routes
Route Stages
Transportation connectio

Route FR0005 Dresden - Northern France

Route Stages

Itin.	C...	Cat	Dep.point	Description	ST	ServcAgent	Distance	Unit	Tot.durat.	Trav.dur.	Work Time	Cal	LegId	SpPt	Sl
1	1		DRESDEN	Dresden, Germany	01										
2	1		SAARBRÜCK.		01										
3			GOUVIEUX	Gouvieux			0,000								

Fig. 2.13 Route definition in SAP ERP

indication of the gross and net weights as well as volume is especially important for logistics processing, because these values are taken into account for the capacity calculation of combined shipments. Volume refers to the volume occupied by a material during transport, not the net contents of a unit of material (for instance, a cardboard box with six 5-L canisters of a cleanser can have 40 L of volume). Fig. 2.14 shows the material master maintenance screen in SAP ERP

In the material master, there is also a *sales view* in which you can define the delivering plant and transportation group as transport-relevant attributes. The *transportation group* is a categorization criterion that allows you to categorize materials having the same defined processing conditions. Examples for values in the transportation group include *palleted goods*, *refrigerated goods* or *dairy products*.

If the material is classified as dangerous goods, you need to create a *dangerous goods master record* for transportation processing. The component *EHS Management* (Environment, Health, and Safety) in SAP ERP lets you save the necessary identifications and definitions for the various norms and carriers. Here, you can store dangerous goods classes and codes, material characteristics, rules for loading together, paper print definitions and other details for dangerous goods definition. A separate dangerous goods master record must be created for each material classified as a dangerous good.

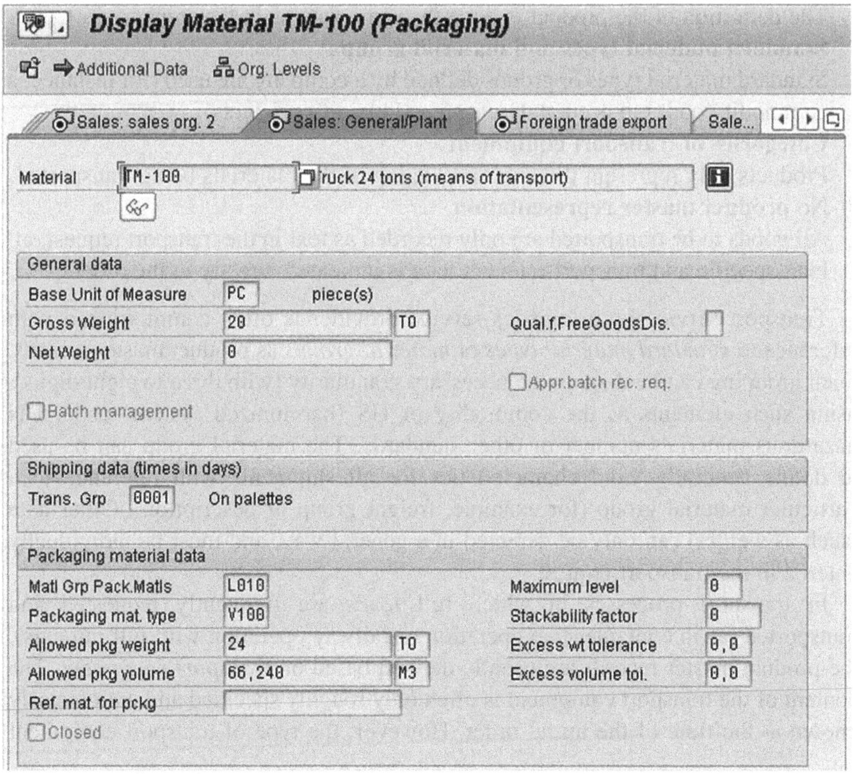

Fig. 2.14 Definition of a packaging material, in this case for a truck, 24 t

2.3.5 Products in SAP SCM

The *product master data* in SAP SCM, which is also be used by SAP TM, is of two basic types:

1. Product master data that maps precisely defined materials that are sold, purchased or transported in connection with a logistics agreement
2. Product master data that constitutes a classification or grouping of various materials or represents a service

The first type is generally used in shipper processing of transports (the exception being contract logistics), and is similar to the material master definition of the ERP system. It is employed in the traditional shipper solution and supported by APO Transportation Planning or SAP TM.

The second type is the product view of a logistics service provider, in which the situation regarding the product master is considerably more multifarious. You can only effectively use this type of product master when employing SAP TM as a service provider solution.

The following possibilities for using the product master exist:

- **Exactly defined products in contract logistics**
  The definition of the product is done the same way as in the shipper view.
- **Standard material types and material groups**
  Standard material types or groups defined by a company are used (for instance, a commodity code) to accurately group and classify products.
- **Categories of transport equipment**
  Products only represent the outer packaging of the materials being transported.
- **No product master representation**
  All goods to be transported are only recorded as text in the transport request; all load-specific and transport-relevant data is indicated directly in the order.

Transport service by a logistics service provider is often commissioned with reference to *standard material types* or *material groups* as product master records. Such grouping can be done in the necessary granularity (with three to eight digits), using such elements as the commodity or HS (harmonized system) code, UN hazardous materials number or other standards. The material group can be used to define generally valid characteristics for all shipments with reference to a particular material group (for example, freight group or description). Other data (such as weight) can only be depicted in a general way, and must be individually entered in the transport request.

In transport processes in which full loads are frequently requested and transported (as in container line operation or railway operation with full rail cars), the product master records are usually defined based on *transport equipment*. The content of the transport equipment is often only roughly specified and not precisely known at the time of the initial order. However, the type of transport equipment must be precisely defined (for instance, a 20-ft standard container or a 67-ft, high-sided gondola). The order then only states the desired number of transport equipment products as the goods to be conveyed. More precise information on the nature of the transported goods is added at a later time.

2.3.6 Transportation Network and Transport Equipment in SAP ERP

The transportation network in SAP ERP is the basis for determining the transportation relevance of shipments and for route determination in ERP Transport. It consists of three major elements: routes, legs and transportation connection points.

The *route* is a fairly detailed, possible transportation route that can consist of one or more legs (see Fig. 2.12, Route *DE_FR_SP*), but which can also be defined without any geographic reference (see "Route *England*"). A route is characterized by its *route identification* and can contain the following attributes, among others:

- Transportation service provider that executes a route
- Shipping type along the route
- Transit duration (total duration including breaks), pure travel duration (not including breaks) and distance
- Permissible total weight
- For dangerous goods shipments, there is the option of including a transit country table.

Route definition without a geographic reference enables you to conduct a pure transit calculation for the shipment without referencing geographic circumstances. For instance, you can define a route called *North Atlantic* in which you define shipments from Europe to the United States with a transit time of 14 days. The ports of departure and arrival remain undefined. If you wish to define ports, you can create one or more legs for that route.

A *leg* is either a connection between two transportation connection points or an individual connection point via which a transportation activity is performed (such as customs clearance). For each leg of a route, you can define a section type (transportation, load transfer or border point), a shipping type, the distance, service provider, travel and transit times as well as details on freight cost relevance. Figure 2.13 shows you a sample route that follows part of Route *DE_FR_SP* from Fig. 2.12.

A *transportation connection point* is a place where goods are dispatched, received, transshipped or processed. Such processes can include activities like customs clearance or railway car cleaning. For each transportation connection point, you can define the type (e.g. load transfer point, airport or seaport), the responsible customs office, calendar and stopover time, as well as a reference to an organizational unit (such as a plant or shipping point), a partner (customer or vendor) or any address.

Transport equipment is generated in SAP ERP as packaging materials (of the material type VERP). The packaging materials define the capacity and characteristics of the means of transport that will subsequently be used in a handling unit for a particular transaction.

2.3.7 Transportation Network and Resources in SAP SCM

The transportation network and resourcesare crucial for executing transportation options. A transportation network represents the geographic circumstances for the transportation of goods, and is modeled using locations, transportation lanes and transportation zones.

Executing a transport is done with one's own or third-party resources which move the goods between locations in a transportation network along transportation lanes. The following categories of *resources* are defined in SAP SCM: vehicles with capacity, tractor trucks, trailers, transportation units (containers, railway cars), and handling resources for goods movement to locations and drivers. *Itineraries*, which are a combination of transportation network and resource, are also significant. Figure 2.15 shows a schematic transportation network with the elements cited.

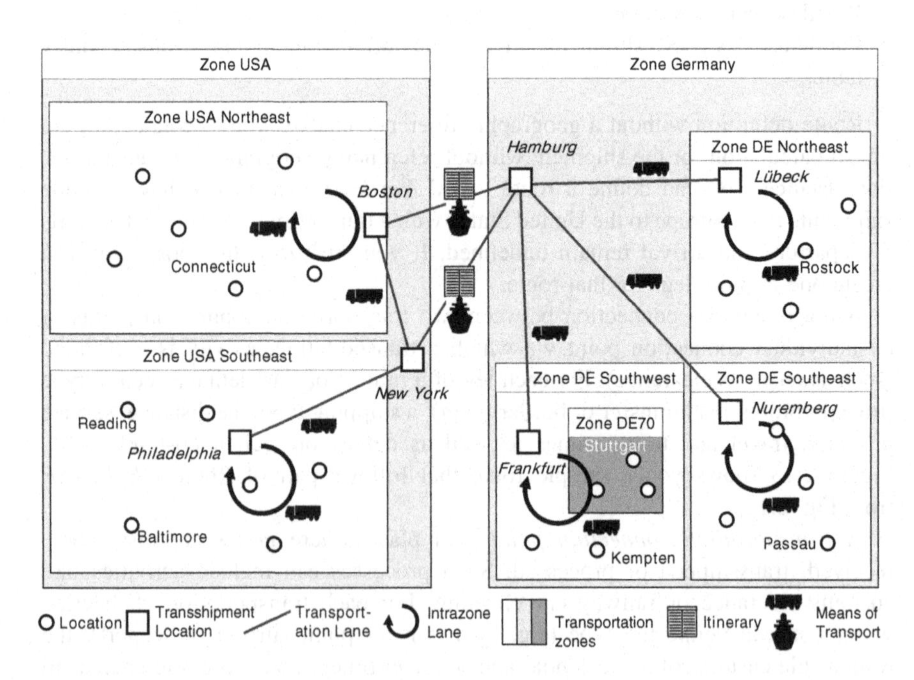

Fig. 2.15 Elements of a transportation network in SAP SCM

Locations are places in which goods are picked up, delivered or transshipped, or in which activities relating to the transportation process are performed (such as customs clearance). Each location is classified by its location type (for instance, a production plant, distribution center, customer, vendor or terminal). In addition to entering the description and address of a location, you have the option of entering further contact details and a reference to a relevant business partner.

A location possesses a *geolocation*, that is, geographic coordinates used by the transportation optimizer to calculate distances. You have the option of linking geocoding software with the SAP SCM system to enable the coordinates of created locations to be automatically determined with the added precision of the geocoder (for instance, down to the town or street number). You can use various geocoding software for a range of countries and regions to fit your transportation needs.

Other location attributes include:

- Minimal transshipment times at locations used for shipment scheduling
- Availability of handling resources such as the number of available forklifts at a hub to load and unload goods
- Alternative identification, with which you can also define the UNLOCODE or IATA airport code for the location

A transportation network generally includes several locations, which can be grouped into *transportation zones*. This is done to aid in the selection process, and to facilitate the definition of transportation lanes. There are three kinds of transportation zones:

- A *direct zone* can be allocated to explicit locations.
- A *postal code zone* contains all locations corresponding to the valid postal code areas of a particular country. You can define these zones with variability (such as ZIP 700xx-729xx and 750xx-753xx).
- A *region zone* is defined by entering a country and region, and contains all locations in that region.

A *mixed zone* is a combination of the above zone types. It is created by entering at least two different zone definitions.

For the selection process, it is possible to arrange the zones in a hierarchy. Figure 2.15 shows the postal code zone *DE-70* as a subzone of the regions *DE-Southwest,* which in turn is a subzone of *Germany*. This hierarchy enables you to process all loads from the areas of the city of Stuttgart, the state of Baden-Württemberg, or all of Germany.

Seen from the perspective of transportation planning, *transshipment locations* are special locations in which goods are allowed to be transshipped from one vehicle to another. A transshipment location can be defined for every location. As a rule, transshipment locations are distribution centers, ports, railway stations, airports or similar places where a change in transit carrier (such as from a truck to a ship) frequently takes place.

Transshipment locations enable the targeted guidance of the flow of goods via defined exit and entry locations. Figure 2.15 illustrates that goods transit from Germany to the United States is either processed through the ports of Hamburg and Boston or Hamburg and New York and that the pick-up and delivery traffic in the southwest of Germany is done through the distribution center in Frankfurt.

Transportation lanes define direct connections within a set of locations and zones. They can be defined between two locations (source and destination), a location and a transportation zone, or between two transportation zones. Important attributes of a transportation lane include a validity period, possible means of transport, duration and distance, cost parameters and data related to transportation service provider selection. Figure 2.15 shows an example of a transportation lane between a distribution center in Frankfurt and the port in Hamburg, and between the ports of Hamburg and Boston.

Of special significance are *intrazone lanes*. They refer to the reachability of a location within a zone from every other location. Thus, it is not necessary to create individual transportation lanes between pairs of locations within a zone; it suffices to define one intrazone lane. For instance, in the transportation network shown in Fig. 2.15, every location in southwestern Germany (within the zone *DE-Southwest*) can be reached by the Frankfurt distribution center. Goods coming from Hamburg, however, cannot reach Stuttgart directly, but must be transshipped in Frankfurt, since Hamburg does not lie in the zone *DE-Southwest*.

As in the case of locations, you can also use an external system to determine distances for transportation lanes. The distance and travel time of a transportation lane between source and location and the selected distance plant is then automatically determined using predefined location coordinates.

Itineraries indicates a predefined sequence of locations that serve as stops. They are used for regular water travel, train travel and in road traffic, such as for recurring trips in retail supply or regular main legs in system transit.

The heading *Resources* within the context of transportation planning in SAP SCM refers to all means of transport provided by a transportation capacity or the ability to move a loaded transportation capacity. Every resource has an identification number and a calendar that defines when it is not available (its downtime). *Downtimes* can occur due to such factors as maintenance or breaks. For transport equipment, you can also define attributes such as means of transport, registration number, owner, provided capacity (that is, how much can be loaded), consumed capacity (that is, how much is utilized when the resource itself is loaded onto another resource, such as when a container is loaded onto a truck), further equipment (loading cranes, accompanying forklifts) and a home location (supply chain unit). Via the means of transport, transportation lanes can be allocated.

Capacities can be maintained in a variety of dimensions, such as weight and volume capacities. Several nondimensional capacities can also be maintained, such as TEU (Twenty-foot Equivalent Unit for containers), loading meters and pallet storing position.

As we have already mentioned, the following types of resources are among the means of transport and transport equipment:

- **Vehicles with their own capacity**
  Vehicles with their own capacity are self-moving means of transport. Examples include a 40-t truck, a container ship, a freighter ship, a container ship with 5,390 TEU or a cargo-model Airbus 340.
- **Tractor trucks**
  Tractor trucks do not have their own loading capacity, but are capable of moving passive transportation capacities (such as trailers).
- **Trailers**
  Trailers have a transportation capacity (to the same extent as a vehicle resource), but they must be combined with a tractor to execute a shipment. If a trailer is available in transportation planning but not a tractor, the planning cannot produce a result.
- **Transportation unit**
  Transportation units (containers, railway cars), like trailers, have a capacity but cannot move autonomously. They have to be loaded onto a means of transport in order to execute a shipment.

Other types of resources include:

- **Handling resources for goods movement in locations**
  Handling resources provide goods movement capacities in a location. Examples include forklift operators, loading cranes, filling stations or traffic congestion workers. If, for example, only one forklift is available at a distribution center having 10 loading ramps, a bottleneck would automatically be created that needs to be considered during planning, since vehicles will have to wait longer for loading and unloading.
- **Drivers**
  Drivers can be allocated to a planned shipment as a resource. As attributes, they possess time availability and qualification credentials such as a driver's license or dangerous goods permits.

In order to further increase the flexibility of transportation resources, you can define *compartments* for every resource into which certain goods can be loaded, such as a truck with a dry and a refrigerated compartment or a tanker trailer with a diesel and a gasoline compartment.

You can create *vehicle combinations* from several means of transport and transport equipment elements. This would, for instance, enable you to combine certain tractor trucks and trailers, which are usually moved together.

You have the option of creating a *means of transport hierarchy*. Special means of transport can thus be made subordinate to generally defined means of transport. Since the characteristics of superior means of transport can be passed on to subordinate means of transport, a hierarchy can be useful for easy description of a transportation network. An example of a means of transport hierarchy would be a 12-t truck and a 40-t truck allocated to the superordinate means of transport "Truck." If you then allocate a transportation lane to "Truck," both 12 and 40-t trucks can be utilized without the need of further definitions.

2.3.8 Organizational Data in SAP ERP and in SAP SCM

The primary organizational element in the SAP ERP transportation solution is the *transportation planning point*. It is defined as an organizational unit of logistics that is responsible for the planning and execution of transportation activities. The transportation planning point organizes the responsibilities of a company into such categories as the type of shipment, carrier or according to regional departments. Thus, there can be a transportation planning point for New England or the Midwestern United States, or in other companies, for truck and ship transit. In addition to its function as a categorical and search criterion for shipment documents, the transportation planning point, via its company code assignment, enables logical allocation to the respective organizational areas in Accounting for freight cost invoicing.

Further organizational units in SAP ERP logistics, such as the sales organization or distribution channel, are only significant to sales and delivery processing, but are not utilized in ERP Transportation.

In SAP TM, the organizational structures are realized via *SAP Organizational Management*. It allows you to set up the organizational structure of your company in a flexible manner. In the simplest case, this means individual employees who perform various functions. In a larger enterprise or a logistics service provider, various organizational areas are needed:

- **Sales organization (logistics service provider-specific)**
  The sales organization structures the sale of logistics services and executes them. It can have several sales groups and sales offices as suborganizations. You can also allocate information to distribution channels and divisions. In SAP TM, the following processes, among others, are related to a sales organization:
  - Quotation generation
  - Order acceptance
  - Contracting of freight transit
  - Invoicing of sold freight services
- **Sales organization**
  The sales organization executes all sales procedures pertaining to logistics services of freight forwarders and carriers. It can have several purchasing groups. The following processes are among those related to sales organizations in SAP TM:
  - The sale and subcontracting of freight services
  - The purchase of freight capacity
  - Tendering of freight services
  - Contracting of freight purchasing
  - The settlement of purchased freight services

- **Planning and execution organization**
  The planning and execution organization arranges the planning and allocation of accepted shipping orders and the shipments to be transported, and performs any necessary activities or monitors them if they are subcontracted. The following procedures are related to planning and execution organization in SAP TM:
  - The distribution of regional and mode-specific planning responsibilities
  - MRP and transportation planning
  - The management of transportation resources

Figure 2.16 shows a comparison of these organizational structures with regard to transportation in SAP ERP and in SAP TM. APO Transportation Planning is a planning tool that is not dependent on an organizational definition.

Because no direct relationship to finance-related classification objects exists in SAP TM (such as to company codes, accounts or internal orders), for purposes of invoicing, organizational data is transferred to the ERP system or a connected account settlement system for financial allocation.

When SAP TM is used with SAP ERP as an account settlement system, you can set up analogous organizational systems in both systems to achieve a coherent categorization of sales and purchasing structures.

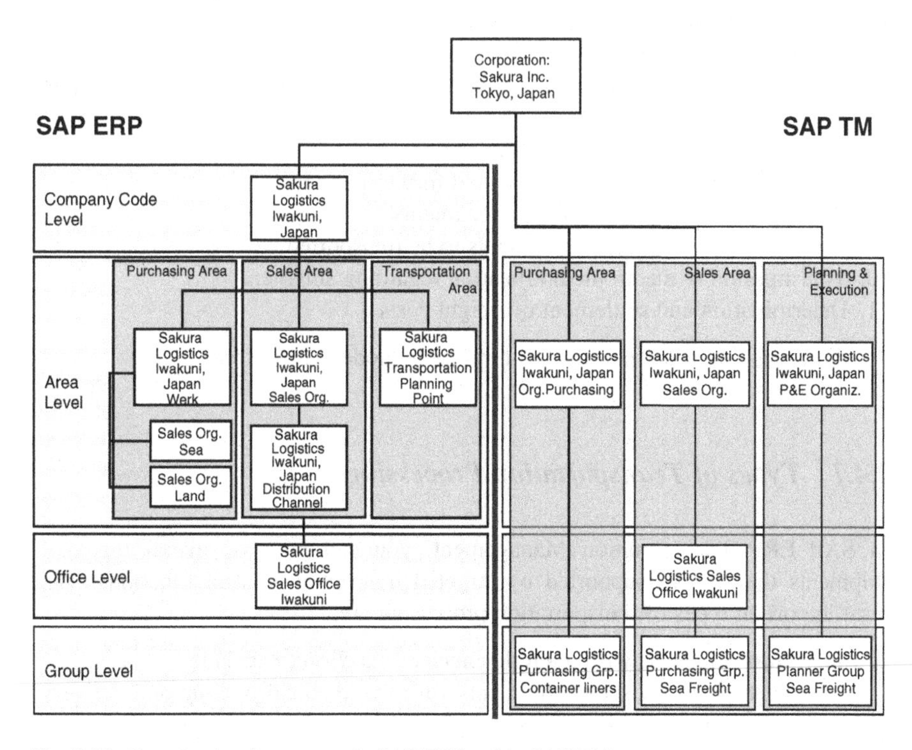

Fig. 2.16 Organizational structures in SAP ERP and in SAP TM

2.4 Transportation Management with SAP ERP

Transportation management in the SAP ERP Logistics Execution System (component LE-TRA) was developed as a traditional shipper solution primarily geared toward the transportation needs of SAP customers who also employ the modules for sales and distribution (SD) and procurement logistics (part of Materials Management, MM) (see Fig. 2.5).

Starting with sales orders in your Sales Department or purchase orders in Procurement Logistics, one or more delivery documents are generated that are to be allocated as transport demand. In ERP Transportation Management, you can now create one or more delivery documents that contain deliveries as shipments. SAP ERP planning tools can provide for efficient processing in this regard. Although optimized planning functionality is not a part of ERP Transportation Management, you can use external planning systems or SAP APO if optimization is desired. After transportation planning, you can create a freight cost document for every shipment document, which allows you to calculate and invoice service provider costs.

The main steps involved in SAP ERP Transportation Management include:

1. Determination of shipment types, carrier and means of transport
2. Execution of transportation planning and delivery allocation
3. Determination of transportation routes and stages
4. Planning of shipping dates
5. Determination of the forwarding agent, invitation of bids to and commissioning of forwarders
6. Definition of transportation packaging
7. Entering of transport details, texts and further partners
8. Printing of shipping and transfer documents
9. Posting of goods issue for shipments to be transported
10. Transmission of electronic notification regarding shipment
11. Determination and settlement of freight costs

All of these steps are possible with multiple modes of transport.

2.4.1 Types of Transportation Processing

In SAP ERP Transportation Management, you can process several types of shipments that are all supported by targeted transaction control functions. The most important types of transportation processing are:

- **Individual shipment with a single carrier as a direct carriage**
  A single shipment or set of shipments with the identical pick-up and delivery locations is transported by one vehicle directly from the pick-up location to the destination.

- **Consolidated shipment with a single carrier**
  Several shipments with various pick-up and destination locations are delivered with one vehicle in a sequence (delivery sequence) from their respective pick-up locations and delivered to their destinations.
- **Transportation chain with several carriers**
  One or more shipments are transported with several carriers sequentially (for example, truck – cargo ship – truck). Separate shipment documents are generated for each carrier, each of which displays the respective carriage identification (pre-carriage, main carriage and on-carriage). The transportation management system uses its process control to make sure a shipment is completely planned and allocated only when a seamless chain of individual shipments has been created in which that particular shipment is contained. Figure 2.17 shows such a transportation chain with two pre-carriages, two on-carriages and one common main carriage.
- **Empty run**
  A vehicle is transported empty from a source location to a destination.
- **Return shipment**
  Return shipment is a special type of processing for the transportation of return deliveries.

2.4.2 *Shipment Documents*

Shipment documents are frequently generated manually in SAP ERP. In every shipment document, there is a set of obligatory data that serves to control processes

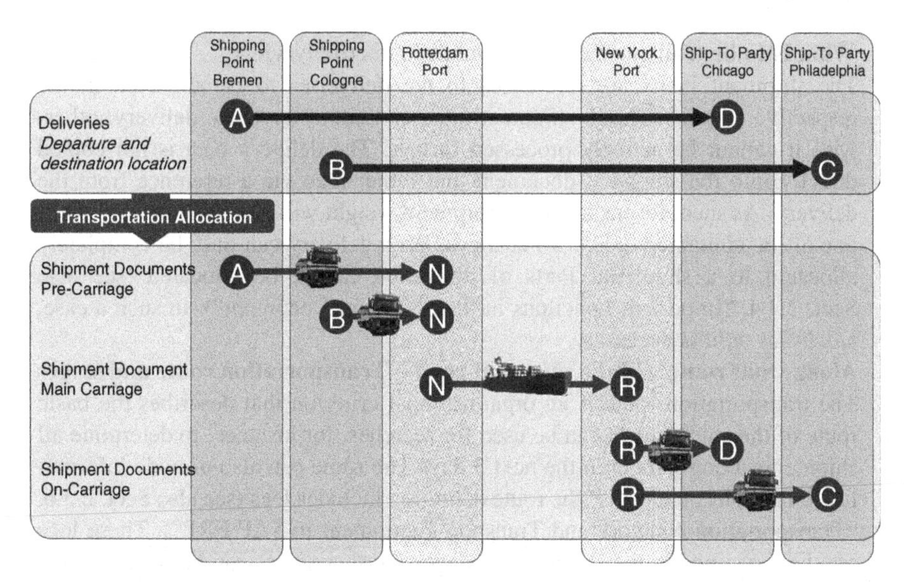

Fig. 2.17 Transportation chain with two pre-carriages, one main carriage and two on-carriages

and authorization behavior. The most important attributes that have to be defined when the document is created include:

- **Transportation planning point**
  Organizational unit responsible for the planning and processing of a shipment. The transportation planning point is allocated to a company code in the ERP system for cost accounting purposes.
- **Shipment type**
  Classification of the shipment with regard to carrier, means of transport and carriage identification (pre-carriage, main carriage, on-carriage and direct carriage). The shipment type as a central control attribute enables the following functions to be controlled in the shipment document, each of which can be configured in Customizing:
  - Number assignment for shipment documents (number range)
  - Text type definition, that is, what kinds of texts can be entered
  - Printed documents and electronic messages
  - The way to determine shipment stages (route determination), setting the way stages are adopted
  - The determination of attributes (such as individual or combined shipments)
  - Loading and packing functions

When you enter these two criteria, a new shipment document opens so that you can continue data entry. However, transportation planning points and means of transport cannot be subsequently changed. Figure 2.18 shows an overview of an ERP shipment document.

Important data that you can define and enter in the shipment document is explained below:

- **What is to be transported? – Shipment items (deliveries)**
  The shipment items are references to the deliveries to be allocated in the respective shipment. Each shipment must contain at least one delivery, otherwise it cannot be actively processed further. The delivery data is not copied directly into the shipment document, but rather read via a reference from the delivery. As such, for instance, the shipment weight will change if the weight of one of the contained deliveries changes. Each delivery can only be completely allocated to a shipment. Parts of deliveries cannot be allocated (see also Sect. 2.4.4, "Important Functions in the Shipment Document"). In such a case, a *delivery split* is necessary.
- **Along what route will the shipment run? – Transportation route and stages**
  The transportation route is an organizational criterion that describes the basic route of the shipment. It can be used for searches, for instance, to determine all shipments along a route in the next 3 days.The route can also provide information on the physical path if the route definition includes legs (see also Sect. 2.3.6, "Transportation Network and Transport Equipment in SAP ERP"). These legs are then assumed in the shipment as stages. Further stages can be defined, such

as for pick-up by the shipper or delivery to the goods recipient (see Fig. 2.19). In every stage, you can indicate the departure and destination location, shipping type, freight forwarder, distance and duration, as well as other data.

- **Who is shipping? – Freight forwarder**
  You can define a primary freight forwarder for the execution of a shipment on the overview screen. The forwarder must be defined as a supplier in the supplier master. Using the copying control, it is possible to copy a freight forwarder from a delivery document, if that delivery is allocated to a particular shipment. Depending on the forwarder selected, freight costs can be subsequently determined. A further use of the freight forwarder field is for tendering. Here, the freight forwarder serves as the recipient of bid invitation information and can either accept or decline the shipment job. For each shipment stage, you can indicate another forwarder for local performance.

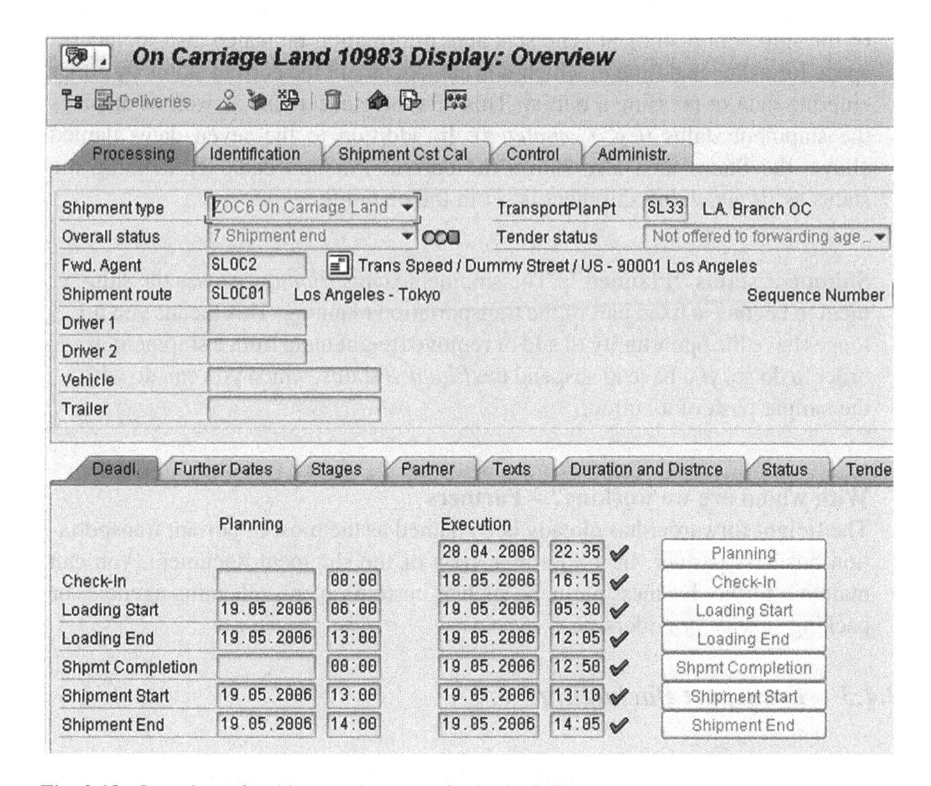

Fig. 2.18 Overview of a shipment document in the SAP ERP logistics execution system

Deadl. | Further Dates | Stages | Partner | Texts | Duration and Distnce | Status | Tender | Addit. data | Dng.G

St...	Departure point	Destination	FwdAgent	Forw.agent	S...	Shippi...	Distance	U...	L	Leg in...	In...	Incoter...
	Tokyo/1000000...	Tokyo/1000004...	SLOC2	Trans Speed	01	Truck	60,836	KM				

Fig. 2.19 Stages in an ERP shipment document

- **When is the shipment to be executed? – Dates**
  On the total shipment level (shipment header), you can define the planned dates and times for the following process steps:

  – Registration (arrival of the means of transport at the place of loading)
  – Loading start
  – Loading end
  – Shipment completion (departure of the loaded means of transport at the first place of loading)
  – Transportation start
  – Transportation end

  You can also enter free dates that can be used to schedule other process steps.
- **Where are we in the shipping process? – Shipment status**
  The shipment status values are closely related to the current dates on the header of the shipment document. The schedule information indicated also includes a space for a date and time in which you can document the current status by either entering data or pressing a button. This sets the current date as well as displays the shipment status (e.g. *Completed*). In addition to the seven dates named above, the Status area also shows the current *Planned* date. By defining the status as *Planned*, the shipment is set in the planning.

Shipment status "Planned". The shipment status *Planned* causes the shipment to become a fixed part of the transportation planning. This means you no longer have the opportunity to add or remove freight units from a shipment. In order to do so, you have to suspend the *Planned* status, which you can do with the simple push of a button.

- **With whom are we working? – Partners**
  The freight forwarder has already been named as the most important transportation business partner. In the Partner view of the shipment document, you can maintain further business partners, such as customs agents, cleaning agencies or packing service providers.

2.4.3 Transport Packaging

Similar to delivery processing, in transportation processing, you also have the option of defining packaging for the goods to be transported. This packaging can be simple or have multiple levels. In contrast to packaging in a delivery document, a shipment document allows you to assign packaging to more than one delivery, that is, several freight units can be assigned to a single transportation packaging.

Packaging in the shipment document is done as it is in the delivery document, with the aid of handling units (see Volume 1, Chap. 6, "Distribution Logistics", and Volume 2, Chap. 3, "Warehouse Logistics and Inventory Management"). If you assign deliveries that are already packed (such as a delivery of three Euro-pallet handling units) to a shipment document, you will see the delivery handling units in the shipment document packing section and can pack these units further.

You can use transport equipment (such as pallets, sacks, crates, pallet cages or containers) as well as means of transport (like trucks, trailers, ships or railway cars) as transportation packaging. Depending on the packaging material types and capacities, you can then pack one into another.

2.4.4 Important Functions in the Shipment Document

In addition to editing the shipment document and its data, you can make use of helpful tools in ERP Transportation Management to support your processing efforts:

- Transportation planning
- Route determination
- Subsequent freight unit split
- Transportation tendering
- Freight cost estimation
- Shipment tracking
- Graphical shipment information system

For your transportation planning needs, you have access to a *planning list* with which you can freely select and allocate deliveries from a delivery due list for existing or new shipment documents (see Fig. 2.20). At that point or a later time, you can select criteria defined by you (such as all unplanned deliveries that must leave Hamburg for northern Germany the next day). In the Planning screen, you can create new shipment documents. The deliveries can either be individually or

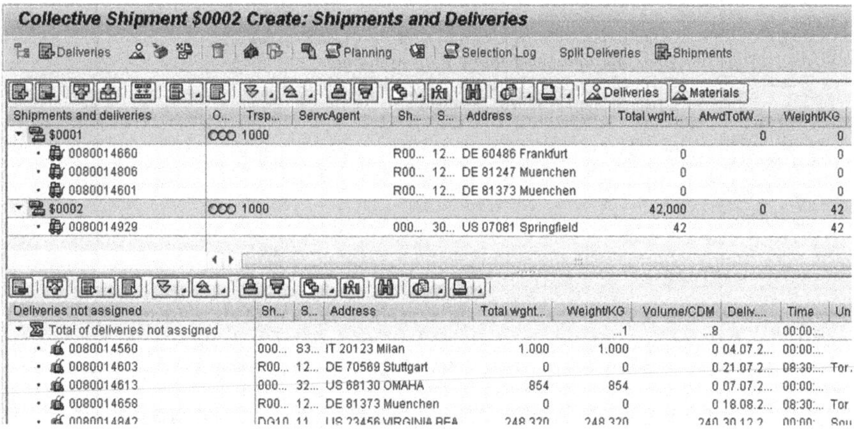

Fig. 2.20 Planning list with delivery allocations to shipments

collectively allocated to the shipment documents using the drag-and-drop function. For every individual shipment document, you can branch to its overview screen and edit details from there.

Interactive transportation planning. In Fig. 2.20, you can see an example of the interactive transportation planning screen. An existing shipment with a delivery was augmented by two other shipments ($0001, $0002), each of which has a delivery allocated to it. The temporary shipment numbers ($) indicate that documents have not yet been saved. In the delivery queue (in the illustration below), there are further deliveries that have not yet been planned.

Route determination is a function that you can start manually or automatically when the status *Planned* is set (see Fig. 2.21).

Starting with the details on the transport route and route determination settings in the shipment type, a route is generated according to the sequence of shipment stages. Subsequently, a stage sequence is generated that serves all pick-up locations

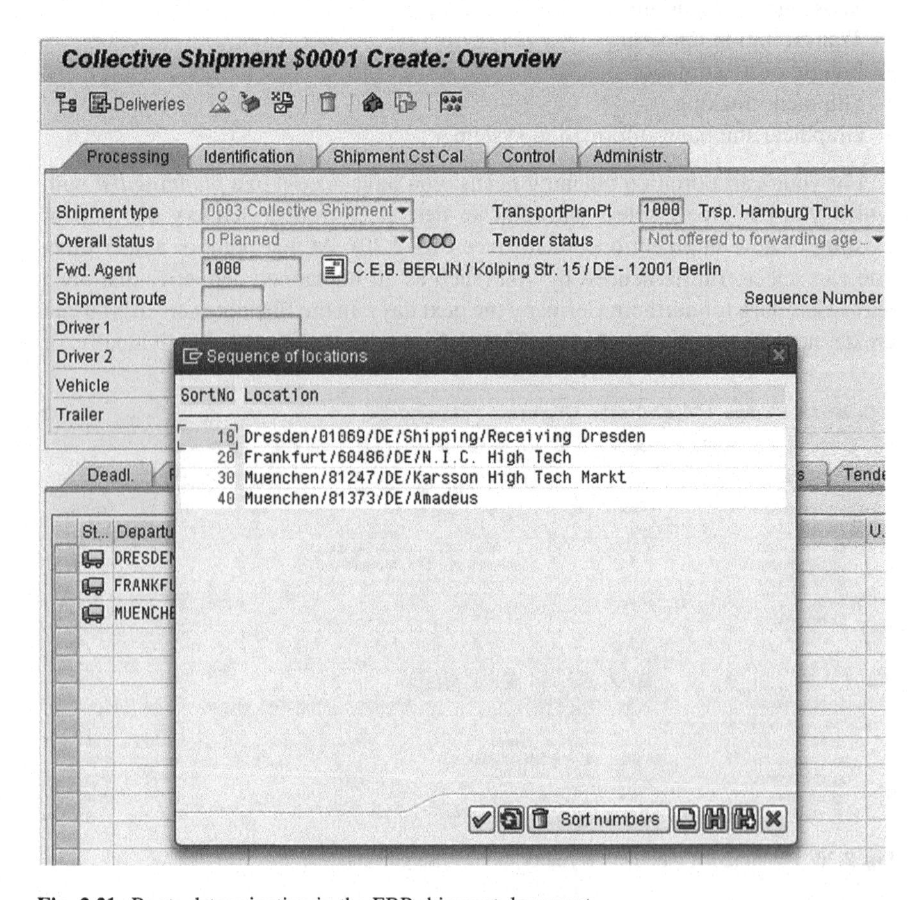

Fig. 2.21 Route determination in the ERP shipment document

before the route is begun. At the end of the route, further stages are added that lead to all ship-to parties. Following the route determination process, you have the opportunity to interactively alter the order of the stops along the route before the actual shipment stages are generated.

Route determination result. Figure 2.21 shows the result of route determination for a shipment of three deliveries (from Dresden to Frankfurt and Munich), for which no legs are defined.

Sometimes a vehicle cannot be loaded in such a way as planned in the transportation plan, because, for example, the dimensions of the vehicle or the load were incorrectly entered. In such a case, you can perform a subsequent freight unit split, to divide a partial delivery into a new freight unit. If, for instance, the rear door of a truck cannot be closed because the last two pallets are hanging out of the back of the truck by 5 cm, you can use the subsequent freight unit split function to reassign those pallets as a new delivery, to which you can allocate another shipment document. Then the two pallets can be unloaded. When the documents are printed out, you will then receive the correct shipping documents and the planning situation in the system corresponds to what has taken place.

Transportation tendering serves to obtain an offer from one or more freight forwarding agents. In order to perform the tendering process, the shipment document must indicate at least one stage and one defined freight forwarder, and have the status *Planned*. You can indicate the tender details and a target price in the tendering data. To determine a target price, you can use the freight cost estimation function, which you can access from the shipment overview screen. The freight cost estimation tool uses the shipment data to perform a cost estimation without, however, generating a freight cost document in the database (see Sect. 2.4.6, "Freight Cost Calculation"). Instead, the costs are temporarily calculated and presented as a decision-making aid in the context of the shipment document.

A bid invitation is then transmitted to the freight forwarder named in the shipment document header, either by sending a bid invitation message or placing it on a tendering portal, where the forwarder has direct access to the bid invitations directed at him. The forwarder can accept the bid invitation, accept it with alterations (such as an altered pick-up time) or reject it. He can also suggest a purchase price. The transportation planner can then check the tendering status in the ERP shipment document and either award the contract or send a bid invitation to another forwarder.

A *freight exchange* can also be used as a forwarding agent. This enables the transport demand to be placed on an open or closed marketplace for transportation services. If the offer is accepted by a forwarding agent or carrier on that market, that offer is assumed in the shipment document with the carrier's name, so that it is obvious who will perform the transport.

The ERP shipment document is linked to an event management process for the *tracking of shipments*. If a shipment document is created and given the status

Planned, the individual milestones and attributes of the shipment are transferred to Event Management, where they are used to generate an event handler. This event handler allows the status of the shipment to be tracked based on status messages that are transmitted via EDI, mobile device, the Internet or manual entry. The status messages can be viewed with a comparison of target/actual values in the tracking view of the shipment document. The standard process generates an event handler for the shipment and monitors all scheduled processing, departure and arrival dates of that shipment. If configured accordingly, you also have the option of tracking individual freight units or packaging units (such as pallets or containers) of the transport. More information on the topic of *Event Management* can be found in Chap. 5, "Controlling and Reporting".

2.4.5 List Processing and Planning Functions

SAP ERP Transportation Management offers a selection of lists and collective functions to ensure an efficient overview and processability of your inventory of transport demands and shipment documents.

The transportation planning list (see Fig. 2.22) allows you to easily obtain an overview of all pending shipments possessing certain criteria. For example, it is possible to have the system display a list of all shipment documents that were generated the day before (date) from the Hamburg distribution center (departure location) as a road shipment (shipment type) but which have not yet been completely planned (do not have the status *Planned*). Using that work list, you can continue processing individual shipments in a targeted manner by following a link in the list that branches to the document editing screen. After editing a document, you can return to the list.

To enable more efficient transportation planning and allocation, you can employ the *transportation planning collective run*. Using it, you can select targeted transport demands (deliveries) for the collective run and configure it, thus automatically generating shipment documents. In addition to the generation of direct shipments, you can also create transportation chains.

The distribution of the transport demands to the individual shipments is done heuristically, and the system attempts to optimally utilize allocated maximum

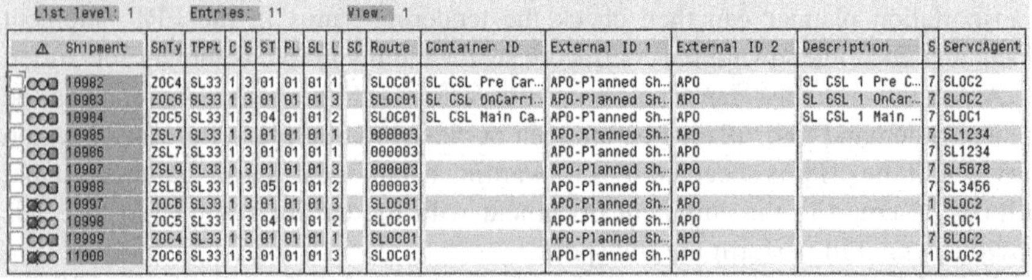

Shipment List: Planning

Mass Change | Delivery | Change Delivery Quantity

List level: 1 Entries: 11 View: 1

Shipment	ShTy	TPPt	C	S	ST	PL	SL	L	SC	Route	Container ID	External ID 1	External ID 2	Description	S	ServcAgent
10982	ZOC4	SL33	1	3	01	01	01	1		SLOC01	SL CSL Pre Car...	APO-Planned Sh...	APO	SL CSL 1 Pre C...	7	SLOC2
10983	ZOC6	SL33	1	3	01	01	01	3		SLOC01	SL CSL OnCarri...	APO-Planned Sh...	APO	SL CSL 1 OnCar...	7	SLOC2
10984	ZOC5	SL33	1	3	04	01	01	2		SLOC01	SL CSL Main Ca...	APO-Planned Sh...	APO	SL CSL 1 Main ...	7	SLOC1
10985	ZSL7	SL33	1	3	01	01	01	1		000003		APO-Planned Sh...	APO		7	SL1234
10986	ZSL7	SL33	1	3	01	01	01	1		000003		APO-Planned Sh...	APO		7	SL1234
10987	ZSL9	SL33	1	3	01	01	01	3		000003		APO-Planned Sh...	APO		7	SL5678
10988	ZSL8	SL33	1	3	05	01	01	2		000003		APO-Planned Sh...	APO		7	SL3456
10997	ZOC6	SL33	1	3	01	01	01	3		SLOC01		APO-Planned Sh...	APO		1	SLOC2
10998	ZOC5	SL33	1	3	04	01	01	2		SLOC01		APO-Planned Sh...	APO		1	SLOC1
10999	ZOC4	SL33	1	3	01	01	01	1		SLOC01		APO-Planned Sh...	APO		7	SLOC2
11000	ZOC6	SL33	1	3	01	01	01	3		SLOC01		APO-Planned Sh...	APO		1	SLOC2

Fig. 2.22 SAP ERP transportation planning list

shipment amounts (such as truck load capacities). However, unlike in SAP TM, optimization does not place. The result of the collective run is presented in a report detailing the individual steps and results.

If you have created several shipment documents and need to make the same targeted changes to a large number of documents (such as changing the freight forwarder), you can use the Mass Maintenance transaction for shipments (see Fig. 2.23).

In Mass Maintenance, as in the case of the planning list, you can initially select shipment documents to be edited that have certain criteria. You can then select the shipments to be changed from a results list and edit them via the various tabs, such as process data, identification, durations and distances, schedules and other data. The changes are then adopted for the shipment documents marked.

Other work lists and overview lists available include a capacity list and a list for available freight room, with which you can evaluate a selection of shipments either based on their utilized or free capacity and process them further, as is done in the planning list.

2.4.6 *Freight Cost Calculation*

Freight cost calculation, which primarily deals with the calculation and settlement of the freight costs of service providers, has long been a component of Transportation Processing. It includes the following basic steps:

- Generation of freight agreements and prices
- Calculation of freight costs
- Settlement of freight costs with service providers

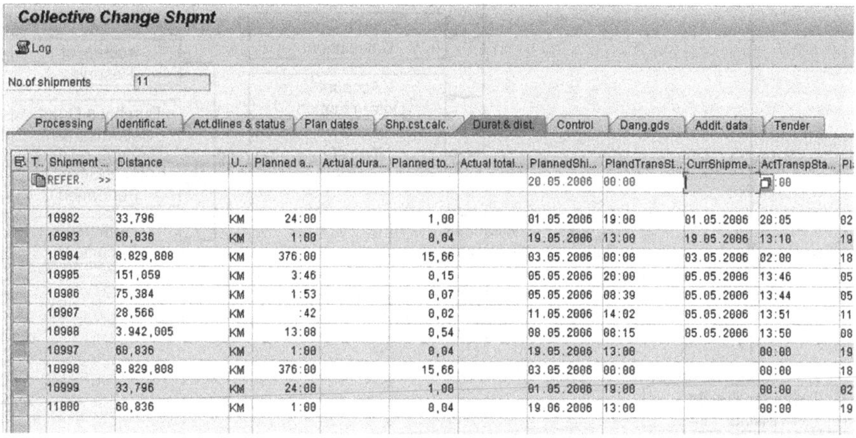

Collective Change Shpmt

Log

No.of shipments 11

Processing | Identificat. | Act.dlines & status | Plan dates | Shp.cst.calc. | Durat.& dist. | Control | Dang.gds | Addit. data | Tender

T..	Shipment ...	Distance	U...	Planned a...	Actual dura...	Planned to...	Actual total...	PlannedShi...	PlandTransSt...	CurrShipme...	ActTranspSta...	Pl
	REFER. >>							20.05.2006	00:00		:00	
	10982	33,796	KM	24:00		1,00		01.05.2006	19:00	01.05.2006	20:05	02
	10983	60,836	KM	1:00		0,04		19.05.2006	13:00	19.05.2006	13:10	19
	10984	8.829,808	KM	376:00		15,66		03.05.2006	00:00	03.05.2006	02:00	18
	10985	151,059	KM	3:46		0,15		05.05.2006	20:00	05.05.2006	13:46	05
	10986	75,384	KM	1:53		0,07		05.05.2006	08:39	05.05.2006	13:44	05
	10987	28,566	KM	:42		0,02		11.05.2006	14:02	05.05.2006	13:51	11
	10988	3.942,005	KM	13:08		0,54		08.05.2006	08:15	05.05.2006	13:50	08
	10997	60,836	KM	1:00		0,04		19.05.2006	13:00		00:00	19
	10998	8.829,808	KM	376:00		15,66		03.05.2006	00:00		00:00	18
	10999	33,796	KM	24:00		1,00		01.05.2006	19:00		00:00	02
	11000	60,836	KM	1:00		0,04		19.06.2006	13:00		00:00	19

Fig. 2.23 Mass maintenance transaction of shipment documents

- Transfer of costs to Financial Accounting
- Transfer of freight costs to customers who generate transport demand (customer freight billing)
- Accrual of reserves for expected freight costs

Figure 2.24 illustrates the respective steps in a graph.

The freight cost documents generated from the shipment documents and their follow-on documents (such as service entry sheets and invoices) are linked with one another via the document flow, such that you can easily link from one shipment document to its freight cost document and other documents. From the document flow screen, you can open the respective documents. Figure 2.25 shows the document flow of a shipment document up to the invoice receipt.

A freight cost document can be generated from every shipment document. Thus, there is always a one-to-one ratio between the two document types. The prerequisites for generating a freight cost document are:

- The shipment must be identified as freight-cost relevant.
- The shipment must have the necessary overall status set in the definition of freight cost type.
- The shipment must at least have the status of *Planned*.
- The shipment must have a service provider.

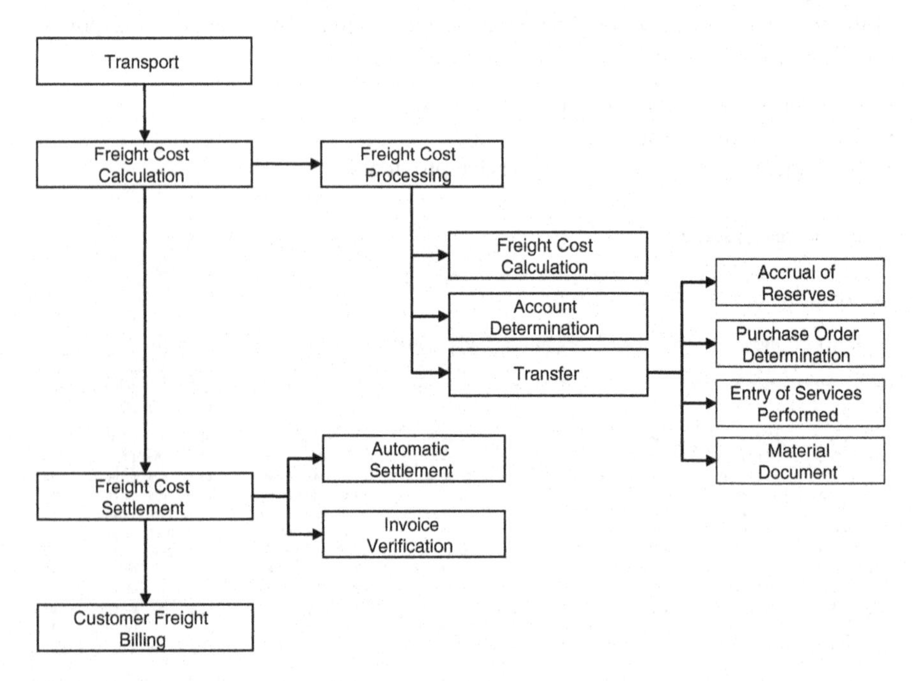

Fig. 2.24 Processing steps for freight cost calculation

The freight cost document contains header information, item information and subitems:

- **Freight cost header**
  In the freight cost header, you will find data pertaining to the freight cost type, document status and pricing date, as well as reference data and the processing report.
- **Freight cost item**
  This contains information on the item category, business partner (freight forwarder), utilized pricing procedure, the pricing date, tax amounts, settlement dates and references to the shipment and the service entry sheet. One freight cost item might, for instance, include costs for the total shipment (document charges, insurance), while another might indicate costs for the pre-carriage or main carriage.
- **Freight cost subitem**
  The freight cost subitems provide information on the calculation basis used and the calculation result, tax rates and references to the deliveries or freight units.

With regard to the calculated costs, the freight cost document offers several views that you can display depending on what information you need:

- An overview of all freight cost items and the calculated costs
- Costs per delivery item
- Costs per shipment stage

Figure 2.26 provides an overview of the freight cost document item view. Figure 2.27 shows the conditions of a freight cost item in detail, which illuminates the pricing procedure and tax determination.

You can define the automated processing of transportation information in freight documents with the respective customizing settings. The following process steps can be automated:

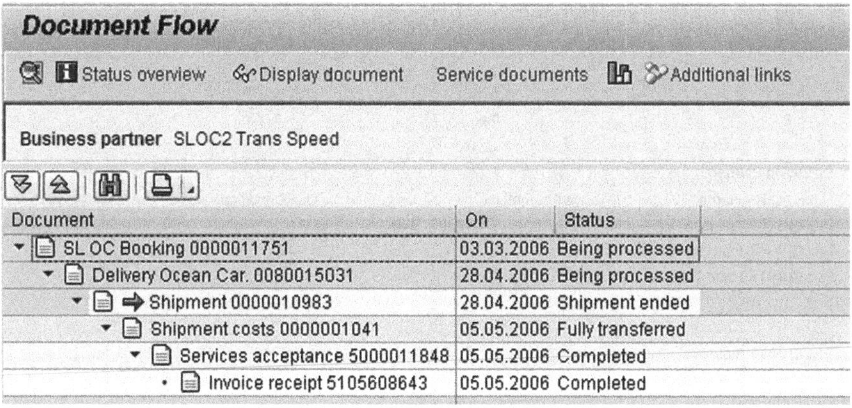

Fig. 2.25 Document flow between the shipment document and the service provider invoice

- The generation of freight cost documents with freight cost items
- The calculation of costs for freight cost items
- Determination of the accounts
- Transfer of costs to Financial Accounting, accrual of reserves

Calculation example for freight costs. In Fig. 2.28, you can see an example for the calculation of freight costs for a combined shipment from San Francisco via Detroit to New York. Two deliveries are transported, but they are only transported together in the first stage.

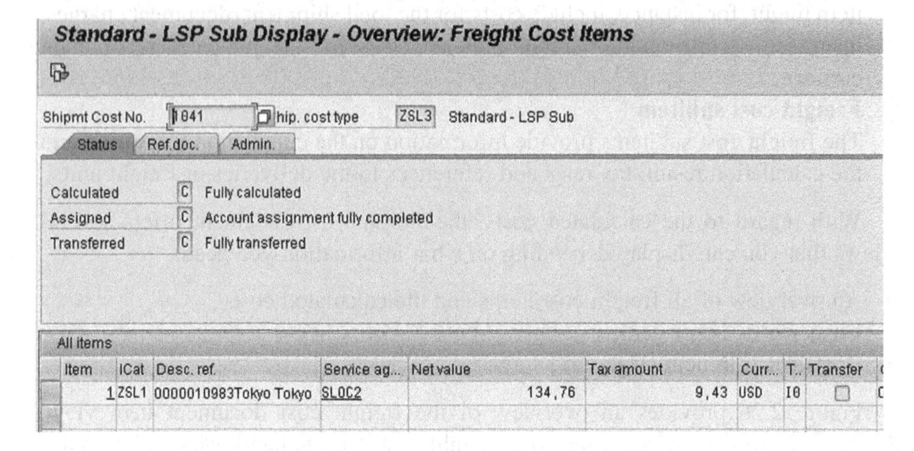

Fig. 2.26 Overview of the freight cost document

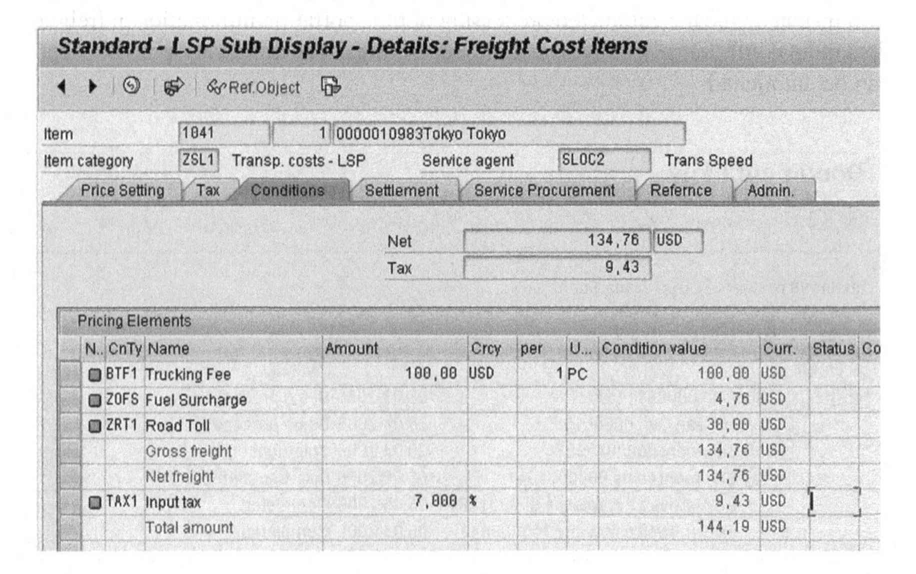

Fig. 2.27 Item and condition detail in the freight cost document

Settlement of costs is conducted with two service providers: a forwarding agent and an insurance company. The insurance costs represent a separate freight cost item with reference to the complete shipment. Two further items were generated for the stages; the costs for the first stage are calculated from the total freight of both shipments (subitems 2.1 and 2.2). The total freight of 10,000 kg is used to determine the scaled value of $0.40/kg.

The calculation, scaled values, applied individual conditions per shipment and other special calculations can be configured using Customizing and master data maintenance. Scales can be created and maintained in multiple dimensions. They can be defined as from-scales, to-scales or with an exact value. Table 2.2 presents an example of a three-dimensional scale table at a price per kilogram with the dimensions "From ZIP code" area and "To ZIP code" area as exact scale values and "Weight" as a to-scale value.

For the calculation procedure and schemes, special forms of freight calculation are also taken into consideration, such as:

- **Shortest main carriage**
  The costs are always calculated such that as short a main carriage as possible (direct carriage) is assumed.
- **Minimum and maximum value for freight cost conditions**
  For example, a freight price of $11.80 per ton with a minimum of $35.
- **Average weight calculation**
  If using the next scale up is cheaper, this value is used (for example, 9 t at $120 = $1,080 is replaced by 10 t at $100 = $1,000).

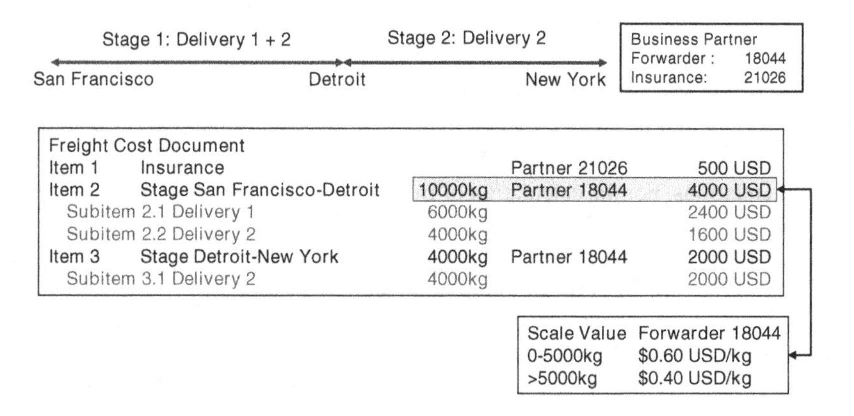

Fig. 2.28 Example for the calculation of freight costs

Table 2.2 Three-dimensional scale table

From ZIP code	To ZIP code	To 100 kg	To 200 kg	To 500 kg
69xxx	80xxx	$1.35	$1.25	$1.08
70xxx	80xxx	$1.14	$1.03	$0.97
20xxx	69xxx	$3.68	$3.32	$3.12

- **Freight comparisons**
  Conditions of a pricing procedure can be summarized in groups. These groups are then each calculated and then compared. The freight calculation can, for example, always select the least expensive group (such as: the freight calculation can either be done according to weight or volume).

Before the actual calculation, freight cost calculation first determines a *pricing procedure* per freight cost item for the calculation, determined in relation to the transportation planning point, service provider, freight cost item and shipping type.

An important criterion for the calculation is the calculation base, that is, the level on which individual freight prices are determined. You can employ the following levels in freight calculation as a calculation base:

- For each shipment stage
- For each shipping unit
- For every shipment
- For every delivery item under consideration of the freight classes (that is, goods type-specific)

Geographic circumstances of a shipment are determined in freight cost calculation by one of two options:

- **Distances**
  You can manually enter the distance into the shipment stage or shipment header, or it is automatically assumed from the leg.
- **Locations and zones**
  A few pieces of information from the address data in the shipment document (such as country, ZIP code, transportation zone), divided into departure and destination locations, can be used for cost determination. Based on this information, a price zone can also be determined, such as one that combines several locations within a ZIP code.

The pricing and billing data can be automatically determined. Using definable scheduling rules, you can set a series of transportation dates that can serve as recommendations. For the calculation of freight costs, you can also have the system automatically determine the respective tax.

In the functional area *Freight invoicing*, work lists are available that can give you an overview of the current operation queue.

The freight cost calculation list is a report that can be used to list freight cost documents for which the calculation has not yet been completed. You have such selection criteria as the status of the calculation, the date of the calculation and the pricing date to limit the selection of freight cost documents.

The list of scales allows you to obtain an overview of the existing scales and their applications. With the freight cost invoicing list (see Fig. 2.29), you have the opportunity to search for freight cost documents that have already been or must soon be settled.

If you as a shipper wish to transfer freight costs that you have paid to a service provider for the transport of goods stemming from a sales order to an ordering party, you have the option of performing *customer freight invoicing*. For this, through special conditions on the invoice of the sales order, the conditions of the freight invoice are accessed, and the individual cost elements can be assumed in the invoice of the purchasing document. This can be done without making any changes, but you can also apply a proportional charge or surcharge. The ordering party then receives an invoice for the freight costs together with the material costs.

List Shipment Costs: Settlement

Choose Save Item Reference Document Purchase Order

ShptCstNo.	SCTy	Sett. date	C	A	T	Ext. ID 1	Ext. ID 2	Created by	Created on	Changed by	Changed on
1000	Z002		C	C	A			MICHALSKY	23.08.2002	MICHALSKY	23.08.2002
1010	ZSL2		C	C	C			PEDERSENJ	13.04.2005	PEDERSENJ	13.04.2005
1011	ZSL1		C	C	C			PEDERSENJ	13.04.2005	MIDTBOELL	15.04.2005
1012	ZSL3		C	C	C			PEDERSENJ	13.04.2005	PEDERSENJ	13.04.2005

ShptCstNo.	Item	ICat	CoCd	C	A	T	Net value	Tax amount	Curr.	Sett. date	Ext. ID 1	Ext. ID 2	Plnt	POrg	PGr	Purch.Doc.	Item	Ser
1000	1	Z002	CPFO	C	C	A	2.075,00	0,00	USD	23.08.2002	APO-gepla..		CPF3	CPF1	001			600
1010	1	ZSL1	SL01	C	C	C	478,40	0,00	USD	12.04.2005	0000001205		SL31	SL31	000	4500016602	1	SL2
1011	1	ZSL1	SL01	C	C	C	184,00	0,00	USD	12.04.2005	0000001206		SL31	SL31	000	4500016605	1	SL1
1012	1	ZSL1	SL01	C	C	C	147,20	0,00	USD	13.04.2005	0000001208		SL31	SL31	000	4500016603	1	SL5

Fig. 2.29 Invoice list for freight costs

2.5 Transportation Planning with SAP APO

As mentioned in the previous section, the ERP transportation solution itself offers no option for optimized transportation planning, since SAP ERP does not include planning tools with an optimization function. In several cases, however, optimized planning is a method with which to improve efficiency and save money.

That is why Supply Chain Management includes the planning and optimization component APO (Advanced Planner and Optimizer). In combination with ERP logistics processes, it offers a few significant advantages for transportation planning:

- **Integration of order processing, global availability check and transportation planning**
  You have the option of optimizing the availability of products in all international divisions of your company as well as the planning of shipments between corporate locations and customers or suppliers in a combined process.
- **Cost-optimized transportation planning**
  The optimization process for shipments offers efficient models and strategies to influence overall transportation costs.

- **Multiple transportation mode planning**
  Optimization can be performed using several modes of transport, that is, you can plan complete transportation chains including transshipment in several distribution centers.

APO Transportation Planning (TP/VS) can be employed for purchasing as well as sales-based processes (see the process flow diagram for the sales side in Fig. 2.7).

2.5.1 Transportation Optimization with SAP APO (TP/VS)

The basic process when working with APO Transportation Planning consists of the following elements:

1. The documents that form the original transport demand (sales orders, purchase orders) are created in the ERP system and subsequently transferred to the APO system.
2. In the event that the *Routing Guide* (dynamic route determination) is used, a planning run is performed during the global availability check before the sales order is saved.
3. *Vehicle Scheduling* (VS) aids you in achieving a consolidation of transport demands and determining an optimal route and delivery sequence and creating the corresponding tours. It considers optimal resource exploitation as well as minimal procedural effort (such as for loading and unloading). Existing solutions can be taken into account for further planning runs and revised based on altered situations.
4. For service provider selection, you have the option of optimizing the allocation of transportation service providers to existing tours according to cost, distribution, quality or quota processing aspects. You can also determine additional shipments, to save money by dispatching two shipments at one time.
5. Collaborative transportation planning with service providers enables you to exchange data relating to expected transportation volume. You can provide service providers with short- and long-term planning data for your transport demands via a collaboration portal, to enable them to prepare enough transport capacity.
6. Shipment tendering serves to communicate planned shipments to service providers and, if necessary, plan for any requests for changes or rejections.
7. Release shipments can be transmitted back to SAP ERP after optimization and service provider allocation. Based on the data, ERP shipment documents and deliveries are then generated.
8. Transportation processing subsequently takes place in ERP Transportation Management.

The Optimizer in APO Transportation Planning is a universal tool for planning comprehensive transport scenarios.

2.5.2 Documents and Transportation Optimization

APO Transportation Planning primarily works with two transaction objects: The *freight unit* and the *shipment document*. Freight units represent transport demands created from sales orders or purchase orders. With the aid of master data objects such as resources, service providers and transportation lanes, freight units are planned in a planning run, consolidated and allocated to shipments. These shipments mainly correspond to tours that service several loading and unloading points and are processed with one resource.

Planning itself can either be performed in batch processing, as a report or as interactive planning with optimization support. Figure 2.30 shows an overview of the *TP/VS planning cockpit*, where you can select various planning views (resource view, shipment view, tabular planning, planning board, etc.).

The *Optimizer* is a software tool integrated into APO Transportation Planning via interfaces. Since the Optimizer was not developed in the SAP programming language ABAP, but rather in C++, it only supports selected operating system platforms (such as Windows and Linux). The Optimizer gets its data and results are provided via the preliminary and subsequent processing of master and movement data in Supply Chain Management.

The Optimizer conducts continual optimization of the transportation scenario under consideration of indicated transport demands, resources, incompatibilities, cost functions and transportation networks, until either a preset optimization duration has been reached or two sequential solutions have not resulted in an improvement in the overall result. As such, the Optimizer can provide a result at a very early stage – even if that result is still suboptimal. The progress of the optimization process can be monitored and checked in the optimization screen (see Fig. 2.31).

After optimization has been performed, the generated shipments can be analyzed and adjusted in the planning cockpit. The system also provides an *overview planning board* to help you evaluate not only the capacity-related but also the

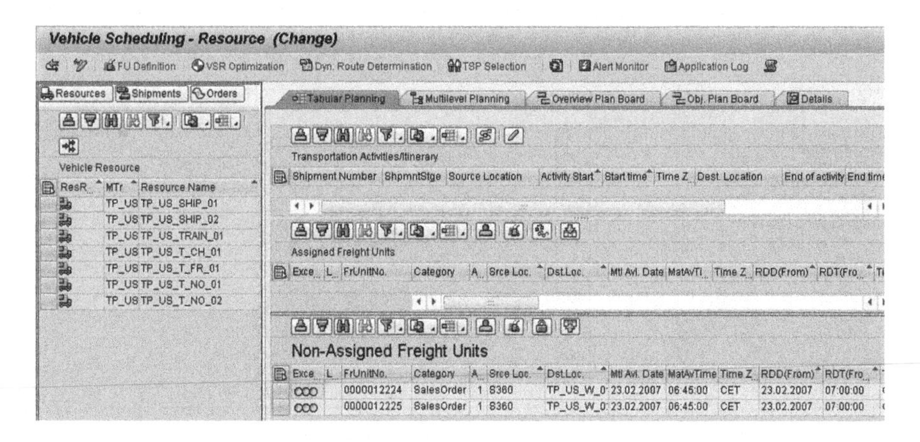

Fig. 2.30 Planning cockpit of SAP APO transportation planning

time-lapse-related aspects. It allows you to display the time lapse of the individual shipments in Gantt diagrams and evaluate them. For this, you have access to an order-based as well as a resource-based time display. Such displays clearly illustrate what resources are available or occupied at what times (see Fig. 2.32).

2.5.3 Scenarios with APO Transportation Planning

Multi-pick and *multi-drop* scenarios involve the optimization of load consolidation with regard to resources. In an overall scenario in which freight is picked up and dropped off at several loading points, transportation costs for resource employment is minimized, that is, the most economical resource selection and load combination is determined.

In order to be able to properly evaluate these networks, you can use the *Supply Chain Cockpit* tool, which provides a graphic display of network components such as plants, customers, suppliers, resources and transportation lanes in map form. Such a network in the Supply Chain Cockpit is shown in Fig. 2.33.

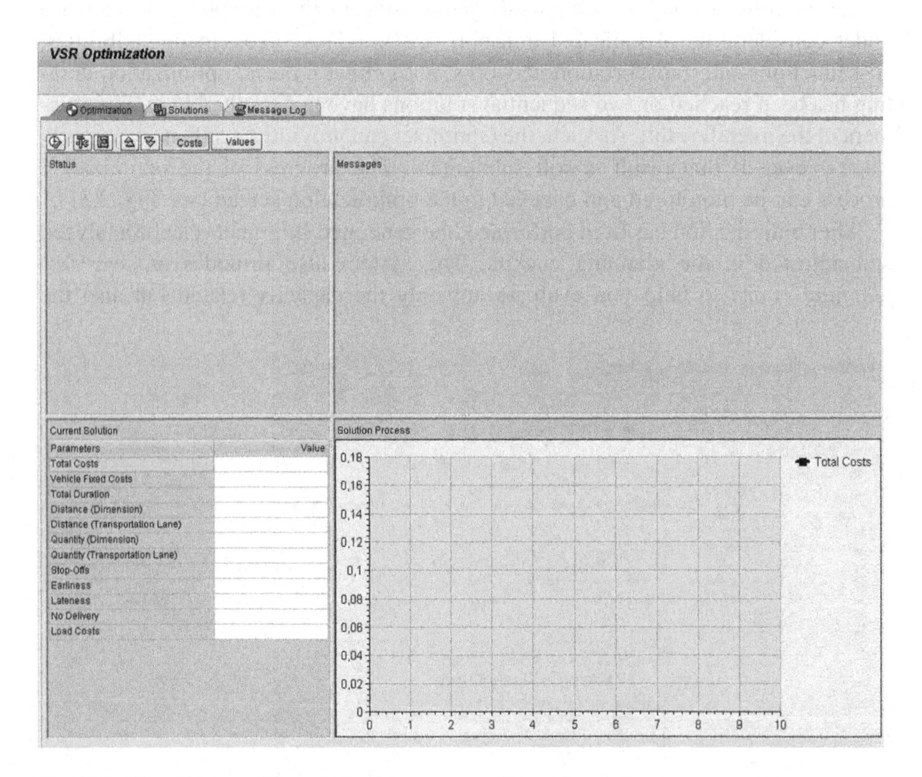

Fig. 2.31 Planning run with optimization progress and result report

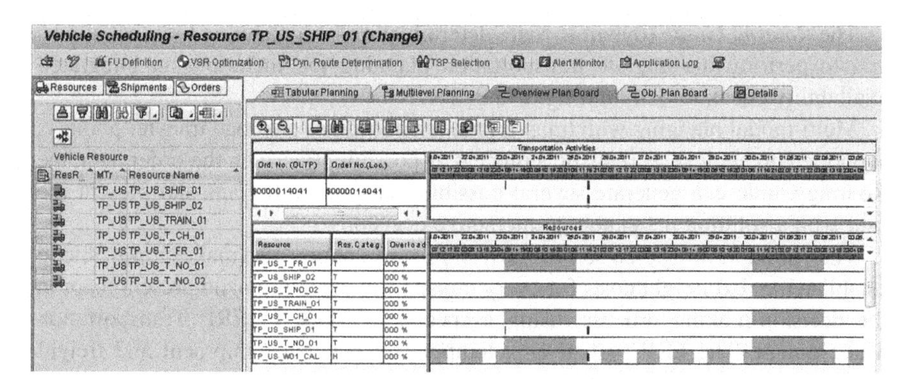

Fig. 2.32 Graphic planning board in SAP APO transportation planning

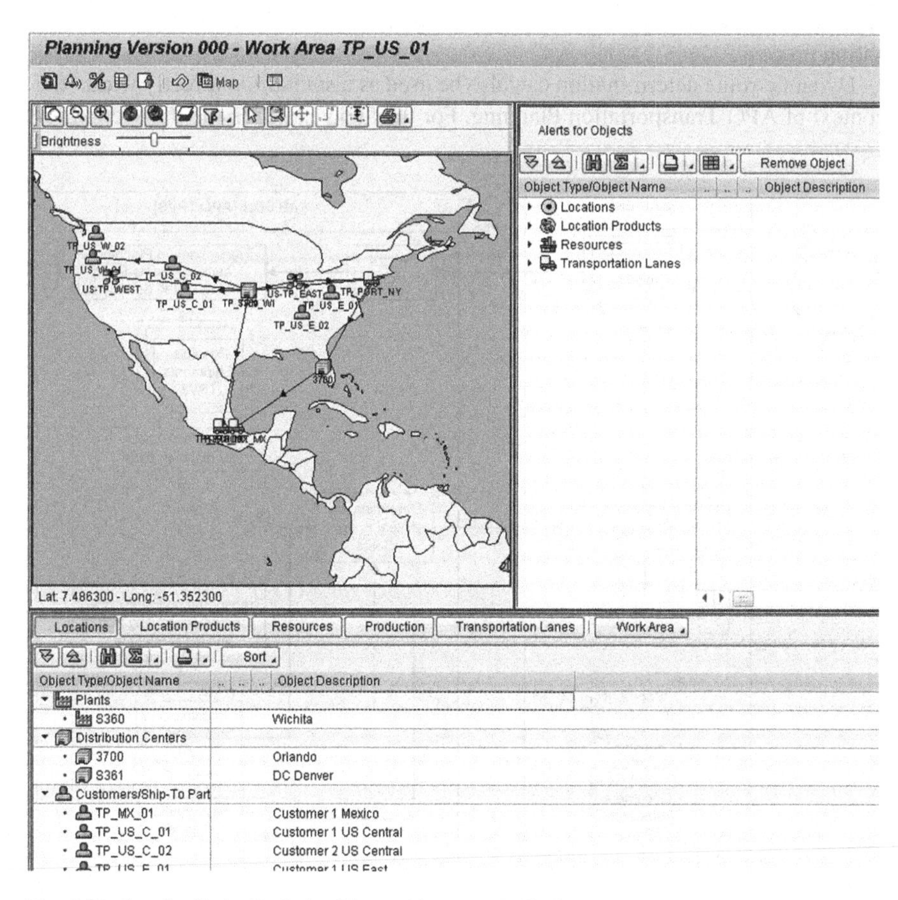

Fig. 2.33 Supply Chain Cockpit with graphic network display

The *Routing Guide* (dynamic route determination) is a planning tool that allows you to perform a complete transportation planning job including an integrated availability check directly during data entry.

Multi-modal planning with transshipment at terminals and load transfer points is supported, to achieve as realistic a process as possible. Based on the order data, the Routing Guide can generate several possible transportation plans and present the user with a solution list enabling interactive decisions.

The option of integrated service provider determination also makes it possible to establish the real freight costs for every transportation solution and present them in the decision-making process, using a special interface to ERP Transportation Management. The ERP system only temporarily generates shipment and freight cost documents to perform the calculation.

After the desired transportation route has been selected and the sales order saved, the shipment documents pertaining to the selected solution are also saved in the APO system. Transportation solutions that are not selected are automatically discarded. Figure 2.34 provides a graphic depiction of the dynamic route determination process.

Dynamic route determination can also be used as a *simulation* directly within the context of APO Transportation Planning. For this, you can open the transaction for

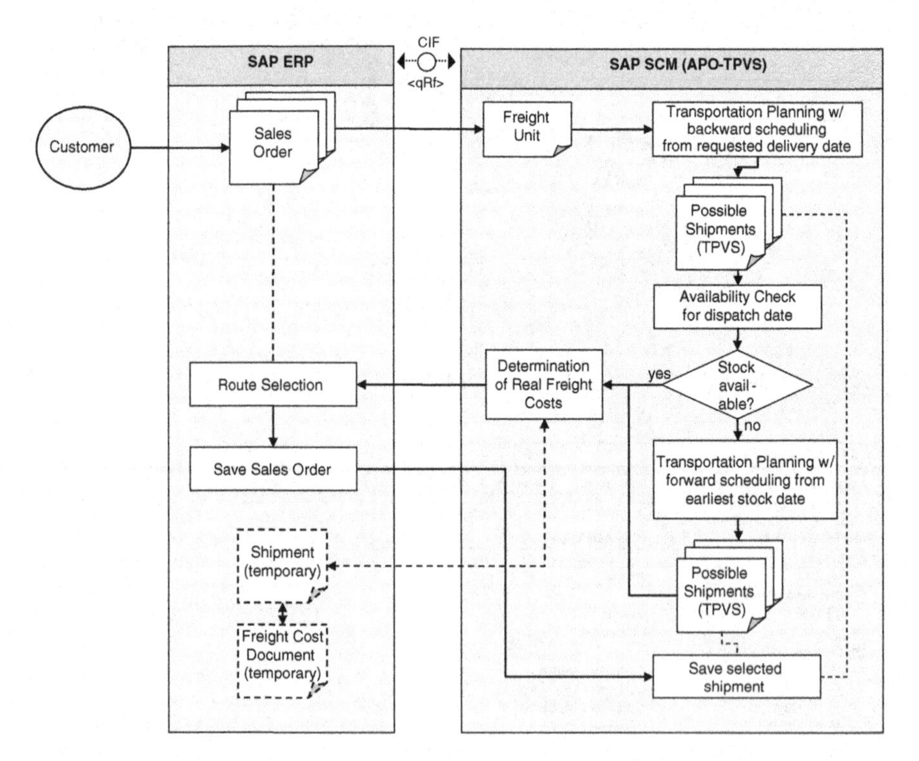

Fig. 2.34 Process in the determination of route recommendations using routing guide (dynamic route determination)

interactive route determination and directly enter the freight unit data. The planning function then immediately presents the desired route recommendation as a result. They are displayed in the transaction as shown in Fig. 2.35.

A *Continuous move* (additional shipment determination) allows you to optimally combine matching individual shipments that you can contract to a single service provider. This enables you to book longer freight routes per service provider and vehicle, often achieving a better freight price. This is especially true if you can offer the service provider return freight from the drop-off location of the first shipment.

Multiple ERP use of APO Transportation Planning. Multiple ERP use of a transportation planning system is a scenario that is necessary from time to time when several organizations of a company use their own ERP systems for order processing but plan transportation with a central planning system. The generated shipments then contain deliveries from several ERP systems but are only mapped in a single ERP system.

Unfortunately, this scenario cannot be executed with ERP Transportation Management and APO Transportation Planning without system modifications and extensions, since the necessary delivery references cannot be consistently mapped in the central ERP transportation system. SAP TM should be used for such scenarios.

2.6 Transportation Management with SAP TM

SAP TM is a software solution with which you can process transportation logistics processes in complex transportation networks. It supports you in basic business processes, ranging from the sales offer and order to planning, subcontracting and

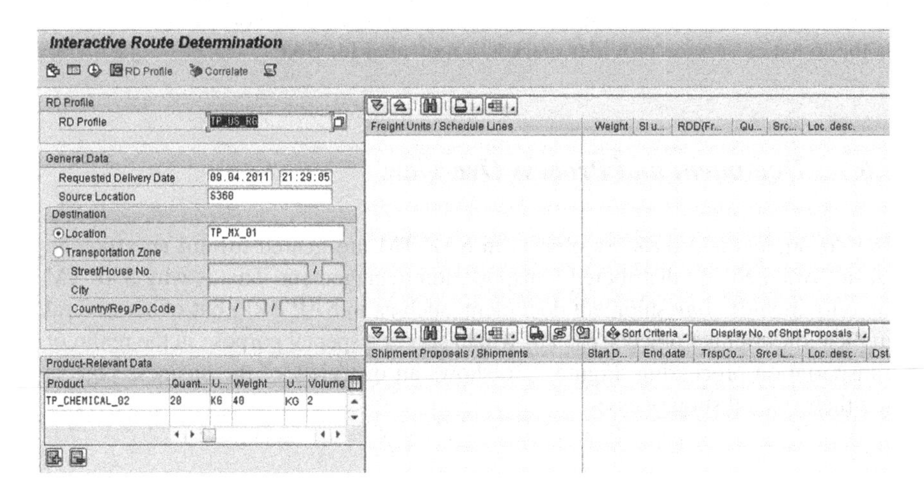

Fig. 2.35 Interactive route determination in SAP APO Transportation Planning

pricing to invoicing and settlement of transportation services. The solution is equipped with algorithms for shipment optimization and route determination

SAP TM is part of the supply chain management solution of SAP and is integrated with SAP ERP for financial settlement. In addition to SAP ERP, SAP Event Management is also employed for various tracking processes.

One significant difference to ERP Transportation Management is a strong emphasis on transportation processing from the service provider view and not only as a partial process of sales and distribution, production or procurement. Transportation is treated as an autonomous process that begins with the tendering procedure and ends with invoicing and settlement. In comparison with SAP ERP, SAP TM offers a considerable amount of additional functions:

- The opportunity to handle transportation processes with or without material master records
- Customer master data is not necessary (however it is helpful for invoicing); supports one-time customers
- Sales and purchase order management for transportation services
- Expanded functions for planning and allocation of complete and partial shipments in complex and also itinerary-supported networks
- Partial processing of transportation chains
- Division of the processing of a transportation order between various participating organizations (import/export view)
- Incoming and outgoing shipments are treated equally, which enables the planning of round trips with drop-offs and pick-ups
- Complex debit- as well as credit-related freight rate calculation and support of internal costs for a company's own fleet
- Calculation of the profitability of orders
- Support of EDI communication
- Complete transportation processing, even in multiple ERP environments

With SAP TM, SAP addresses existing markets, such as the shipper market, for which transportation processing had previously been the realm of SAP ERP, as well as the logistics service provider market, a new area for SAP.

2.6.1 Document and Process Overview

The documents and business objects in SAP TM are geared toward an efficient, service-oriented and distributed transportation management. That is why SAP TM not only includes *one* shipment document, such as in SAP ERP, but rather several application-oriented objects, each of which serves a special purpose in the realm of transportation processing. Figure 2.36 shows an overview of the business objects and their related subprocesses.

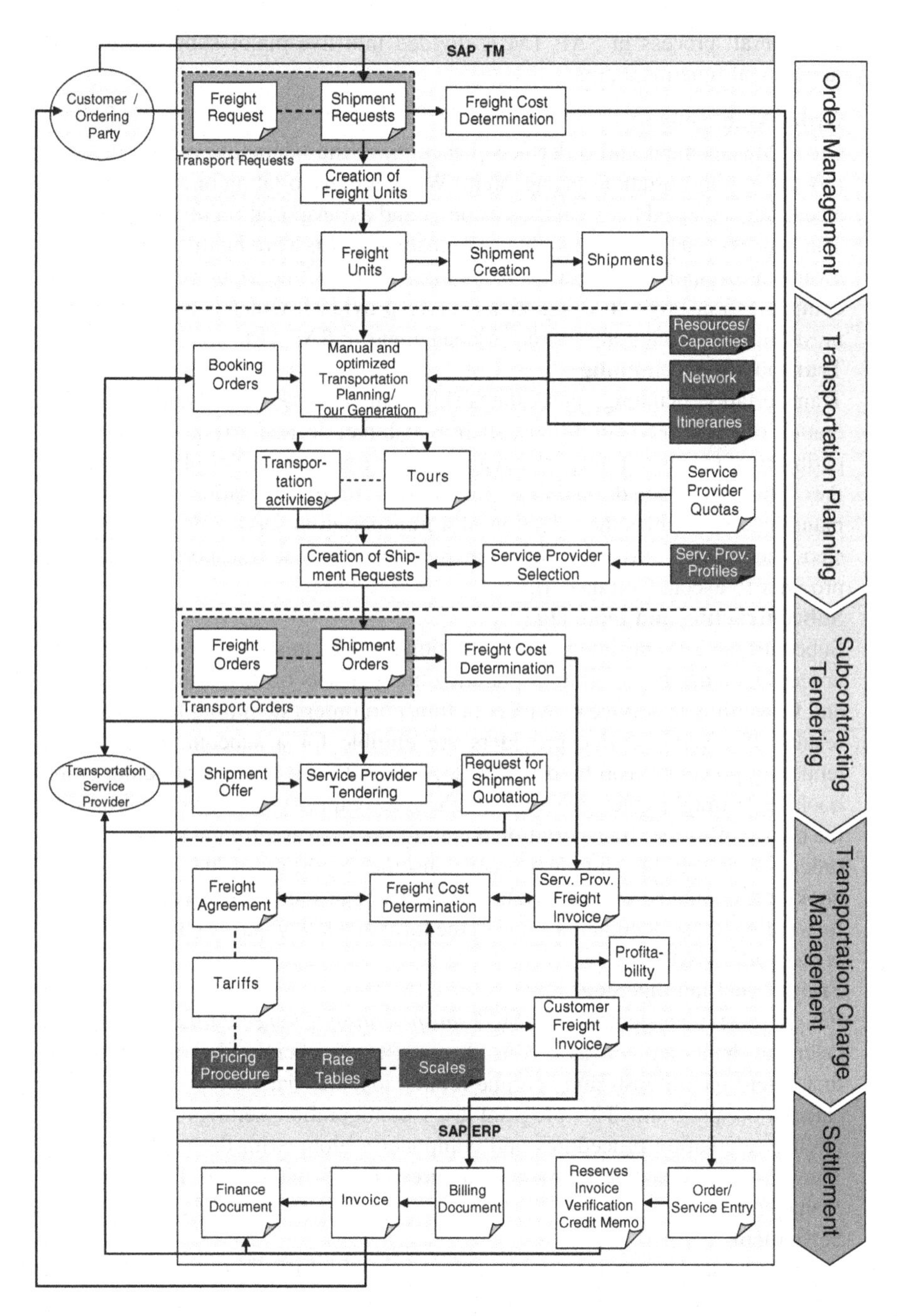

Fig. 2.36 Overview of the SAP TM object model

The overall process in SAP TM is divided into five major subprocesses and organizational functional areas:

- **Order Management**
  Order Management and order acceptance mark the beginning of an operative process in transportation management. When a transportation order is awarded, a contract is created between the customer and a transportation service provider. This process can take place within a company (such as a producing company that awards a transportation contract to its logistics department) as well as between companies (an order to a logistics service provider). Order acceptance is the most important function of Order Management in SAP TM.
- **Transportation planning**
  Transportation planning consolidates shipments into cargo loads under consideration of predefined conditions, such as volume, desired arrival time or compatibility of means of transport and goods to be transported. The planner also has the option of manual planning. In addition to transportation optimization, planning also enables the selection of a transportation service provider. In this step, the system aids in finding the most economical transportation service provider to execute a transport.
- **Subcontracting and tendering**
  Subcontracting is consigning transportation services to a shipper or carrier. The process includes transportation planning and consists of the further processing and forwarding to service providers of transport orders stemming from planning results. If several service providers are eligible for a subcontracting job, a tendering procedure can be performed. A second form of subcontracting is the booking of freight space, wherein freight space capacity is reserved and used in the course of further transportation planning. As soon as a freight space booking order is received, the shipment order consumes a portion of the volume of the booking.Booking as well as freight orders can be printed. In the realm of freight forwarding, these documents come in the form of a *bill of lading* (B/L) or *master air waybill* (MAWB).
- **Freight cost management**
  Freight cost management (TCM, *Transportation Charge Management*) can calculate debits as well as credits. It consists of several subcomponents that first determine all cost- and revenue-related logistics data and provide it to the calculation program. The program uses configuration settings to determine pricing calculation procedures and components along with the relative scales. Thus, even complex installment structures can be mapped in Transportation Management.
- **Settlement**
  Freight cost settlement links freight calculation with ERP Financing on the debit side and ERP Purchasing on the credit side.

Within these components, the system determines to which accounts, cost centers, etc., the freight order is to be booked. It can also perform allocation duties, such as distributing costs among all participating cost centers.

2.6.2 Cross-Divisional Functions

One significant characteristic of SAP TM is the option of distributing a general process among several processing employees. This enables various people in different organizations to perform transportation processing steps.

Two functions in SAP TM are used to do this:

- The *Personal Object Worklist* (Personal Object Worklist, POWL)
- SAP authorization control

The *Personal Object Worklist* is the central element for role-specific user access in SAP TM. Its basic elements are configurable queries and work lists. Depending on functional area, the Personal Object Worklist offers you access to the business objects and their directly subsequent business objects, such as:

- Personal Object Worklist for sales orders
- Personal Object Worklist for shipment requests (transport orders)
- Personal Object Worklist for tours
- Personal Object Worklist for shipment orders
- Personal Object Worklist for customer freight invoice requests (pro-forma customer invoice)

Depending on the business object and area, you have the opportunity to display, create, edit or delete objects.

You can configure POWLs for each user or for a user group through an administrator. This allows every user access to exactly those work lists with the objects that he or she needs upon entering the work area (such as order management). Figure 2.37 shows an example of the Personal Object Worklist for shipment orders.

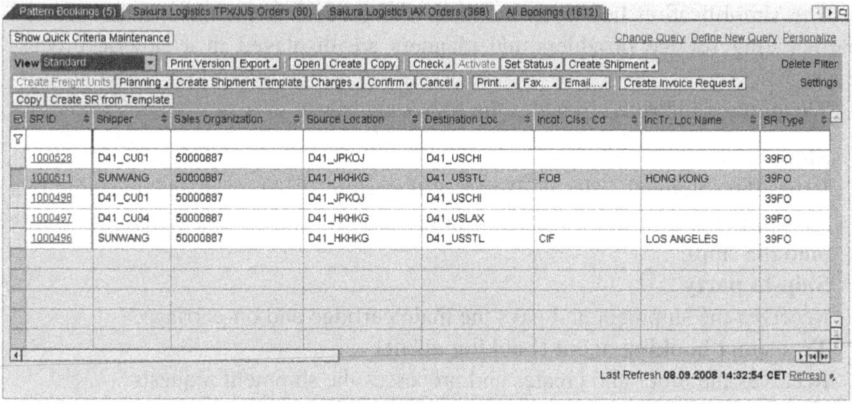

Fig. 2.37 Personal Object Worklist of a booking agent

The Personal Object Worklist can be configured by users in several ways and customized to task-specific needs. You have the following customizing options:

- Order of display and layout of the query view
- Selection of the table display
- Number and order of columns and number of lines in the table display
- Column order and selection in the table display
- Sorting, calculating and filtering in the table display

With the *authorization profile*, you have the opportunity to limit the access of certain users or user groups to business objects or those with certain semantic content. Using the respective settings, for instance, you can configure the following authorizations:

- A booking agent in a call center in Hamburg only has full access to sea freight orders dispatched from Germany. He only has reading access to other orders if they are processed via Germany.
- An import employee in a Singapore office only has access to freight orders that arrive in Singapore via air freight and on- and preliminary carriages within Singapore by truck.

2.6.3 Business Objects and Functions Illustrated in a Sample Process

We would like to highlight the business objects and detail functions of SAP TM with the aid of a sample process. This process, divided among several employees, involves full container processing for sea freight (*Full Container Load*, FCL), representing a standard process of a logistics service provider. A simplified process flow diagram can be seen in Fig. 2.38.

The simplification lies in the summary of process steps and the fact that all participating service providers and planners are displayed in a generic process track.

The process indicated includes the following roles:

- **Shipper**
  Issues the shipment order FOB port of departure (*Free on Board*, meaning that the shipper pays the pre-carriage and terminal fees up to and including loading onto the ship).
- **Ship-to party**
  Receives the shipment and pays the main carriage and on-carriage.
- **Transport booking agent (booking agent)**
  Receives the order and creates and processes the shipment requests.
- **Transportation planners (import/export)**

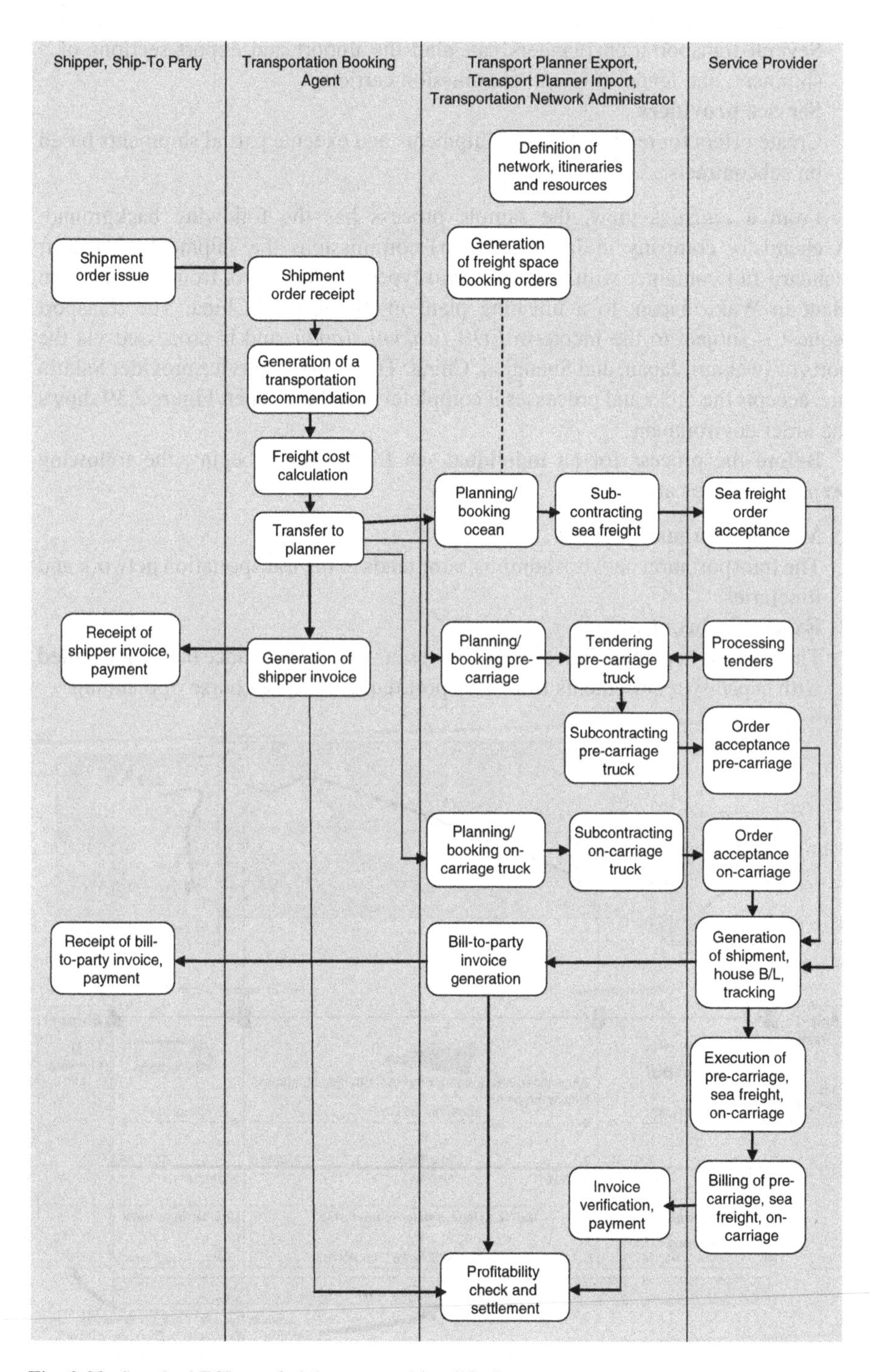

Fig. 2.38 Standard FCL sea freight process (simplified)

Several transportation planners can plan the import and export sections of a shipment, and tender it to and commission carriers.

- **Service providers**
  Create offers for tendered partial shipments and execute partial shipments based on subcontracts.

From a logistics view, the sample process has the following background: A chemistry company in Japan (shipper) commissions the shipment of a 20-ft standard full container with barrels of two types of elastomers from a production plant in Waki, Japan, to a finishing plant in Zhongshan, China. The transport request is subject to the Incoterm *FOB Iwakuni, Japan*, and is processed via the ports of Iwakuni, Japan, and Shanghai, China. The logistics service provider Sakura Inc. accepts the order and processes it completely for the shipper. Figure 2.39 shows the order environment.

Before the process for an individual sea freight order begins, the following preparatory steps are taken:

1. **Master data maintenance**
   The transportation network administrator tends to the transportation network and itineraries.
2. **Execute subcontracting**
   The sea freight planner books capacity on sea vessels in advance that will be filled with capacity requirements from transport requests in the course of planning.

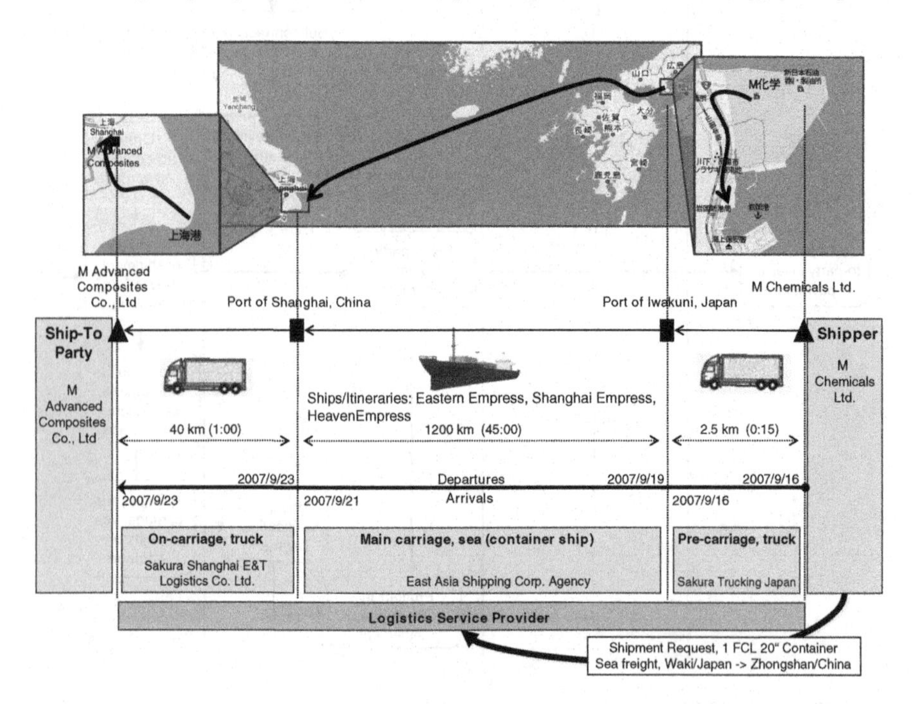

Fig. 2.39 Logistics view of the sample process

The main steps in transportation processing are listed below (see the total object model Fig. 2.36):

1. **Order management – Shipper**
   The shipper commissions transportation via telephone and would also like to receive a transportation proposal and pricing information.
2. **Order management – Transport booking agent**
   The transport booking agent enters the order in the system (business object *shipment request*). Then he activates the shipment request, creates three route recommendations using the transportation proposal function, clarifies the desired transportation process with the shipper, and calculates the freight prices for shipper and recipient. This activation triggers the system to automatically generate transport demands (freight units).
3. **Transportation planning – Main carriage**
   The planners are automatically informed of the new freight units via their Personal Object Worklists. First, the sea freight planner books a container on the ship requested by the customer, consuming a portion of the previously reserved capacity (business object *booking order*) in the process. At the appropriate time, the sea freight planner generates a subcontract for the sea vessel (business object *shipment request*), which serves as the *master B/L*. Here, of course, consolidation with other shipments may also take place, but this is not the case for our sample process.
4. **Transportation planning – Pre-carriage/on-carriage**
   After the sea freight booking, the import and export planners can independently plan the pre- and on-carriage shipments in their work list and generate subcontracts for them (these are also business objects of the type *shipment request*).
5. **Freight cost management and settlement**
   The transport booking agent can already create and send the invoice to the shipper (*FOB prepaid*, from the pick-up location to on board the ship). For this, he generates a pro-forma invoice (business object *customer freight invoice request*) and forwards it to SAP ERP FI/CO.
6. **Subcontracting and tendering – Pre-carriage**
   The pre-carriage planner can use the service provider selection and tendering functions to submit inquiries to several service providers regarding price quotations and pre-carriage processing.
7. **Quotation – Pre-carriage**
   The contacted service providers can submit quotations for the processing of the pre-carriage transport via the collaboration portal.
8. **Service provider selection – Pre-carriage**
   After the tendering period has expired, the pre-carriage planner can select a service provider and grant the transport job by forwarding the pre-carriage shipment request.
9. **Subcontracting and tendering – On-carriage**
   Like the pre-carriage procedure, the import planner can commission on-carriage with the aid of the on-carriage shipment request.

10. **Generation of a house bill of lading**
    Before the process is executed, the export planner creates the shipment, and can generate the house bill of lading from it as well as initiate shipment tracking.
11. **Freight cost management and invoicing the recipient**
    At any time after cost calculation in the shipment request, the invoice for the recipient can also be initiated. Here, too, a customer freight invoice request is generated and sent to SAP ERP FI/CO.
12. **Freight cost management and payment of service provider invoice**
    The supplier freight invoice request is generated from the shipment orders for pre-carriage, main carriage and on-carriage and also forwarded to SAP ERP Materials Management. This makes them available for verification of incoming carrier invoices.
13. **Profitability calculation**
    After receipt of the supplier invoices and, if applicable, adjustment of invoice amounts, the shipment request can be subjected to a profitability check. Thereafter, the shipment request is marked as *completed*.

2.6.4 Order Management

Order Management provides functions necessary for the acceptance of a shipment order. A variety of business objects are available for this. Table 2.3 provides an overview of the application areas of order business objects.

Table 2.3 Order business objects in SAP TM

Business object	Application
Shipment request (SE)	Agreement between a transportation service provider and an ordering party with relation to the shipment of goods or transportation equipment from a supplying location to a receiving partner or location in accordance with stipulated conditions
Freight request (FA)	Request from an ordering party to transport goods from one or more supplying partners or locations to one or more receiving partners or locations. A freight request is a combination of shipment requests
Quotation	Offer from a transportation service provider (supplier) to an ordering party (customer) for the transportation of goods at desired conditions
Template for shipment request	Partially filled-out shipment request that can be used as a master for regularly recurring, similar shipment requests
Shipment (SN)	Contractual document in logistics that the transportation service provider sends to an ordering party. It contains information on goods that are transported together in one or more means of transport throughout the entire transportation chain or during the main carriage
Freight unit (FE)	Combination of goods that are transported together throughout the entire transportation chain. A freight unit can include transportation restrictions for transportation planning

The shipment request is the central order business object with which all major verification and processing steps in order management are performed. A shipment request can be created as a shipment order in SAP TM either through electronic transmission (EDI) or manual entry.

Manual creation can occur in the system in several ways:

- A shipment request is newly created and generated in SAP TM with the order data provided by the ordering party.
- An order is created from which a shipment request is generated through copying and, if necessary, editing.
- For frequently recurring, similar order processes, a template can be created for a shipment request. It includes the recurring data (such as shipper, ship-to party, goods description and transportation conditions). Through copying and editing, a new shipment request is created from the template.
- An existing shipment request can be used as a master for a new shipment request.

Figure 2.40 diagrams the possible ways of generating a shipment request.

Role and type codes in SAP Transportation Management. In SAP TM, data fields with role codes and type codes are used in many business object nodes, in order to map categories and characteristics of data. This enables you to define as many characteristics of certain data fields as you like. In contrast to explicit data storage (in which, for example, data fields are defined in the business object

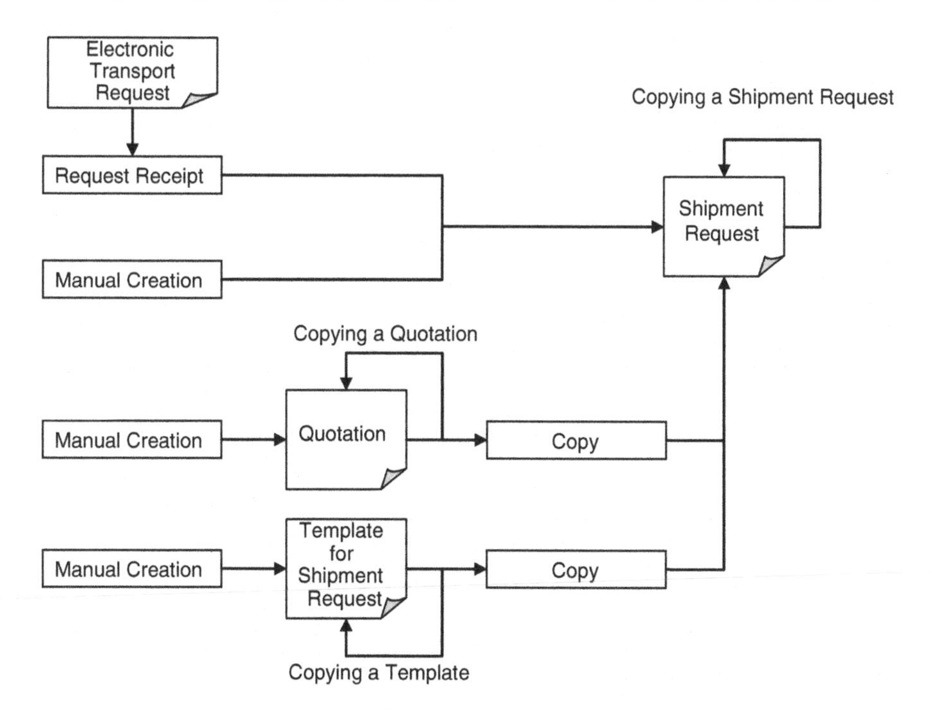

Fig. 2.40 Alternatives for the generation of shipment requests

for business partners like *Shipper* and *Ship-to Party*), in role code-based definition, there is a table with the columns *Role* and *Business Partner*. This lets you enter information for the roles Shipper and Ship-to Party and the respective business partners in addition to as many other user-defined business partners as you wish (such as *Packing Service Provider*, *Customs Agent*). You are not able to indicate these in the explicit definition, since the respective fields are missing.

In our sample process, the shipper grants the transportation order via telephone. The employee processing the transportation order accepts the order and adds a new shipment request using the *Create* function taken from his personal work list for shipment requests.

In the shipment request template, the shipper enters the shipment request type " Full container sea freight house-to-house". Inferable information, such as the sales organization, can be automatically set as a personal default value of the transportation order processor (user settings for the corresponding data fields).

Transportation order for an FCL sea freight shipment. Figure 2.41 shows the basic view of the shipment request screen in which you can enter the most important data on a single screen. The Details button switches to a more detailed display, which enables you to access all available data areas.

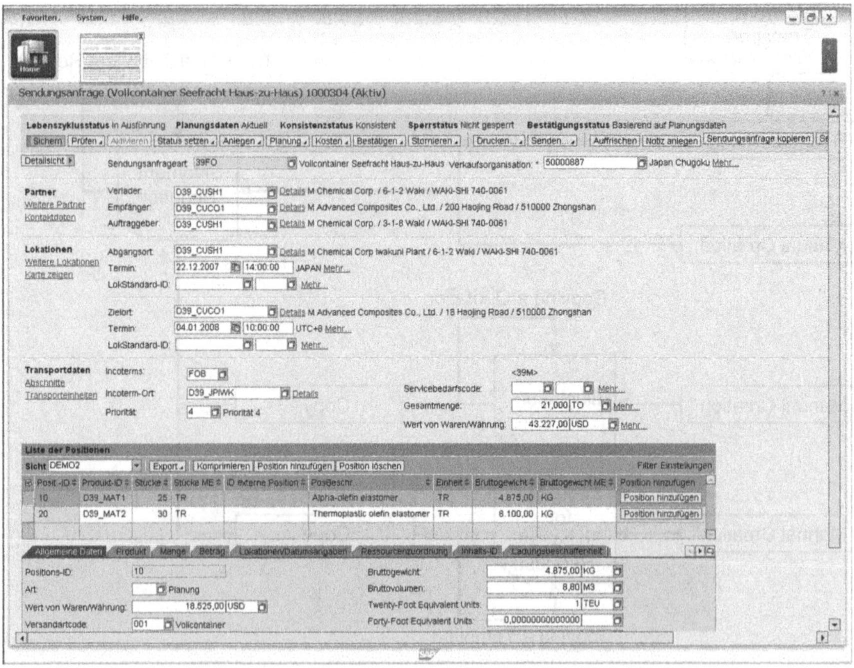

Fig. 2.41 Entry of a shipment request in SAP TM (basic view)

Now we will look at the individual data areas for shipment requests that are filled out by a transportation order processor. The most important data areas are:

- **Shipment request header data and transportation data**
  For categorization of a shipment request in various order types (such as sea freight, air freight and general truck cargo) and allocation to organizational units (like a sales organization or sales office). The header data also includes status information regarding the shipment request that applies to the entire order. The transportation data includes Incoterms, load composition, service definitions and total dimensions and values.
- **Business partners**
  Business partner data includes the obligatory information about the shipper, ship-to party and ordering party, as well as other parties involved in the shipping processes. They may include a predefined service provider (for instance, one a customer has requested), agent or bill-to party. Using the role concept, you can assign as many business partners as you wish.
- **Item data**
  Used to define details of the goods to be shipped. Here, you can indicate the type of goods, goods description, Material numbers, selection numbers, etc. In further subnodes, you can enter the departure and arrival location and dates for each item. You can also enter any necessary measurements (weights, volumes, count, dimension), values, and information regarding dangerous goods, customs and packaging. All fields, with the exception of the goods description and amount fields, are optional.
- **Packaging**
  In the packaging node, you can enter information on the individual packages necessary for the shipment order. A package is defined as shipping material (such as a pallet, cardboard box or pallet cage) with reference to the shipping items or parts thereof. A shipping item with 20 household appliances can, for instance, be packed on five pallets.
- **Resources**
  Resources are subdivided into transportation unit and vehicle resources. You can either use these business object nodes to define resources provided by the customer (such as an ordering party who picks up goods that have already been packed in a container), or you can indicate special resource requirements of the ordering party.
- **Shipment stages**
  In the shipment stages, the individual partial shipment segments are defined. The stages can be multimodal, and can be used for the calculation of distance and mode-dependent transportation costs.
- **Memos**
  Within a shipment request, any number of language-dependent memos can be entered. Notes are categorized free texts (such as of the category *Shipping Memo*) that can be used to forward information in the process chain as well as for printout and communication purposes.
- **Dossier**
  The dossier offers the opportunity to create document references and information on necessary documents and data attachments (such as scanned

documents related to the shipment request) and make them available to people to edit them.

- **Costs and payment methods**
  These subnodes enable the recording of costs and invoice information for shipment requests. Here, the calculated costs are saved for which the shipper or ship-to party will later be invoiced. In addition, information regarding the feasibility of the cost calculation is stored here, and can be accessed to analyze the calculated prices.
- **Official requirements**
  These can be stored under the nodes for official requirements that are relevant for performance of export and dangerous goods checks.

When the transportation order processor enters the shipper and ship-to parties, the ordering party data and the departure and destination location are automatically filled in. Afterward, the processing employee can enter the desired departure and delivery dates. He enters the Incoterms (FOB Iwakuni) into the transportation data, as well as the load characteristics (chemicals, flammable), priority, shipping type code and service requirement code. The total amounts and values are automatically calculated.

The materials loaded into the full container are entered in the shipment request items. There are 25 and 30 barrels with various elastomers. In addition to the gross weight and volume, the gross and net weights and volumes of each shipping unit are recorded. The total value of each item is also entered, as is the value per shipping unit. This data is used later to generate a packing list and pro-forma invoice for customs. Figure 2.42 shows an overview of the amounts and values of a shipment request item.

The container is entered in the shipment request as a transportation unit resource. The individual container numbers are entered in the Registration Number field. Depending on the category of transportation unit resource, a corresponding check of the registration number takes place. The previously entered items can then be allocated to containers to enable labeling of packages.

When a shipment request is created electronically (such as via EDI) or manually, it is initially inactive. This state is part of its *life cycle* and characterizes the initial shipment order of a customer – the customer's wish, so to speak. Through its activation, which can be done by an administrative employee or through program control, a working copy of the shipment order is generated that a planner can use for further processing. This ensures that the customer's wish is preserved in the system in its original form and can be referenced as needed. The various life cycle statuses of the shipment request include *In Planning*, *Planned*, *Confirmed*, *Ready for Execution*, *In Execution*, *Executed*, *Completed* and *Canceled*.

Following activation, the booking processor can have the system issue a transportation proposal (see Fig. 2.43).

Depending on the configuration, SAP TM generates one or more recommendations on how the ordered shipment can be transported, taking into account dates, shipment characteristics, available transportation capacities and routes, and transportation costs.

Transportation proposal for an FCL sea freight shipment. Figure 2.43 shows a transportation proposal for our sample shipment from Japan to China. Three possibilities with varying costs were determined. The most economical is displayed in detail with the individual stages. The desired recommendation can be selected from the presented transportation proposals and stored in the transportation stages of the shipment request.

When the transportation proposal function is accessed, SAP TM automatically generates *freight units* according to defined rules. A freight unit is a shipping volume set by a shipper that is moved collectively through the transportation chain and forms the basis for further transportation planning and optimization. The rules used to create the freight unit define amount restrictions and the granularity of the constructed freight units, such as:

List of Items

View DEMO2 | Export | Collapse | Add Item | Delete Item

Item ID	Product ID	Pcs	Pcs UOM	External Item ID	Itm.Descr.
10	D39_MAT1	25	DR		Alpha-olefin elastomer
20	D39_MAT2	30	DR		Thermoplastic olefin elasto

General Data | Product | Quantity | Amount | Locations / Dates | Resource Assign

View DEMO2 | Export | Add Quantity | Delete Quantity | Filter Settings

Qty TC	Quantity	UoM	Type Code
UNIT_VOL	0,35	M3	Unit Volume
GROSS_VOL	8,80	M3	Gross Volume
GROSS_WT	4.875,00	KG	Gross Weight
PIECES	25	DR	Number of Pieces
UNIT_NWGT	180,00	KG	Unit Net Weight
UNIT_GWGT	195,00	KG	Unit Gross Weight
HEIGHT	2.350	MM	Height
LENGTH	5.895	MM	Length
TEU	1	TEU	20' Equivalent Units
WIDTH	2.392	MM	Breadth

Fig. 2.42 Shipment request items with amounts and values

- Freight unit per container, pallet or other packaging
- Freight unit per shipment or shipment item
- Freight unit per 100 kg of an item
- Freight unit per 1 cubic meter of a shipment order

Figure 2.44 shows the principle of freight unit generation using a shipment request with two items.

Important attributes of the freight unit include the shipper, ship-to party, items and their amounts and material characteristics, and transportation restrictions that can be used to regulate the transit route of a freight unit. Transportation restrictions provide load transfer points and periods for a freight unit that will subsequently be taken into account by Transportation Planning. In addition to the shipment stages of the shipment request, the transportation proposal function also generates transportation restrictions for a particular freight unit.

In our example, the freight unit is constructed on the basis of containers, meaning that the shipment request leads to a freight unit corresponding to a 20-ft container with its contents. Based on the transportation proposal, the ports of

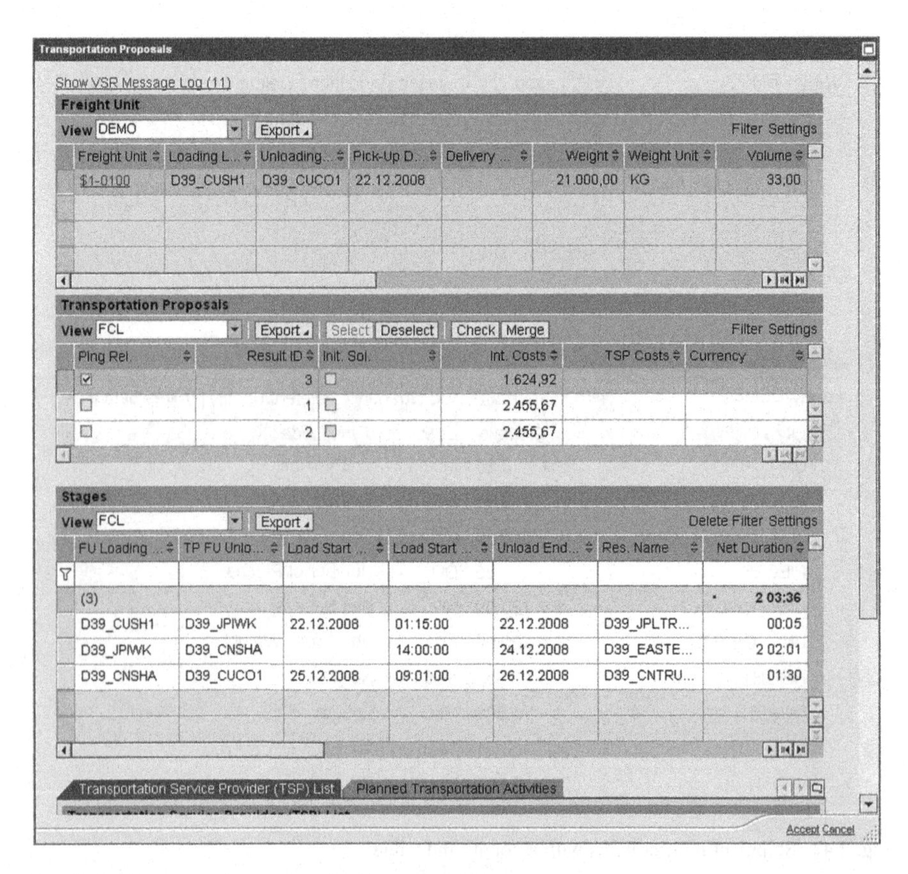

Fig. 2.43 Transportation proposal with route details

Iwakuni in Japan and Shanghai in China are set as load transfer points. This instructs Transportation Planning to avoid routing the freight through Hong Kong, for instance.

At a later time, the planner can, if necessary, create a shipment from the freight units of a shipper through allocation or consolidation that can serve as a basis for the generation of the bill of lading (see Fig. 2.44).

Based on the data in the shipment request (header, items, stages, resources), a booking employee can execute a freight sales price calculation in the shipment request, in order to confirm the freight price for the customer. Determination of freight costs is done using the SAP TM component *Transportation Charge Management*, which is accessed from the shipment request (see also the overview of the TM object model in Fig. 2.36). In Transportation Charge Management, the list of freight price components is generated with the aid of existing freight contracts, tariffs, price calculation schemes and installment structures, and the individual prices are calculated. The result is then stored in the freight cost details of the shipment request. In Fig. 2.45, you can see an overview of the price components that are billed to the ship-to party in our sample process.

To print documents (see the example in Fig. 2.46) in SAP TM, the *Post Processing Framework* (PPF) is used, which represents a further development of the SAP ERP printout control. In the PPF, you can define flexible printing times and requirements, print routes, used forms and their contents for SAP TM standard documents as well as your own documents. Document definition itself is based on Adobe Interactive Forms, with which you can create and edit documents in a graphical editor.

Print document generation using the example of a pro-forma customs invoice. Figure 2.46 shows a sample document (*Customs Invoice*) based on a shipment request with the allocated data segments from which the field values

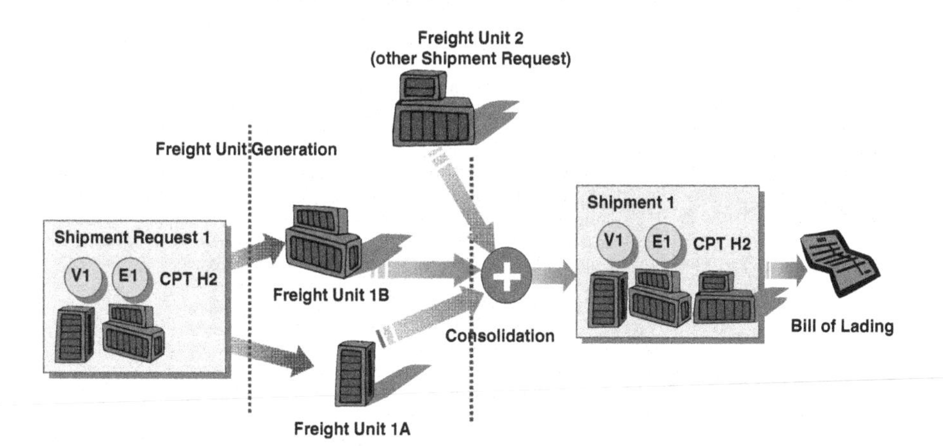

Fig. 2.44 Creation of freight units and shipments

are filled in. A variety of standard document types are available in SAP TM, such as bills of lading based on various norms, shipment request confirmations or transportation instructions.

2.6.5 *Transportation Planning and Optimization*

Transportation planning and optimization operate on several levels, with the initial transport demand being more concretely transformed into transport requests with each level. Figure 2.47 provides an overview of the planning levels.

1. Shipment requests represent the initial transport demand as defined in the request overview, which generally only indicates departure and destination locations (A, B, C, D).
2. After the transportation proposal, a detailed itinerary is determined as a binding requirement for the transportation restrictions of the generated freight units. For cost calculation purposes, respective shipment stages are generated in the shipment request. Additional load transfer points (N, R) are added, which will be taken into account in subsequent optimization.
3. Transportation planning and optimization considers the individual transport demands in the stages (A-N, B-N, N-R, R-D and R-C) and, under consideration of available capacity, schedules, freight costs, transshipment times, service providers and incompatibilities (such as refrigerated goods in a nonrefrigerated container), generates a cost-optimized solution. The planning run results in tours

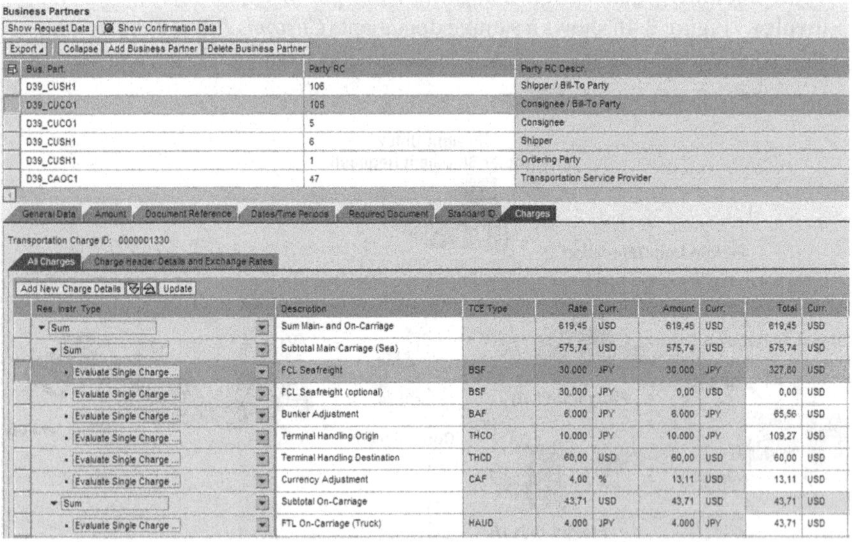

Business Partners

Show Request Data | Show Confirmation Data

Export | Collapse | Add Business Partner | Delete Business Partner

Bus. Part.	Party RC	Party RC Descr.
D39_CUSH1	106	Shipper / Bill-To Party
D39_CUCO1	105	Consignee / Bill-To Party
D39_CUCO1	5	Consignee
D39_CUSH1	6	Shipper
D39_CUSH1	1	Ordering Party
D39_CAOC1	47	Transportation Service Provider

General Data | Amount | Document Reference | Dates/Time Periods | Required Document | Standard ID | Charges

Transportation Charge ID: 0000001330

All Charges | Charge Header Details and Exchange Rates

Add New Charge Details | Update

Res. Instr. Type	Description	TCE Type	Rate	Curr.	Amount	Curr.	Total	Curr.
Sum	Sum Main- and On-Carriage		619,45	USD	619,45	USD	619,45	USD
Sum	Subtotal Main Carriage (Sea)		575,74	USD	575,74	USD	575,74	USD
Evaluate Single Charge ...	FCL Seafreight	BSF	30.000	JPY	30.000	JPY	327,80	USD
Evaluate Single Charge ...	FCL Seafreight (optional)	BSF	30.000	JPY	0,00	USD	0,00	USD
Evaluate Single Charge ...	Bunker Adjustment	BAF	6.000	JPY	6.000	JPY	65,56	USD
Evaluate Single Charge ...	Terminal Handling Origin	THCO	10.000	JPY	10.000	JPY	109,27	USD
Evaluate Single Charge ...	Terminal Handling Destination	THCD	60,00	USD	60,00	USD	60,00	USD
Evaluate Single Charge ...	Currency Adjustment	CAF	4,00	%	13,11	USD	13,11	USD
Sum	Subtotal On-Carriage		43,71	USD	43,71	USD	43,71	USD
Evaluate Single Charge ...	FTL On-Carriage (Truck)	HAUD	4.000	JPY	4.000	JPY	43,71	USD

Fig. 2.45 Freight cost calculation (here for the ship-to party with the Incoterm FOB)

with allocated means of transport and service providers with whom the freight units are moved between the load transfer points.

4. Generation of the shipment order creates shipment orders for each tour and their freight units that can be forwarded to the service provider. In the example presented in Fig. 2.47, three shipment orders were generated:
   - Pickup of goods from A and B with a collective truck tour and delivery to Port N
   - Shipping of the goods in a collective sea transport from N to R (master B/L)
   - Collective truck transport from destination port R to goods recipients C and D

In the FCL business process example, the following planning steps are performed:

1. The shipment booking employee has already issued a transportation proposal upon entering the shipment request. This generates shipment stages in the shipment request as well as transportation restrictions in the freight unit. In our example, they are the segments A-N (Waki to Iwakuni port), B-N (Hiroshima to Iwakuni port), N-R (Iwakuni port to Shanghai port), R-D (Shanghai port to Zhongshan) and R-C (Shanghai port to Shanghai).
2. The individually generated freight unit transportation restrictions now automatically appear in the Personal Object Worklist of the import/export planner.

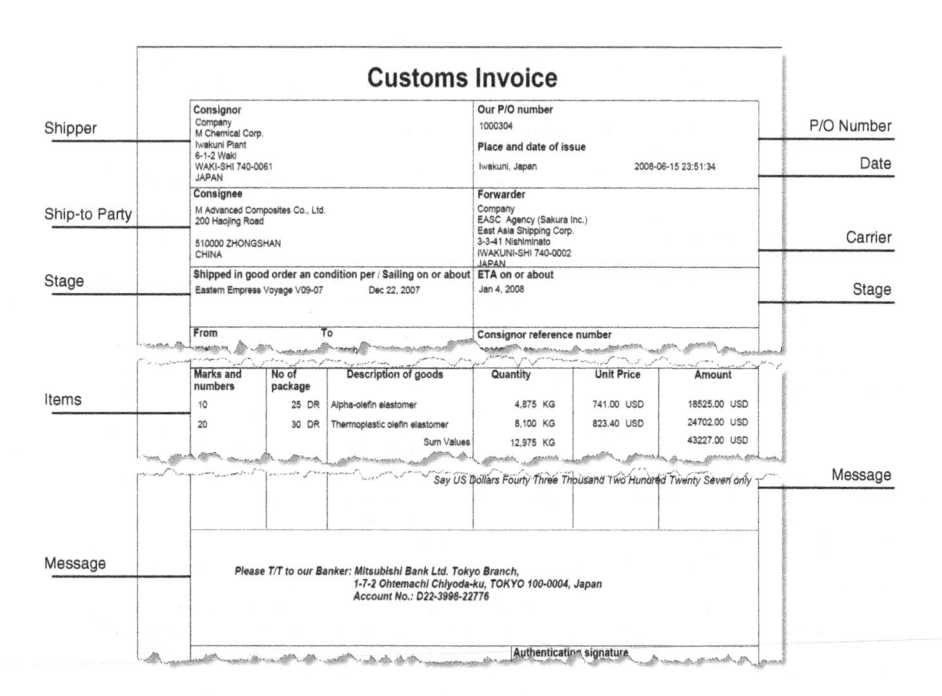

Customs Invoice

Consignor	Our P/O number
Company M Chemical Corp. Iwakuni Plant 6-1-2 Waki WAKI-SHI 740-0061 JAPAN	1000304 **Place and date of issue** Iwakuni, Japan 2008-06-15 23:51:34
Consignee	**Forwarder**
M Advanced Composites Co., Ltd. 200 Haojing Road 510000 ZHONGSHAN CHINA	Company EASC Agency (Sakura Inc.) East Asia Shipping Corp. 3-3-41 Nishiminato IWAKUNI-SHI 740-0002 JAPAN
Shipped in good order an condition per / Sailing on or about	**ETA on or about**
Eastern Empress Voyage V09-07 Dec 22, 2007	Jan 4, 2008
From **To**	**Consignor reference number**

Marks and numbers	No of package	Description of goods	Quantity	Unit Price	Amount
10	25 DR	Alpha-olefin elastomer	4,875 KG	741.00 USD	18525.00 USD
20	30 DR	Thermoplastic olefin elastomer	8,100 KG	823.40 USD	24702.00 USD
		Sum Values	12,975 KG		43227.00 USD

Say US Dollars Fourty Three Thousand Two Hundred Twenty Seven only

Please T/T to our Banker: Mitsubishi Bank Ltd. Tokyo Branch,
1-7-2 Ohtemachi Chiyoda-ku, TOKYO 100-0004, Japan
Account No.: D22-3998-22776

Authenticating signature

Fig. 2.46 Example of a print document having a data origin (Customs Invoice, not part of the SAP TM standard collection)

3. Initially, the sea freight export employee in Japan plans the two transportation restrictions of the freight units intended for sea transport based on booking orders or itinerary-based resources, thus determining the sea route of the freight.
4. Subsequently, the land freight export planner in Japan can plan the road transport from the shippers to the port in Iwakuni. Depending on the resource situation, costs and time restrictions, load consolidation may take place (as shown in Fig. 2.47, both freights are picked up sequentially by the same vehicle on a single tour and brought to the port), or separate tours are planned. Once the truck tour is complete, the planner can create the shipment order.
5. When the deadline for accepting sea shipments has arrived, the sea freight planner can create and release the shipment order for the sea freight service provider.
6. As soon as it is decided, based on the deadline, which freight units are to be transported on the ship and unloaded in Shanghai, the import planner in the Chinese organization of the logistics service provider can perform planning for transportation restrictions in China. As in the export portion, here, too, circumstances may require consolidated road transport. The planned tours are then used as the basis for generating shipment orders for the on-carriages.

User entry in the Transportation Planning realm is done by selecting two profiles that will largely control the course of planning:

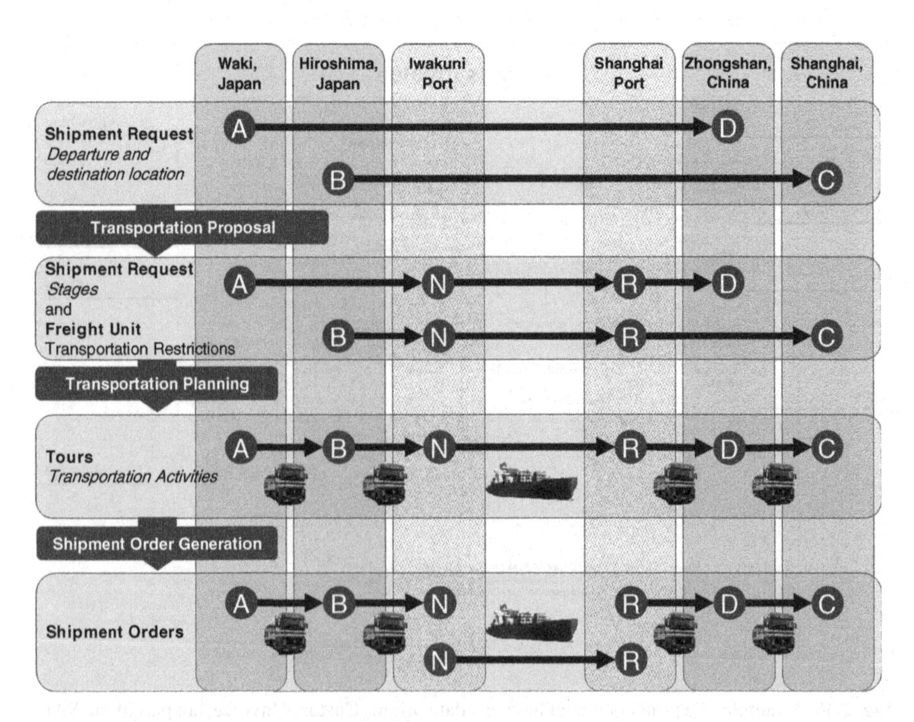

Fig. 2.47 Planning levels in SAP TM

- **Requirement profile**
  The requirement profile defines which freight units are to be selected for planning. For this, you can set geographic criteria (departure and destination locations or zones), temporal criteria (such as delivery within the next 3 days) and other conditions (such as only refrigerated goods). The selected freight units are then integrated in the planner's Personal Object Worklist.
- **Planning profile**
  The planning profile controls the functions of planning and optimization. In the planning profile, you can configure elements in several subareas such as cost evaluation, use of queue and loading time definitions, quality of the optimization results and individual planning steps (optimization, tour generation, shipment order generation or service provider determination).

By selecting those two profiles, a user can determine his work area when entering the planning realm. Examples for work areas pertaining to our sample process displayed in Fig. 2.38 include:

- **Sea freight planning**
  Planning of sea freight shipments from Japanese to Chinese ports for the coming 2 weeks, including tour generation but without creating a shipment order.
- **Land freight planning for Japan**
  Planning and pick-up of FCL freight in the area of western Japan with the destination of western Japanese sea ports within the coming 3 days, including the generation of tours and creation of shipment orders.
- **Completion of sea freight planning**
  Shipment order creation and release for the planned sea freight tours from Japan to China.
- **Land freight planning for China**
  Planning of a delivery of FCL freight from the Shanghai port with the destination of the Shanghai metropolitan area within the next 3 days, including the tour generation and shipment order creation.

After selecting the profiles, an interactive planning screen opens. On this screen, the user can select the freight units to be planned and the resources to be used. Then he can either start manual planning or an optimization run with the selected units. Of course, it is also possible to include all selected freight units and resources in a common planning run.

You can also make use of screens with a tour overview, graphic map display of the transport course or a resource overview. As a result of the planning, the individual transportation activities (loading, transport, unloading, trailer coupling and uncoupling) are stored in SAP TM. If tours have already been created, they are stored as a summary of several activities.

SAP TM Transportation Planning enables you to plan multimodal scenarios, that is, you can also plan the transport in our sample process in a planning run that determines and schedules land as well as sea freight stages together. Figure 2.48 shows a sample network.

Transportation planning and the optimizer used in SAP TM allow you to employ multifaceted planning parameters and optimization strategies and goals. Below, we cite the most important elements that can influence the optimization result:

- **Multidimensional load capacity and time-dependent capacity consumption for resources and booking orders**
  Resources (vehicles, trailers) and booking orders have a certain capacity that can be defined via several dimensions (such as a trailer with a 25-t payload, 100 m shipping volume, 24 pallet storage spaces and 16 loading meters). In the course of time, freight units are loaded and unloaded, such that a consumption profile results for every capacity dimension that serves as the basis for any further loading. If the trailer mentioned above has already been loaded with 25 t but only 50 cubic meters in volume, no further loading can take place.
- **Fixed costs per vehicle**
  Fixed costs can be allocated to every vehicle for optimization purposes.
- **Use of freight cost-relevant elements**
  Cost-relevant elements, such as the distance of a transportation route, number of loading devices, loading weight and volume, transportation duration-dependent costs or costs incurred through intermediate stops can be taken into consideration during optimization. For that process, either optimization results with real freight costs can be calculated and compared, or imaginary optimization costs can be used.
- **Restrictions for transportation duration, number of intermediate stops or total distance**
  You can configure certain settings to limit such things as the maximum transportation duration. This is advisable if a pick-up vehicle can only travel for a maximum of 8 h. Optimization can prevent a 10-h tour from being scheduled.
- **Load compartments**
  For vehicles and trailers, you can define fixed or variable load compartments that can then be planned independently of one another (for instance, a tractor-trailer

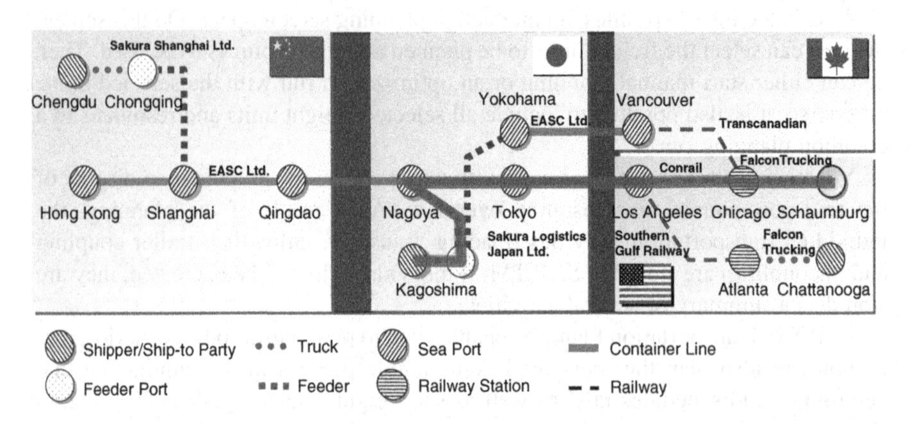

Fig. 2.48 Planning in complex, multimodal networks

with dry and refrigerated compartments). Through an incompatibility definition, the system can be forced to only plan compatible loads for the respective compartments (milk in the refrigerated compartment, rolled oats in the dry compartment).

- **Decreasing capacities in variable load compartments**
  If load compartments are used for individual separation of the freight of individual shippers or recipients, an increasing number of compartments can lead to a decrease in the total capacity of a vehicle in the case of incomplete capacity utilization. This factor is taken into account through modeling of decreasing capacities.
- **Availability and capacity utilization of handling resources in load transfer points**
  If shipments are carried out via load transfer points, the number of available handling resources at these points are decisive for loading lead time at the load transfer point. For example, if a distribution center only has one forklift, wait times will inevitably occur if three trucks need to be unloaded at once. This is considered in optimization planning.
- **Opening hours**
  Different opening hours can be defined for each location and are considered during planning. Freight can only be picked up or dropped off during the opening hours.
- **Duration of loading and unloading**
  Depending on products and locations, variable loading and unloading durations can be modeled. Palleted goods can be unloaded much faster than goods only in cardboard boxes, meaning a product-dependent variance occurs for the durations. This is taken into account during planning.
- **Incompatibilities**
  Incompatibilities represent a means of preventing optimization solutions from having a certain combination of planning element attributes or characteristics. Planning elements that are taken into account include locations, resources, products, business partners, means of transport, service providers or load compartments. Examples of using incompatibilities include:

  - Customer A does not want to be supplied by Carrier X.
  - Shipper B has a ramp that can only be accessed by trucks with a maximum of 12 t.
  - Milk is not to be transported in unrefrigerated compartments.
  - Certain chemicals may not be transported together.
  - Concrete is not to be filled into tanker trucks.
  - Alcohol is not to be transported through Saudi Arabia.

- **Minimum and maximum stopover durations at load transfer points**
  The attributes defined at the locations regarding the minimum and maximum waiting times are used by the optimizer to generate a realistic time lapse when scheduling transports.

- **Depot locations**
  You can define depot locations for vehicles, that is, a location to which a vehicle can return after successful delivery or pick-up. The trip to the depot location is automatically included in the tour by Planning. Vehicles without a depot location can begin a new tour directly from their last loading/unloading point.
- **Waiting and stopover durations (location- and product-specific)**
  In addition to the waiting times defined at the locations, special waiting and stopover durations can be set, and these can be added for a specific product. For example, you can schedule a regular loading time of one hour, but add an additional hour for products that are especially difficult to load.
- **Itineraries**
  Itinerary-based resources (such as liner ships, airplanes, trains, transit system networks) can be supplied with itineraries having regular or irregular departures. The departure times are taken into account during planning.
- **Tractors and trailers, possible combinations**
  Vehicles can be defined as active resources (with loading capacity, such as a truck, or without loading capacity, such as a tractor) or as passive resources (trailers). You can establish predefined combinations of active and passive resources that you can use to model tractor-trailer or railway car combinations. These vehicle combinations are taken into account for the optimization process. It is also possible to unhitch trailers during a tour (leaving them at a depot) and take new trailers.
- **Penalties for early or late delivery or non-delivery**
  Optimization not only takes into account costs related to freight, legs, vehicle and time, but also anticipated penalties that occur due to early or late pick-up or delivery, or because of a failure to deliver.

When several optimization runs are preformed sequentially, a particular evaluation of transportation costs and penalties may lead to planned freight units being removed from the transportation plans. Penalty costs can depend on the significance of a customer and the priority of his transportation contract. Non-delivery may occur if, for instance, individual freight units are removed from planning based on insufficient vehicle capacity or other, more highly prioritized, freight units. An example for such a decision can be seen in Fig. 2.49, which assumes loading a ship with 10 container loading spots.

The loading process depicted in Fig. 2.49 results in the freight unit types with various priorities listed in Table 2.4:

On the *first booking day*, the capacity situation is not yet serious; all of the freight units resulting from orders can be transported as requested.

On the *second booking day*, overbooking occurs, which is accommodated for by removing two of the freight units having a lower priority from the plan (the firm's own containers that are to be repositioned).

On the *third booking day*, extreme overbooking takes place, which leads to customer freight units not being able to be transported. Units with a lower priority are taken out of the plan and shipped with subsequent transports. The decision of which freight units not to transport is made on the basis of which ones would produce the lowest penalties.

2.6.6 Booking Freight Space

A *freight space booking* is a reservation of freight space on a ship, airplane, train or truck for which it may not yet be clear what is to be transported at the time of the booking. For instance, in sea transit, connections from Asia to Europe or to the United States are often used to capacity, because many goods are produced in Asia but consumed in the rest of the world. That is why there is a very unbalanced flow of goods from West to East compared with that moving East to West, a phenomenon which also affects freight prices.

In order to secure transportation of the goods with regard to available capacity, you can use a booking order to book freight space in advance and receive a booking confirmation. As is the case with a shipment request, you can enter the following data in a booking order:

- Departure und destination location: generally a departure and destination harbor, airport or railway station

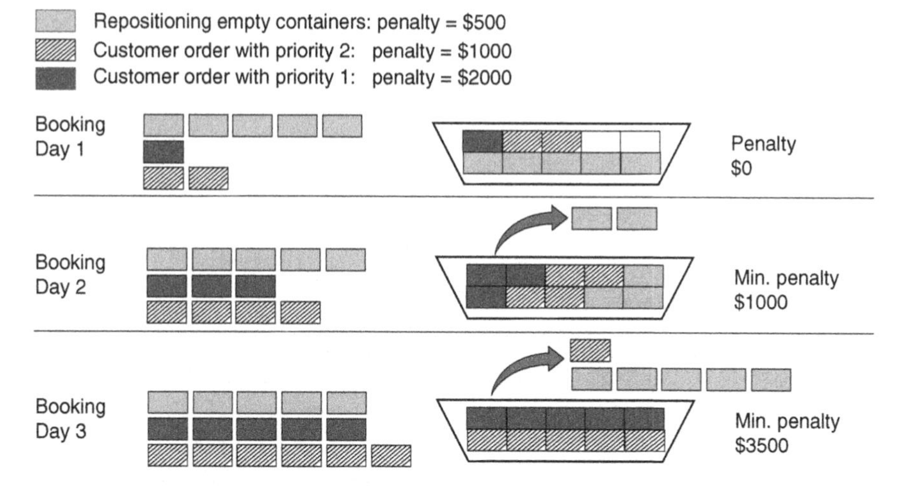

Fig. 2.49 Optimization situation with a minimization of penalties

Table 2.4 Penalties for non-fulfillment of shipments (sample values)

Freight unit type	Priority	Planned penalties for non-fulfillment
Repositioning empty containers	3	$500
Freight units of standard customers	2	$1,000
Freight units of preferred customers	1	$2,000

- Departure and arrival dates
- Reserved freight space capacity with capacity type (such as ten 40-ft containers, 8 loading meters, 3 t)
- Reserved transportation units
- Service provider that executes transport
- Identification of the flight, ship, cruise or train

Booking order for a container ship in the FCL process. Figure 2.50 shows a booking order for sea freight transport in our sample process. Here, freight space for five 20-ft containers is reserved. Thus, the booking order also has a consolidating function.

Confirmation of a booking order can either be done by the service provider electronically or by one of your employees. After the booking order has been confirmed, the confirmed capacity is available for transportation planning, as is the case for the itinerary resource. You can then assign freight units as transport demand to the booking order as transportation capacity. All of the transportation activities allocated to a booking order are used to create a tour. A shipment order can then be created from that tour. In accounting terms, the shipment order represents the master bill of lading (Master B/L, MAWB).

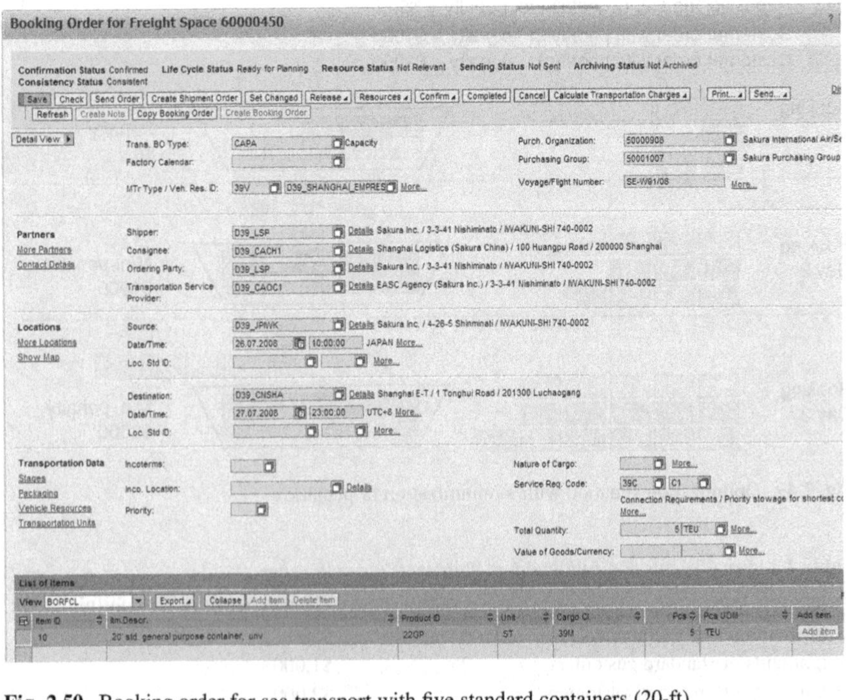

Fig. 2.50 Booking order for sea transport with five standard containers (20-ft)

2.6.7 Subcontracting

The subprocess *Subcontracting and Tendering* enables you to create, manage and settle transport requests to third parties, such as carriers, other logistics service providers or one's own fleet. The central business object in this area is the *shipment order* (see Fig. 2.36). You can also use a freight order to consolidate several shipment orders for a single service provider.

Shipment orders are generally the result of manual or optimized transportation planning after the created tours have been processed by shipment order generation. The same is true for *freight orders*, either generated as a result of consolidation or from a connected transportation planning effort (*Continuous Move*), for which a series of tours is allocated to a service provider to achieve higher cost efficiency.

You can also manually create shipment orders to request a shipment that has either nothing or only little to do with actual transportation planning. Such orders can include the following cases:

- Provisioning or pick-up for empty containers
- Work orders to service companies, such as for packing, aeration, measuring, declaring or loading freight

Shipment orders are connected to the original shipment request via the transportation activities and freight units referenced therein. This enables SAP TM to relate the sales side (shipment requests) with the purchasing side (shipment orders) and perform a profitability calculation for transportation orders (see Fig. 2.36).

In terms of their content, *shipment orders* and *shipment requests* are very similar, since they represent an outgoing transportation order (that is, an order to a transportation service provider) and an incoming transportation order (meaning a transportation order from a shipper). Thus, you can naturally enter the same data in a shipment order as in a shipment request. The structure of the messages used for electronic communication is also the same. It corresponds to the EDIFACT format IFTMIN. Figure 2.51 illustrates order development in SAP TM and the subsequent communication with service provider systems (here, including SAP TM).

Based on the original shipment requests (on the left of the illustration), freight units are formed that, with the aid of Transportation Planning, are allocated to tours and shipment orders. Several shipment orders can be assigned to a freight unit (for example, a pre-carriage, main carriage and on-carriage) that are then communicated electronically to various service providers. If these service providers also use SAP TM, the incoming messages are converted to shipment requests that each service provider can individually plan.

The shipment order user interface is designed in the same way as that of the shipment request, with the exception that there is no basic view (see Fig. 2.41), since manual entry is only done in exceptional cases. Figure 2.52 shows the detail view of a pre-carriage shipment order.

Unlike the shipment request, in which the shipper and recipient of goods are entered as locations, the shipment order cites the freight station of the logistics service provider at the port of dispatch as a destination location. The cost view in the shipment order shows the billing items that are invoiced by the transportation service provider. Because the shipment order can be a consolidation of several freight units – including from different shippers – it is also the business object that represents the master bill of lading in SAP TM.

The *transportation service provider selection* is a function that is executed based on shipment orders. You can manually allocate a shipment order to one or more possible transportation service providers, or this can be done with the support of the

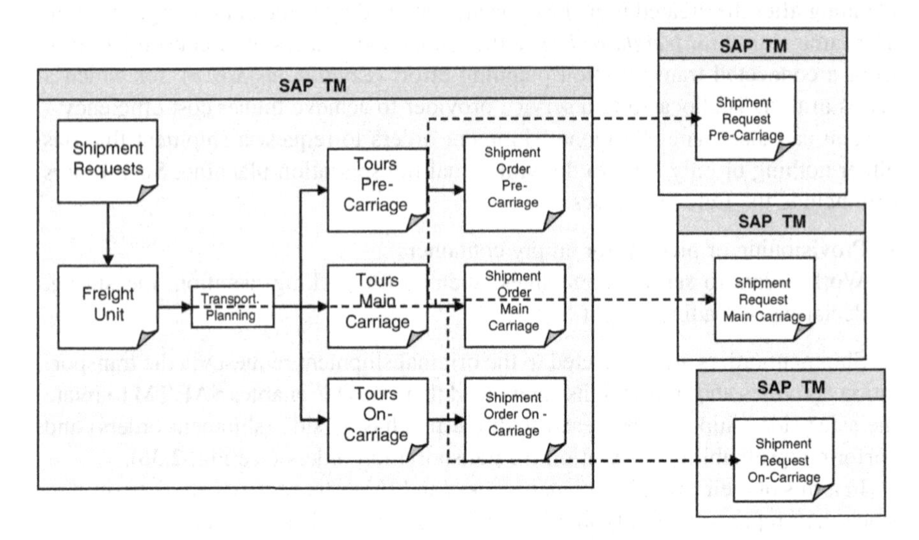

Fig. 2.51 Communication of shipment orders to service provider systems (as a transportation order or shipment request)

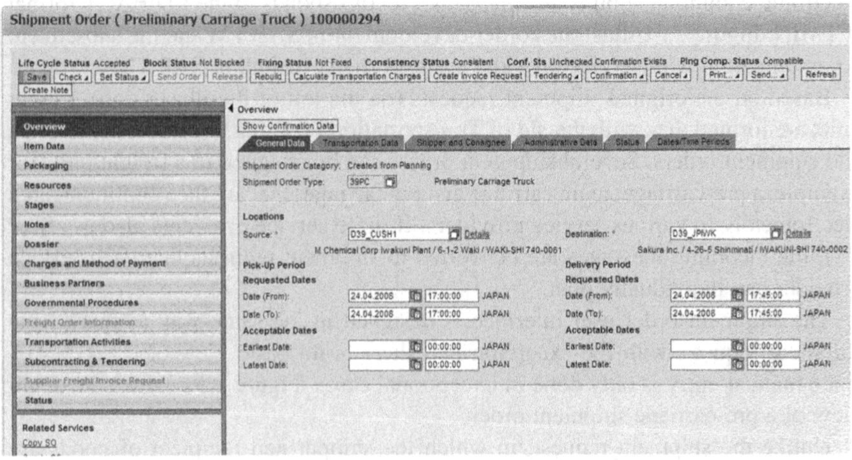

Fig. 2.52 Detail view of a pre-carriage shipment order

system. Automatic transportation provider selection is a good idea in the following instances:

- When several transportation service providers are possible for certain shipments, and the one that is most inexpensive, reliable or has the highest priority is to be selected.
- When certain service providers serve particular regions or only execute certain types of transports (for instance, some will not transport dangerous goods or only ship within New England), and the allocation of a suitable service provider is to be done automatically.
- Outline agreements are stipulated with various service providers regarding percentage-based or absolute quotas in a transport job that must be fulfilled to receive special shipping conditions and prices (such as a minimum of 500 TEUs on the Hamburg-to-Singapore container line or 20% of all truck transports in New England).

As described above, you can allocate a shipment order to one or more service providers. The shipment order is always only given to one service provider. However, the indication of several service providers is required for the transportation *tendering* function. Tendering serves to select service providers from the list who:

- Ensure/confirm available transportation capacity and performance
- Offer the lowest price (spot quote)

SAP TM offers three procedures for the execution of the transportation tendering. The tendering procedures are each controlled by a process triggered in SAP Event Management. It ends the current process step when a service provider has responded or the tendering period has expired, and begins the next tendering step or presents the tendering result:

- **Peer-to-peer tendering**
  Tendering is performed sequentially and for each individual service provider, in order of priority. This means that the service provider with the highest priority is contacted. If he answers with an acceptance of the tender and the price, tendering is complete. If he does not respond, rejects the bid or offers insufficient conditions, tendering goes to the next service provider.
- **Broadcast tendering**
  Tendering is sent simultaneously to all service providers. Upon expiration of the tendering period, the offers are compared and the best one selected.
- **Open tendering**
  Open tendering works in much the same way as broadcast tendering; however, not only preselected transportation service providers are contacted that have already been allocated to a shipment order, but all service providers maintained in the system that correspond to certain selection criteria (such as truck transportation in New England).

Tendering can either be performed via electronic data communication, meaning the service provider receives a request to submit a quote via EDI, or you can give service providers access to a *collaboration portal*, where they can view tenders directed at them and respond to them.

After the tendering process has been completed, the selected service provider is commissioned via electronic data communication, fax or email.

After the service provider has been selected, you can calculate the freight purchase price in the shipment order just as was done in the shipment request. After the calculation, in the corresponding views of the shipment request, you will find a detailed list of the individual cost items (see Fig. 2.45). They can be manually adjusted and subsequently used in the invoice verification process for the service provider and in the credit memo procedure to settle open invoices.

2.6.8 Transportation Charge Management

Transportation Charge Management is a powerful tool in SAP TM with which you can calculate all costs occurring in transportation processes:

- Earnings from the sale of transportation services
- Costs for the purchase of transportation services
- Internal costs for use of a company's fleet
- Internal settlement between organizational units of a logistics service provider

The realm of Transportation Charge Management is primarily composed of the following elements:

1. Contract, tariff and freight rate data (master data)
2. Operative cost structures in the order documents
3. Settlement-related cost structures in the settlement documents
4. Incoming and outgoing invoices and internal clearing
5. Booking in Cost Accounting and Financial Accounting

As depicted in Fig. 2.36, you will find elements 1, 2 and 3 in SAP TM, and elements 4 and 5 in SAP ERP.

The *master data* of Transportation Charge Management is constructed in a multilevel way, with the elements of the central and lower levels being reusable (see Fig. 2.53).

The top level of the definition is represented by *freight agreements*, which stipulate a contractual relationship between parties with the goal of determining transportation prices for freight purchasing or sales. In addition to the contractual partners and application (purchasing or sales), other details such as validity periods, currency definitions and cost limits can also be defined.

Every freight agreement contains one or more tariffs that can be used within the context of the freight agreement. A tariff defines the way in which transportation charges are to be calculated for certain transportation processes. The tariff is also defined by determining conditions for its application; in the example shown in Fig. 2.54, these include the shipment type *sea freight* and the departure and destination ports of *Hong Kong* and *Los Angeles*. Further conditions can include stipulations for types of goods, transportation zones or service conditions.

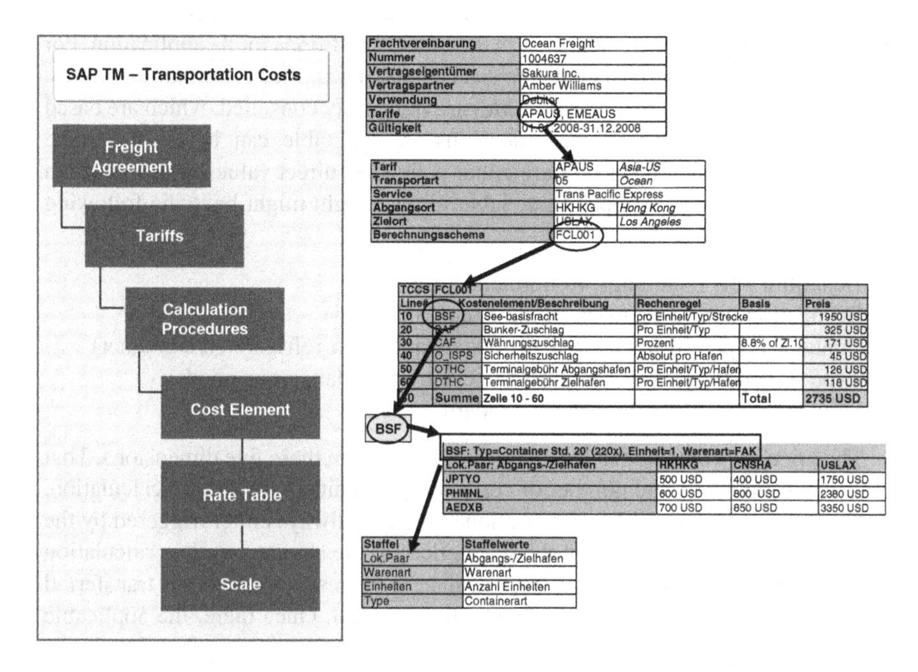

Fig. 2.53 Structure of the charge management master data

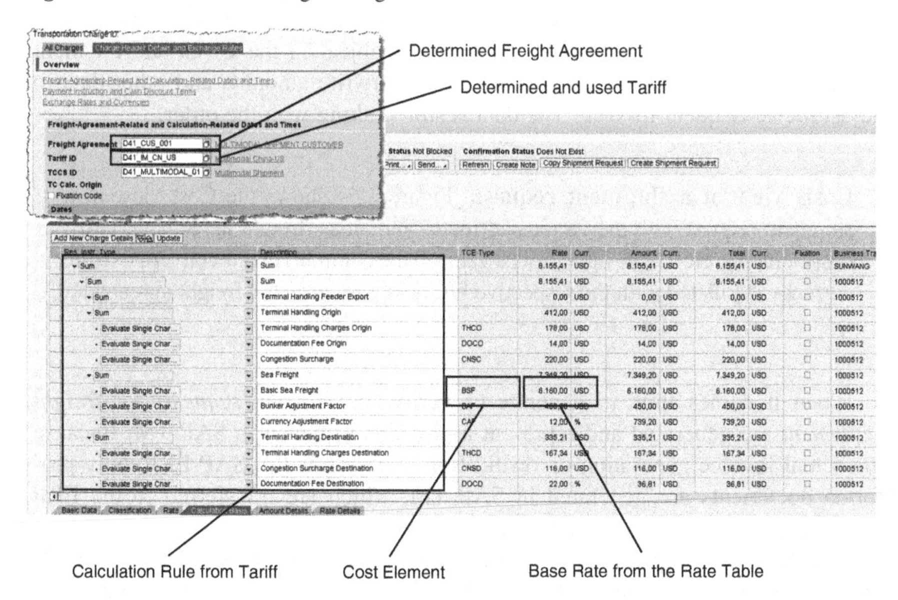

Fig. 2.54 Structure and allocation of transportation costs in the agreement documents (shipment request, shipment order)

A *transportation charge calculation* is maintained in every tariff, which defines individual cost elements and their relationships. A transportation charge calculation can be a list of individual cost elements (such as the ocean base freight, bunker surcharge, or terminal fees), defined either with absolute values, as percentages with reference to other cost elements or with reference to a *calculation rule* (such as a rate table). Every cost element can, in turn, have conditions for its application. For instance, you can define cost elements that can only be applied to certain Incoterms.

To calculate cost elements, *rate tables* are frequently consulted, which are based on scales representing various dimensions. A rate table can have up to nine dimensions, for which the scales are either defined as direct values or as minimum and maximum values. A typical rate table in sea freight might have the following dimensions:

- Departure port (Hamburg, Rotterdam, Antwerp)
- Destination port (Singapore, Hong Kong, Shanghai)
- Container type (20-ft standard, 40-ft standard, 40-ft refrigerated container)
- Type of goods (general freight/FAK, electronics, dangerous goods)
- Service degree of the transport (standard, priority)

The price per container is then defined depending on these five dimensions. That price is multiplied by the number of corresponding units in the charge calculation.

Operative calculation of transportation charges is always either triggered by the shipment request (calculation of the sales price) or the shipment order (calculation of the purchase price) (see Fig. 2.36). The data of the respective order is transferred to freight cost determination in a standardized format. Once there, the applicable freight agreement and tariff are determined. This in effect determines the charge calculation. After each cost element is calculated (under consideration of the condition of applicability) the finished list is returned to the shipment request or order, where it is displayed with the business object in the Costs and Payment Method (see Fig. 2.54). Here, you can also overwrite amounts and other data to adjust the subsequent invoice. The data is stored along with the order.

Costs view of a shipment request. Figure 2.54 shows the Cost view of a shipment request with the various attributes and areas (freight agreement, tariff, cost elements, etc.) high-lighted. The value in the column Total is converted to the currency stipulated with the respective business partner in the freight agreement.

When it comes time to generate or verify the invoice, *requests for freight settlement* are generated and represent a pro-forma invoice in SAP TM. Because the actual invoice generation or verification is performed in SAP ERP, only pro-forma documents are generated in SAP TM, which are transferred to the ERP system after their check and release.

In the case of a shipment request, a customer freight invoice request is generated and directly transferred to SAP ERP, where it leads to an invoice document. This document can be directly used for customer freight settlement and is automatically integrated into Financial Accounting.

In the case of a shipment order, a service provider freight invoice request is generated for settlement. This is initially transferred to SAP ERP to accrue reserves for the expected service provider invoice. If a pro-forma invoice is then transferred, a purchasing document for services is generated with a service entry sheet. Based on this, the data is integrated into Financial Accounting and incoming invoice verification is subsequently conducted.

2.6.9 *Integration with SAP Event Management*

For purposes of process control and tracking the status of transportation procedures, SAP TM is integrated with SAP Event Management. Relevant processes that must be monitored by Event Management are automatically initiated:

- Tracking of the processing status of a shipment request
- Shipment tracking
- Tour tracking (itinerary monitoring)
- Tracking of a resource as an asset (containers and railway car tracking)
- Status and life cycle management for tendered shipment requests

Status values are reported to Event Management either from the current process in SAP TM or, more frequently, from outside via electronic data communication (EDI) or interactive feedback (such as a vehicle computer or onboard unit, OBU). You can then display the status history in SAP TM, in the context of the respective business object. Figure 2.55 shows the view of the status history of a shipment from our sample process.

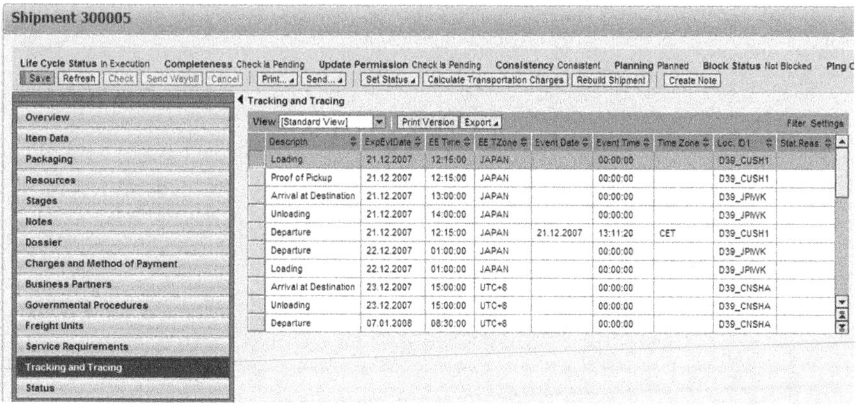

Descriptn	ExpEvtDate	EE Time	EE TZone	Event Date	Event Time	Time Zone	Loc. ID1	Stat.Reas.
Loading	21.12.2007	12:15:00	JAPAN		00:00:00		D39_CUSH1	
Proof of Pickup	21.12.2007	12:15:00	JAPAN		00:00:00		D39_CUSH1	
Arrival at Destination	21.12.2007	13:00:00	JAPAN		00:00:00		D39_JPIWK	
Unloading	21.12.2007	14:00:00	JAPAN		00:00:00		D39_JPIWK	
Departure	21.12.2007	12:15:00	JAPAN	21.12.2007	13:11:20	CET	D39_CUSH1	
Departure	22.12.2007	01:00:00	JAPAN		00:00:00		D39_JPIWK	
Loading	22.12.2007	01:00:00	JAPAN		00:00:00		D39_JPIWK	
Arrival at Destination	23.12.2007	15:00:00	UTC+8		00:00:00		D39_CNSHA	
Unloading	23.12.2007	15:00:00	UTC+8		00:00:00		D39_CNSHA	
Departure	07.01.2008	08:30:00	UTC+8		00:00:00		D39_CNSHA	

Fig. 2.55 Status tracking/history of a shipment

2.7 Summary

In the world of SAP solutions, each transportation management component has a special significance. As the most modern of the components, *SAP Transportation Management* (SAP TM) offers the widest range of functions and the greatest degree of flexibility.

For simple scenarios, however, and considering TCO (*Total Cost of Ownership*) aspects, the use of SAP ERP Transportation Management can also be beneficial; with it, you do not need any additional system for transportation management.

If you have demands in the realm of transportation optimization, you can either integrate an external transportation planning system into ERP Transportation Management, or select APO Transportation Optimization, which also gives you the integration of an availability check.

In the next chapter, we will examine warehouse logistics and inventory management with SAP.

Chapter 3
Warehouse Logistics and Inventory Management

A warehouse is defined as a "structural unit with all resources and organizational provisions necessary for the execution of processes connected to inventory and warehouse management, including the organizational units involved with goods receipt and shipping" (Pfohl 2004, pp. 124–146). Unlike this subjective description, the proverbial saying "The best warehouse is no warehouse" implies an evaluation: It characterizes a warehouse and warehousing as something negative, to be avoided wherever possible.

The latter statement holds a kernel of truth, since inventory (raw materials and supplies, semi-finished and finished products) and warehouse buildings incur costs. The statement is aimed at an improvement in cost-effectiveness – which, of course, is a significant aspect of warehousing.

In addition to improving cost-effectiveness through lowering costs, the functional processing of all goods movement that causes a change in inventory is the primary task of software designed to control warehouse logistics and maintain inventories.

We also see the warehouse as a link between internal and external logistics. In this chapter, we will provide a detailed overview of the functions offered by SAP and their integration in the logistics core processes of procurement and distribution. We will illustrate this with several examples for inbound as well as outbound processes, and for WM (Warehouse Management) as well as SAP EWM (Extended Warehouse Management).

3.1 The Fundamentals of Warehouse Logistics

In this section, we will look at the fundamentals of warehouse logistics and inventory management, explain their significance to industrial management, and examine the systems and requirements for mapping logistics processes.

J. Kappauf et al., *Logistic Core Operations with SAP*,
DOI 10.1007/978-3-642-18202-0_3, © Springer-Verlag Berlin Heidelberg 2012

3.1.1 Management-Related Significance

Warehouse logistics involves the storage and maintenance of goods in warehouses. Warehouse management, or *warehousing*, is a realm of materials management. From a business and management point of view, the warehouse equipment of such a "structural unit" takes up space and ties up fixed assets. In addition, warehouse inventory ties up liquid asset capital. Nevertheless, there are several management-related reasons for creating inventory and maintaining it accordingly (see Table 3.1).

As a rule, a logistics chain contains several warehouses between a raw materials source and the end customer. Thus, on each individual level of the logistics network, stock is established to strike a balance between fluctuating requirements and inward goods movement times, or due to insecurity about future requirements in order to secure material availability.

With the aid of stock procurement, a disconnection between materials procurement and materials use in production or distribution processes is achieved. Stockkeeping fulfills the primary goal of ensuring material availability and represents a reaction to any insecurities in the materials procurement and production planning processes. From a production logistics point of view, stockkeeping enables a streamlining of production by building up warehouse stock in times of low demand and reducing it in times of increased demand.

Against the backdrop of the logistics core function of a warehouse – bridging time – we can identify two central reasons for stockkeeping:

- The establishment of inventory to guarantee ability to deliver
- To offset fluctuations in delivery and demand

Securing the ability to deliver is thus often achieved by maintaining a higher stock level and, consequently, higher capital commitment. Especially with a view to cost pressure, in addition to operative warehouse functions and the modern tasks of warehouse logistics, it is important to reduce a significant driver of inventory – lead times. Among the major factors for success are modern, IT-supported warehouse management systems.

Table 3.1 Reasons for stockkeeping

	Procurement logistics	Distribution logistics
Largely warehouse-free concepts	Procurement from case to case, which triggers the procurement procedure through the occurrence of a requirement	Order-related production for which a customer order triggers production
	Production- and usage-synchronous procurement: "Just in time"	Just-in-time production
Stockkeeping	Stock procurement (for the receiving storage location)	Make-to-stock production (sales stock)

3.1.2 Systems and Applications

Early software for warehouse logistics and inventory management appeared in the early 1970s. In those days, people around the world referred to systems for warehouse management in their native tongues. It was only in the course of increased functionality and the implementation of optimization algorithms that the English term *warehouse management system* found widespread usage.

Originally, warehouse systems were pure *stock management systems*, whose goal it was to maintain quantities and locations within a warehouse and their relation to one another. The quantity refers to the amount of materials stored, while the location means the respective bin location. Additional functions can include the maintenance of transportation systems.

In contrast, modern *warehouse management systems* (WMS) are capable of not only supervising complex warehouse and distribution centers, but also controlling and – with regard to the reduction of internal stock – optimizing them. *Warehouse management* in general thus refers to the control, supervision and optimization of complex warehouse and distribution systems. In addition to the elementary functions of warehouse management, such as quantity and location management, conveyance control and planning, warehouse management also includes comprehensive means and methods for supervising system conditions (with software and automation technology) and a selection of business and optimization strategies. The task of a WMS is thus the operation and optimization of internal warehouse systems.

In the meantime, we not only expect a high-performance warehouse management system to accomplish the operative maintenance of materials and their bin locations, but to also provide continual optimizing control and supervision of material flow, equipment and staff, from goods receipt through all warehouse and processing steps, right up to goods issue. In this regard, most companies need to master the following core processes with the support of a warehouse management system:

- Unloading the means of transport and furnishing the materials in the goods receipt area
- Goods receipt monitoring
- Deconsolidation through the dissolution of loading units and formation of storage units
- Provision for placement into stock
- Putaway, storage, withdrawal
- Picking
- Packing, formation of loading units and loading

In business practice, especially due to its rationalization and optimization potential, the strategic significance of the warehouse has been steadily increasing. As a result, the processes and tasks taking place in a warehouse are also changing. In the past, it often sufficed to manually post goods receipts and issues and place goods, usually in a chaotic fashion. Today, the tasks of warehouse management are considerably more complex and rely on the support of an intelligent IT system. RFID technology (*radio frequency identification*), automated rack systems and operation technology, route optimization, pick-by-voice and robotics are trends that extend far beyond traditional warehouse processes and must be supported by a modern warehouse system.

The performance and efficiency of a logistics chain depend largely on the optimal interplay of the ERP and warehouse management systems employed. That is why warehouse management systems are often integrated or closely linked to an existing ERP system.

In an SAP system environment, we differentiate between pure *inventory management* and *warehouse management*. Inventory management and evaluation is generally the task of an ERP system. Within materials management, inventory management provides information on what total quantity of a material is in stock. It enables the precise indication of the exact location of a certain material quantity in the warehouse, and also tells us whether this amount is idle and situated in its location or in movement.

Inventory management and warehouse management

An SAP system differentiates between inventory and warehouse management.

- **Inventory management**
  The primary task of inventory management is the management of stock with regard to its quantity and value. Inventory management is executed regardless of the warehouse management utilized in SAP ERP.
- **Warehouse management**
  Warehouse management concerns the spatial division of a warehouse, the assignment of warehouse bin locations and the processes within a warehouse. Warehouse management is either executed with integration to SAP ERP or decentrally, either on the basis of an ERP system (SAP LES) or using SAP Extended Warehouse Management (EWM), an SAP SCM component.

The administration of and transparency regarding existing materials is essential to making precise statements about the availability of a material. Goods movements are usually caused by procurement, distribution and the associated goods receipts and issues, or through stock transfer.

SAP has been providing warehouse management functions since the release of SAP R/3 2.0. Thus, it can look back on more than 16 years of experience in warehouse management and countless successful implementations. Ever since the first SAP R/3-based versions and leading right up to the current SAP SCM-based systems, functionality has been continually expanded and adjusted accordingly to customer demands. In addition to *Warehouse Management* (WM) as part of SAP ERP, in 2005 the considerably more efficient *SAP Extended Warehouse*

Management (SAP EWM) was introduced, which is based on SAP Supply Chain Management (SAP SCM).

Originally a part of SAP Spare Parts Management, today SAP EWM is an independent application that can be used in any warehouse environment. SAP EWM was developed for complex warehouse and distribution centers with several different products and a high document volume, and in contrast to WM, it offers many new and expanded functions. SAP EWM does not replace ERP-based warehouse management, but supplements it with a decentralized warehouse system.

In this chapter, we will thus discuss WM (SAP ERP) as well as SAP EWM (SAP SCM), focusing on the respective mapping and integration of logistics core processes. Since both SAP systems partially overlap with regard to existing functions, we will describe the corresponding processes for the system in which the last functional expansion took place (generally SAP EWM). This especially applies to auxiliary logistics services, such as cross-docking. Although both are also possible in SAP ERP, EWM offers considerably more functions.

The selection of a warehouse management system and the associated system architecture is a central issue in every project pertaining to warehouse management. While discussion in recent years has concentrated on whether or not a central or decentral system should be implemented and how to integrate it into an existing warehouse automation system, the focus is now shifting.

Economic issues have not only changed due to the introduction of an EWM system, which is vastly expanded in comparison to an ERP-based management system. While in the past primarily technical influencing factors were examined, such as degree of automation and system performance, it is currently economic factors that have come under scrutiny and are decisive when selecting a product. A sweeping answer is not available for finding solutions to these issues, yet there are certain points of reference and relationships that can indicate a certain direction and possibly exclude certain architectures. Degree of automation continues to be a very important technical as well as economic influencing factor that greatly affects such a decision.

If the connection of warehouse and conveyor technology is a central issue for a project, the new SAP EWM has considerable advantages over the old solution. Starting from Release 5.1, EWM offers in its component *Material Flow System* (MFS) a fully integrated material flow calculator with which warehouse and conveyor elements can be connected and controlled in real time. SAP EWM also allows you to connect *programmable logic controllers* (PLCs) as well as to map checkpoints, rack feeders or conveyor segments using a standard configuration. In addition to the degree of automation, there is a series of other influencing factors for determining the system architecture, each of which must be adjusted to the specific demands of a company.

3.1.3 Organizational Structures and Master Data

Organizational structures play a significant role in the control of logistics processes and the operative administration of material and its corresponding quantities.

Furthermore, technical system integration of decentral warehouse management with SAP EWM is based on the allocation of SAP ERP and SAP SCM organizational units, that is, on whether or not the decentral system is responsible for warehouse execution.

A central organizational element is the warehouse number. The number, based on subordinate warehouse structures and, as the technical and organizational unit of a warehouse management system, represents a warehouse's complex spatial structure and circumstances. In operational practice, the warehouse number corresponds to a warehouse complex or an individual warehouse building. It also serves as the central element under which warehouse-specific material master data, such as information on palleting, picking and placing, is stored.

Warehouse numbers can be assigned to a specific plant/storage location combination (see Fig. 3.1). This assignment represents a link between inventory management and warehouse management, and enables the use of warehouse management functions with SAP ERP or SAP EWM (see also Volume 1, Chap. 3, "Organizational Structures and Master Data").

Organizational structures and allocation

Figure 3.1 shows the plant structure of a company with two production sites: New York and Philadelphia. The plants are allocated to different storage locations, which in turn are assigned a specific WM warehouse number. The subordinate warehouse structures of the warehouse numbers *Central Warehouse 1* and *Central Warehouse 2* are mapped in SAP ERP. The warehouse *Airport*, which is allocated to storage locations *NY-South* and *PH-North*, is a decentral warehouse that is used by both production sites. The warehouse number *Airport* is identified as a decentral warehouse and linked to a SAP EWM warehouse number.

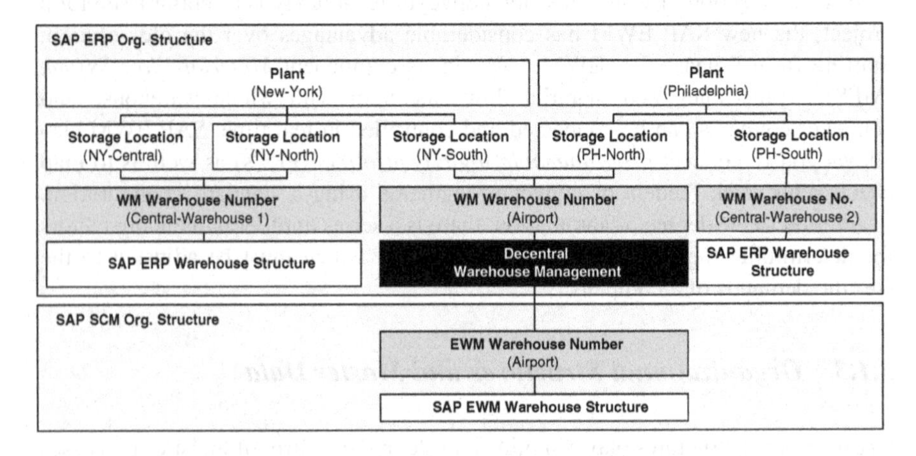

Fig. 3.1 Organizational structures in warehouse management

The materials stored in an EWM warehouse are maintained with reference to quantity at the warehouse level. To achieve this, SAP EWM is seamlessly integrated with the Inventory Management of SAP ERP. Integration is done via the direct assignment of an ERP warehouse to an EWM warehouse number and a system setting that specifically identifies the assigned ERP number as a decentral warehouse. The SAP EWM warehouse number has a similar function as those used in SAP ERP. Although EWM is not functionally related to *SAP ERP Warehouse Management*, the EWM warehouse number also indicates the physical location in which material is stored and maintained. The actual structuring of the warehouse and the allocation of warehouse numbers to warehouse structure elements depend on the respective warehouse management system, and will be explained in the sections below along with the system-specific warehouse processes and their integration.

Both WM and EWM use *SAP ERP Inventory Management* (IM) to maintain quantity- and value-based inventory. Thus, we will initially discuss the basics of Inventory Management in SAP ERP before turning to the most important differences between warehouse structuring and management with SAP ERP and SAP EWM.

3.2 Inventory Management

Inventory Management is a part of Materials Management. As the latter's central area, it is seamlessly integrated with all logistics processes that trigger a change in inventory (see Fig. 3.2). From a logistics standpoint, this integration refers to the moved quantities, and from an accounting standpoint, to the moved values. In all processes, Inventory Management accesses the respective master data and movement data.

Material requirements determination uses Inventory Management, its physical stock, and planned inflow and outflow for the planning and determination of requirements. Requirements are either procured internally or externally. External procurement is generally performed by purchasing from an external vendor. Internal procurement is achieved via in-house production or by internally procuring materials from another plant via stock transfer.

In the case of in-house production, Inventory Management governs the provision of components and records the receipt of finished products into stock. Finished products are maintained and stored in stock until they are shipped to customers or used for internal purposes.

The external procurement of material is executed in Purchasing by placing an order with an external vendor. The integration of inventory management takes place through goods receipt with reference to the purchase order or the delivery. Goods receipts generally lead to an increase in stock. Inventory Management documents the actual quantities received, making it possible to check and verify whether the quantities and values of the order and goods receipt match the figures on the invoice.

Retrieving items from stock is not only done for production, but also for the supply of sales orders. As early as the sales order processing stage, the system can verify whether the required material will be available in a sufficient quantity on the desired delivery date. When doing so, the availability check considers the requirement quantities of planned inflow and other outflow for already-confirmed sales orders. When deliveries are generated, the quantity to be delivered to the customer is updated and subtracted from the total stock when the goods issue is posted.

Goods movement leads to changes in stock. Actual goods movement is controlled in an ERP system by so-called *movement types*. The movement type, a three-digit code, identifies goods movement and controls how it is to be executed and what effects it will have on the system. In addition to quantity and value updates, these effects influence messages that can be issued by the system during goods movement as well as the type of stock that can be posted.

For inventory management, it is not only important to know what quantity of a material is situated in a particular bin location or factory, but also what type of stock it represents (see Fig. 3.3). The type of stock governs what stock type can be posted for each goods movement, as well as whether it is unrestricted or is subject to a block. It is vital to know this detail in order to be able to provide precise information about the availability of a material. If the usability of a material changes, a posting change can be carried out. For this type of posting change, a physical goods movement to another

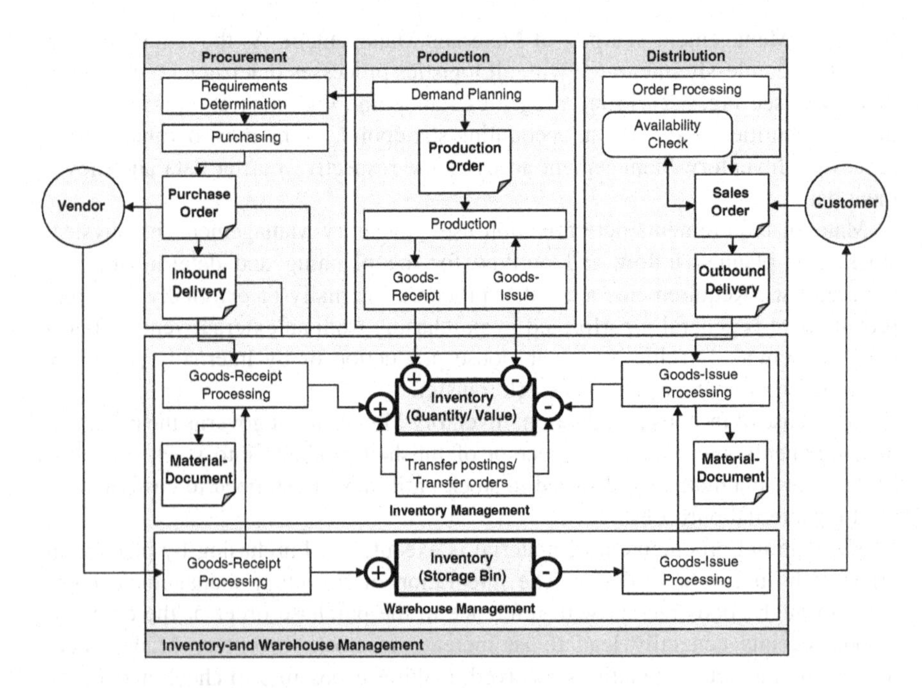

Fig. 3.2 Overview of inventory and warehouse management

storage location can be recorded. Stock types differ in terms of the application and usability of a material, as well as whether it is special stock.

Stock quantities are indicated for the various stock types, and the availability of a material in various departments is also mapped. For instance, the total open purchase order quantity is available to cover requirements in Planning but cannot be freely used in Materials Management, since no goods receipt took place (see Fig. 3.3).

Stock types in goods receipt

In business operations, goods receipts from an external vendor are not immediately posted to unrestricted-use stock, but first are subjected to a quality inspection. Depending on the result of this inspection, the material is either posted (and thus booked as unrestricted-use stock) or blocked. Blocked stock is subsequently sent back to the vendor (see Fig. 3.3).

The following types of stock indicate material availability:

- **Consignment stock**
  Material that a vendor makes available in your warehouse but that does not become the property of your company until it is issued or transferred and invoiced.
- **Unrestricted-use stock**
  Only material from this type of stock can be issued to Production or Sales and Distribution.

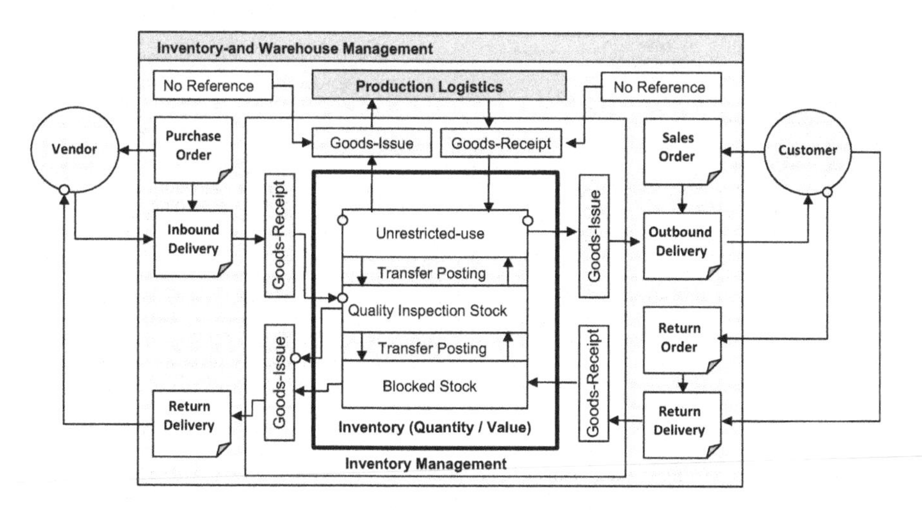

Fig. 3.3 Stock types and goods movement

- **Blocked stock**
  Material that is not to be issued.
- **Quality inspection stock**
  Material intended for use only after an inspection.
- **Goods receipt blocked stock**
  Material that can only be accepted on certain conditions.

These general types of stock are primarily differentiated on the basis of their usability. In addition, the system distinguishes between valuated and non-valuated stock. For instance, the quality inspection stock can be valuated stock that is subject to usability limitations. A blocked stock return is blocked stock with material that has been returned by a customer and accepted conditionally. Until a final decision is made, it is thus neither valuated nor available for unrestricted use. Due to its significance for inventory management, inventory valuation will be explained in more detail in Sect. 3.2.2, "Inventory Valuation".

Transfer posts between stock types

Figure 3.4 shows a transfer post in SAP ERP. Material 100-110 is transferred from unrestricted-use stock to quality inspection stock within the same storage location. This posting is controlled by Movement Type 322.

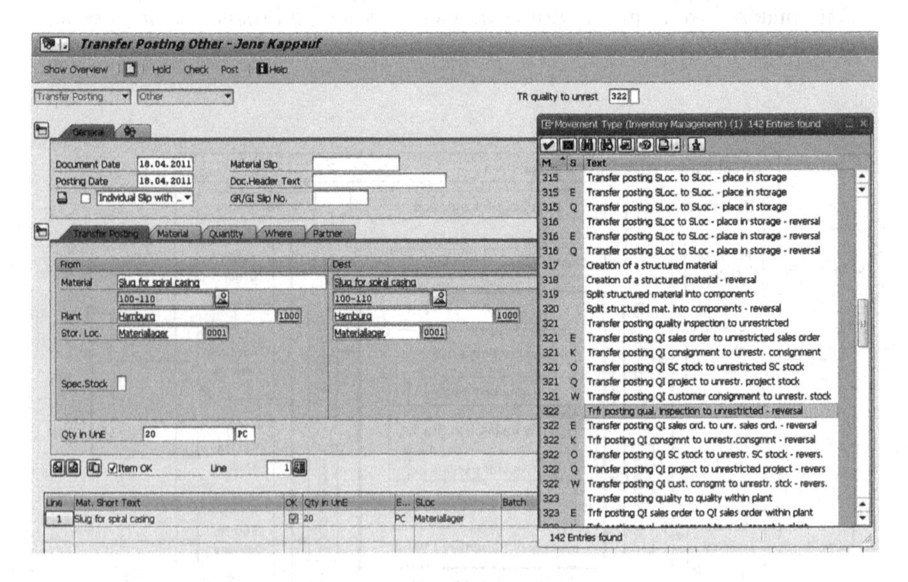

Fig. 3.4 Transfer posting to quality inspection stock

Stock overview after transfer posting

Figure 3.5 shows the stock overview for Material 100-110 after the transfer posting. The unrestricted-use plant and storage location stock amounts to 568 units, composed of the following: 252 units from Plant 1000 in Hamburg and 316 units from Plant 3000 in New York. Twenty units of the material are in quality inspection stock, which is restricted in use.

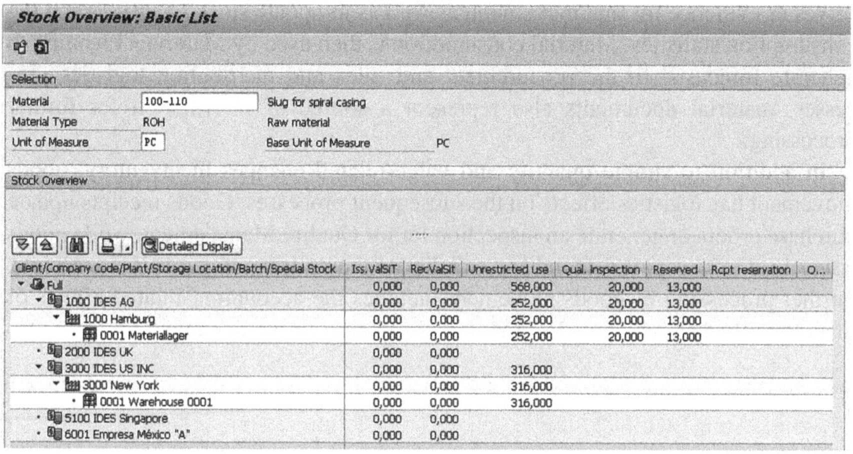

Stock Overview: Basic List

Selection

Material	100-110	Slug for spiral casing	
Material Type	ROH	Raw material	
Unit of Measure	PC	Base Unit of Measure	PC

Stock Overview

Detailed Display

Client/Company Code/Plant/Storage Location/Batch/Special Stock	Iss.ValSIT	RecValSit	Unrestricted use	Qual. inspection	Reserved	Rcpt reservation	O...
Full	0,000	0,000	568,000	20,000	13,000		
1000 IDES AG	0,000	0,000	252,000	20,000	13,000		
1000 Hamburg	0,000	0,000	252,000	20,000	13,000		
0001 Materiallager	0,000	0,000	252,000	20,000	13,000		
2000 IDES UK	0,000	0,000					
3000 IDES US INC	0,000	0,000	316,000				
3000 New York	0,000	0,000	316,000				
0001 Warehouse 0001	0,000	0,000	316,000				
5100 IDES Singapore	0,000	0,000					
6001 Empresa México "A"	0,000	0,000					

Fig. 3.5 Stock overview

3.2.1 *Goods Movement*

The purpose of inventory management is to map the physical stock by recording all activity that changes inventory. Such quantity-based inventory management is done in real time, documented by so-called *material documents*, which form the basis of updating quantities and values and serve as proof of stock movement. In keeping with the motto "No posting without a document", any updating of inventory through goods movement leads to a material document.

Goods movement includes *external* and *internal* procedures. External procedures are processes of procurement and sales and distribution logistics in which material is purchased or sold. Internal procedures are logistics processes in which material movement occurs on the basis of internal stock transfers or production retrievals.

Figure 3.3 depicts the integration of inventory management with core logistics processes, its delimitations to warehouse management and major goods movement types: goods receipt, goods issue, returns, reservations and stock transfers and transfer postings.

3.2.1.1 Effects of Goods Movement

Goods movement represents logistics-related procedures that are recorded in the system. They always result in a change in inventory and lead to an updating in real time that allows an overview of the current inventory situation at any time. In addition to the pure stock overview of plant and storage location stock, it also includes reserved stock and any quantities that are in quality inspection or have already been ordered but have not yet arrived.

Goods movement is documented in *material documents* (Fig. 3.6).

Every incidence of goods movement thus leads to a material document that serves as evidence for that movement, as well as to update stock quantities and the consumption statistics. Material consumption is then used by Materials Planning to generate forecasts. In the procurement and sales and distribution logistics processes, material documents also represent a source of information for further processing.

In addition to simple quantity and value-related changes in inventory, goods movement has logistics effects on the subsequent processes. Goods receipts update purchase orders or generate an inspection lot for Quality Management, while goods issues lead to transport demand and can dictate the printing of goods issue receipts. Further processing of goods movement includes the accounting-related effects of inventory change, which is documented via an accounting document.

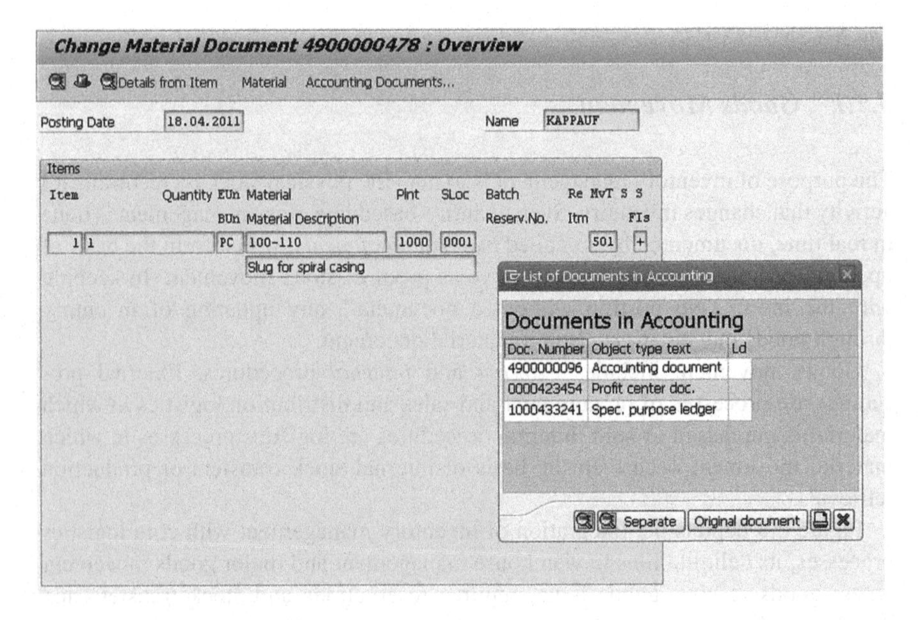

Fig. 3.6 Material document

Material document

Figure 3.6 shows a material document for Material 101-100. Goods movement, in this case a goods receipt of one unit to non-restricted stock of Storage Location 0001 in Plant 1000, was executed with Movement Type 501. Material 101-100 is maintained in stock on a valuated basis. Thus, the system generates an accounting document with the number 4900000096 as a follow-on document of the material document. The accounting document is discussed in Sect. 3.2.2, "Inventory Valuation", and illustrated in Fig. 3.8.

Goods receipts are goods movements that lead to an increase in stock. Goods receipts result either from an external procurement procedure, the receipt of goods from an external supplier or from the inflow of finished products from Production. We can differentiate between the following types of goods receipts:

Goods receipts for a *purchase order* are done via an external procurement procedure with reference to an order placed with a vendor. The goods receipt for this purchasing process enables verification of whether the delivered items correspond to those ordered. In addition, the reference to an order document enables a check of the permissible over- and underdeliveries and allows an evaluation of the goods receipt based on the order and invoice price, since the vendor invoice is also checked and compared to the ordered and delivered quantities.

Production orders from *Production* are another trigger of goods receipts. Production orders trigger production processes not within the scope of this book, but they also serve as an important planning and monitoring instrument for Materials Planning and Inventory Management. The produced materials can either be posted to a certain type of stock or immediately consumed. Goods receipts from Production generally refer to a production order thus determine the storage location where the stock is posted when the finished product enters the warehouse. The type of stock, that is, whether it is unrestricted, restricted or quality inspection stock, depends on the respective system setting in the ERP system.

Inventory management allows you to post goods receipts that do not relate to a preceding document. Such receipts include those initially entered as part of the data being migrated from an older system. You can also create internal as well as external goods receipts without reference to a document. Such instances include free deliveries from a supplier not based on the submission of a purchase order, customer returns without a return order and return deliveries.

A goods issue is generally a goods movement that reduces stock and with which a materials withdrawal or consumption is booked. The withdrawal of materials is usually done in connection with a shipment of goods to a customer or with material consumption related to production processes. Thus, we can differentiate between the following types of goods issue:

The withdrawal, picking and shipment of ordered goods to an external customer comprise the core logistics process of *sales and distribution (SD) logistics*. Goods issue that is done with reference to a delivery represents its effect, from a materials management standpoint, and leads to a reduction of stock. In the case of especially urgent material requisitions for which a preceding document might not be available, goods issue to a customer can be executed without dispatch handling.

Goods issues for Production generally take the form of *withdrawals* of raw materials and supplies for production. We differentiate between planned and unplanned withdrawals. *Planned* withdrawals are done with reference to a production order that has triggered a materials reservation for the planned components. Actual goods issue is related to the production order or the reservation produced by such an order. *Unplanned* goods issues are withdrawals during the production process in which it is determined that, for a certain production order, additional material or a deviating quantity is required in addition to components already withdrawn. Withdrawal is thus unplanned and without reference to a reservation, a so-called *goods issue without reference*.

This type of unplanned goods issue can also be performed for other purposes, such as for the withdrawal of samples and consumption to a cost center or settlement for an internal order.

From the perspective of inventory management, scrapping or withdrawing a sample is considered material withdrawal and thus a goods issue without reference. Scrapping is done when a particular material is no longer needed or has lost quality.

Goods issues for returns are return deliveries to suppliers if the delivered goods for some reason do not correspond to the required type or quantity. Return delivery is done either immediately upon goods receipt, with reference to the purchase order or after the goods receipt has been posted. The material document of the goods receipt or purchase order related to the goods receipt serves as the reference. Actual return delivery is picked and packed, after which a good issue is posted and the shipping documents printed. If the supplier sends a replacement delivery upon receipt of the return delivery, the new goods receipt can reference the return delivery.

A *reservation* represents a request to the warehouse to make a certain quantity of materials available for later withdrawal. This withdrawal is usually earmarked and is taken into account by Materials Planning to ensure that the required material can be procured in a timely manner. Reservations aid in the preliminary planning of future goods movements and can be generated automatically by the system for goods receipts and issues, or be created manually. They are used for planned and unplanned goods issues, or to plan stock transfers. An example of an earmarked goods issue is the reservation of a certain quantity of a specific material for a sales order.

Reservations are used to plan future goods movement. The result of this planning is the reservation document, which increases the reserved inventory of a material without changing the total inventory or unrestricted-use stock. From the viewpoint of material requirements planning, on the other hand, available stock decreases by the reserved quantity (see Fig. 3.7).

Through the specific reference to various account assignment objects (orders, plants or cost centers), reservations simplify and accelerate goods receipts as well as goods issues. The actual reservation is mapped in the ERP system as a reservation document, which consists of a header and item data, and contains planning-relevant information. The header data includes information pertaining to the movement type and accounting object. The item data contains detailed information on the reservation process. It shows what material is required at what time and in what quantity, which plant or storage location is to provide it and where the material is to be moved on the required date.

Stock and requirements list for a material

Figure 3.7 shows the stock and requirements list for Material 101-100 in Plant 1000. The available quantity of 273 units is reduced by planned outflow and increased by planned inflow. Materials planning elements indicate activity such as reservations that reduce inventory and procurement orders that increase stock quantities.

Generally, stock is not exclusively managed from a central site, but rather often decentrally from several locations. In such cases, it may be necessary to transfer materials in one specific location to another storage location or to alter its accounting stock attributes due to internal reasons.

A *stock transfer* refers to a two-step procedure in which material is taken from one storage location and moved to another one. Stock transfers can occur within the same plant as well as between two (or more, but always in pairs) plants or company codes. Stock transfers between storage locations within one plant lead to an update

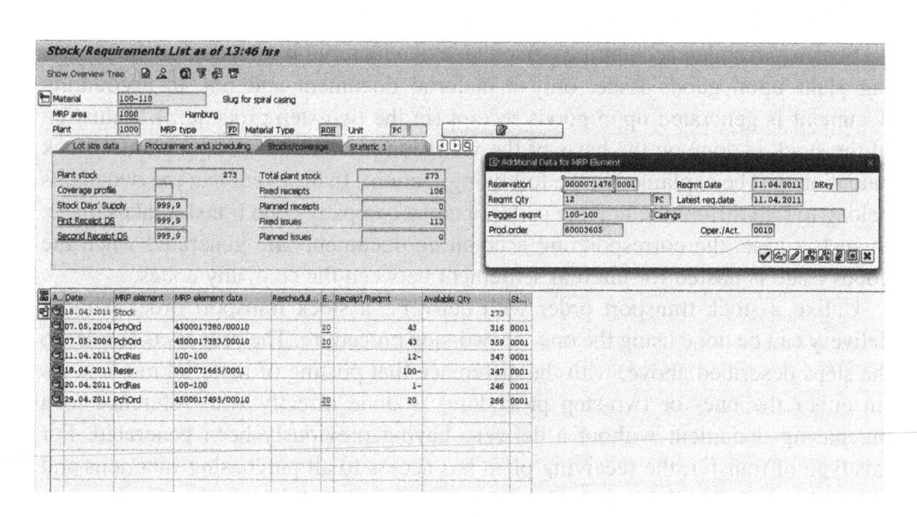

Fig. 3.7 Stock/requirements list

of stock quantities in both storage locations – a reduction in the outflow storage location and an increase in the inflow location. In addition to stock changes, transfers between plants can also influence accounting if both plants belong to different company codes.

Stock transfers can either be done in one or two steps. The *one-step procedure* consists of a goods issue from the issuing storage location and a goods receipt at the receiving location. In this process, only a single material document is generated in the system. The *two-step procedure* also enables the monitoring of the transferred stock through the initial posting of that stock into *transfer stock* after the goods issue posting that stock is maintained in the system with reference to the receiving location. In a second step, the stock is then further posted in the receiving storage location. Another stock transfer option is offered by a *stock transport order*, which can be done with or without deliveries, in a one- or two-step procedure, and either internally or for more than one company code.

Stock transport orders can be executed with or without the *generation of a delivery*. Stock transport orders with delivery are carried out exclusively with the two-step procedure and with two separate goods movements. First, the stock transport order is created in the receiving plant, including any necessary delivery costs. The issuing plant then generates a so-called replenishment delivery with reference to the stock transport order and posts the goods issue. From a management standpoint, the goods issue posting represents a value transfer, because the material is allocated to the receiving plant for accounting. The actual posting is documented through the material document and the accounting document, which is also visible in addition to the stock transport order and replenishment delivery. In order to monitor the quantities being transferred between the two plants, the quantity posted for transfer is first maintained as being stock in transit for the receiving plant. Only after the goods receipt is posted is the quantity booked to the unrestricted-use stock of the receiving plant. The stock in transit is thus diminished and the unrestricted-use stock increased.

Because, from an accounting perspective, the material is allocated to the receiving plant upon goods issue, only a material document and not an accounting document is generated upon goods receipt (in the two-step procedure). Valuation of the stock is done on the basis of the valuation price in the issuing plant. (Stock valuation will be explained in the following section.) In the event that the two plants belong to two different company codes, a cross-company code transfer takes place. In such a case, the corresponding accounting documents are generated when the goods issue is posted for internal settlement between the company codes.

Unlike a stock transport order with delivery, a stock transport order without delivery can be done using the one- or two-step procedure. The process is similar to the steps described above, with the difference that posting of material movements (in either the one- or two-step procedure) is done directly with reference to a purchasing document without a delivery having previously been generated. For this type of transfer, the receiving plant has access to all purchasing functions and can perform such tasks as record delivery costs for a purchase order. Goods issue from the issuing plant is done in Inventory Management without using the SAP ERP Sales and Distribution component SD. In the receiving plant, the quantities posted for transfer are first booked as stock in transit. When the goods receipt is posted, the status of that stock is changed to unrestricted use.

Stock transfers between two plants belonging to different company codes can be done using either the one-step or two-step procedure. Since the material can be valuated differently in each of the company codes, the valuation price in this type of stock transfer is not based on the issuing plant, but rather the condition values stored in the respective company code. Here, too, stock transfer is initially executed with a stock transport order in the receiving plant. Because a vendor is indicated in the stock transport order to which a delivering plant (issuing plant) is allocated in the vendor master record, the system recognizes that the order is a stock transport order with delivery and billing documents, and performs pricing in Purchasing. Pricing can be done based on a maintained purchase info record (see Volume 1, Chap. 4, "Procurement Logistics"). If the order is due for delivery, the issuing plant generates a replenishment delivery either with or without an availability check, depending on the system settings. The goods issue is posted in a one- or two-step procedure. Unlike the types of stock transport order mentioned above, the stock in this case is not posted as being in transit. In the stock overview of the issuing plant, the issued quantity is taken from the unrestricted-use stock and posted to delivery stock.

From the accounting perspective, when there is a stock transport order with delivery and billing, the stock account of the supplying company code is updated with the valuation price of the material in the delivering plant. For internal settlement, the replenishment delivery is invoiced in relation to the delivery. Billing is handled in the issuing organization with pricing according to the regular procedure, based on the condition technique. The invoice is usually automatically forwarded to the receiving plant with a payment block. If the stock transfer is executed using the two-step method, the receiving plant posts the goods receipt with reference to the delivery. Accounting posts the transferred quantity to the unrestricted-use stock, which increases that stock. From an inventory management perspective, the transferred quantity is then allocated to the company code of the receiving plant. Goods movement thus generates an accounting document with the procurement price taken from the stock transport order. After the delivered quantity is checked, a logistics invoice verification is performed. The payment block is removed, and the invoice is released for payment (see also Volume 1,, Chap. 4, Sect. 4.5.3, "Invoice Verification and Handling of Payments").

In contrast to stock transfers, transfer postings not only indicate pure physical changes in stock through movement, but also a change in the stock identification and qualification of a material. For instance, transfer postings can be done when consignment material is received and put into a company's own stock or when blocked stock is released following a quality check.

3.2.2 *Inventory Valuation*

Because of high capital commitment and warehouse maintenance costs, inventory should be maintained such that everyone responsible can receive precise information at any time regarding quantities and values. Inventory valuation serves the task of exactly recording the capital tied up in inventory.

The materials to be valuated are not only examined with regard to quantity (such as in units or kilograms), but also with regard to value (for instance, in the currency of a particular country). This means that the valuation of materials according to trade and tax laws is designed to produce verification of the whereabouts of materials kept in a warehouse. It can also keep track of the inflow and outflow of materials and inventory for accounting, cost accounting, costing and company statistics. Since materials are purchased or manufactured at various times and their prices often change daily, materials are subject to price fluctuations. The receipt and consumption of materials often happen at separate times. Because prices can change, it is necessary to valuate the materials at a fixed point in time and according to officially approved valuation procedures (within trade and tax laws). These valuation procedures for financial valuation of inventory (lowest value determination, LIFO and FIFO procedures) will not be discussed in any detail here. Instead, we refer to the further recommended reading listed in the appendix of this book.

Within SAP ERP, material valuation is not an independent work area, since several functions for the valuation of materials occur automatically, and the tasks that are to be carried out manually are performed either in Inventory Management or Invoice Verification, depending on the respective organizational structure. Due to the integrative quality of the ERP system, materials movement establishes a connection between Materials Management (MM) and Financial Accounting (FI) that accesses and updates the G/L accounts (see Fig. 3.8).

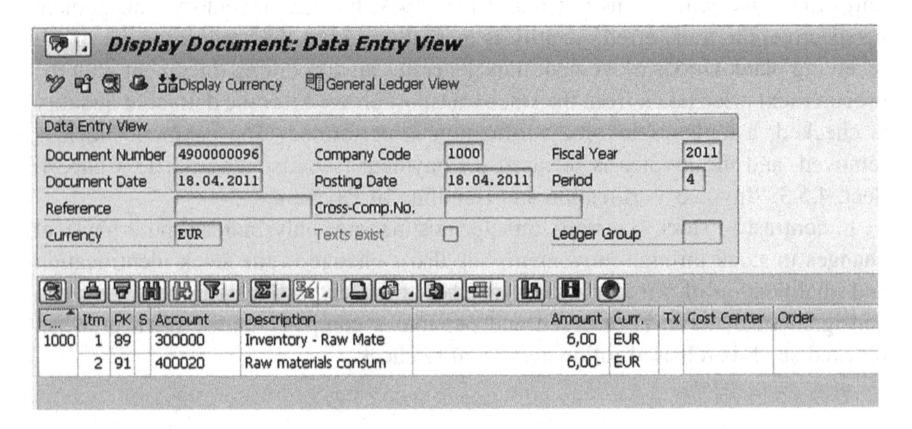

Fig. 3.8 Accounting document for goods movement

Accounting document and stock value

The accounting document in Fig. 3.8 was generated as a document in Accounting for the material document in Fig. 3.6. The subject of the material movement was a goods receipt into unrestricted-use stock. The posting record of this accounting document thus refers to the increase in stock of an unfinished product, and posts the material value in the amount of €6.00 on the debit side of Account 300000. The offsetting entry is posted to the credit side of Account 400020, a stock change account without a cost element. The material value of €6.00 was automatically determined from the material master record of the posted material and rounded off.

The value-based management of material inventory is done automatically through the updating of stock whenever goods are moved. In addition to quantity- and value-based updating of stock, account assignment can also be performed in Cost Accounting. Value-based inventory management is done on the *valuation area* level. The valuation area is an organizational level that can correspond to a plant or company code. The company code, in turn, is derived from the plant via the assigned valuation area responsible for quantity-based inventory management. Generally, valuation of inventory is performed on the plant level.

In Volume 1, Chap. 4, "Procurement Logistics", we briefly looked at material valuation and account assignment in relation to consumable material. Updating of quantity and value (see Fig. 3.8) is based on system settings in addition to the *material type* of a material (see also Volume 1, Chap. 3, Sect. 3.2.2). The system settings control the basic parameters of the valuation: In what level, plant or company code are the materials evaluated? What goods movements are relevant to valuation and what accounts are updated for which procedures? The material type is a fixed parameter that controls whether a material is to be managed in a value- and stock-based manner. In the *accounting views*, the material master determines whether, for instance, various partial stock of a material is to be valuated differently (split valuation) and what price is to be used for the valuation.

The control of the valuation price of valuated stock requires a material type that allows inventory management and inventory valuation. The actual valuation, that is, the valuation price of the material, is based on price control settings in the material master.

According to these settings, stock can be valuated with a constant price (*standard price*) or, adjusting to fluctuations in the procurement price, with a moving *average price*. Moving average prices are advisable for raw materials and externally procured material for which price fluctuations are to be recorded for stock when they happen. Standard prices are constant and suitable for finished and semi-finished products, especially for products manufactured in-house.

Non-valuated stock is maintained for materials for which quantity-based but not value-based inventory management is intended. Such stock includes consumable materials whose valuation and quantity update are based on the settings of the respective material type. Because no stock values are updated for these non-valuated materials, there is no need to maintain accounting views in the material master. The procurement of these materials is thus done only with account assignment.

In the case of materials that are maintained in stock on a quantity as well as a value basis, goods receipt leads to an account-assigned, neutral stock order. The procured quantity is posted to consumption and the value to a consumption account, where the respective account assignment object is debited (see also the procurement of consumption material in Volume 1, Chap. 4, "Procurement Logistics"). Goods receipts for *non-valuated materials* that are managed on a quantity basis lead to an increase in stock and an update of the material master record. The material value is posted to the consumption account, and the account-assigned object is debited with the costs. Non-stock materials represent a special case.

Non-stock materials refer to materials for which neither inventory management nor valuation is to take place. As in the case of non-valuated materials, accounting data is not maintained and procurement is solely done with reference to an account assignment object. When a goods receipt of such materials occurs, neither a quantity nor a value update is undertaken, since the procured quantity is immediately posted as consumption, and the account assignment object is debited with the costs.

In contrast to the uniform valuation of materials on a plant or company-code level, *split valuation* enables a differentiation between the partial stock of a material with regard to various criteria. Such criteria allow you to evaluate certain partial stock with differing values within a plant. For instance, split evaluation is done when a material is simultaneously produced in-house and externally procured, because material stock from in-house production usually has a different valuation price than external production. A further criterion and thus a reason for split valuation is a difference in material qualities and conditions, or batch stock with varying valuation prices.

Split valuation is performed in the following situations:

- Batches with varying characteristics
- Materials of varying sources
- Materials from various deliveries
- Materials of varying quality

The differing valuation of a material is done for each *partial stock*. Thus, you must indicate which partial stock is concerned regarding logistics and value for every valuation-relevant procedure, every goods movement, every invoice receipt and every inventory. Any change in value only applies to that particular stock and is based on the parameters in the accounting view of the material master record that determine whether and how a split valuation is to be performed for the material.

Partial stock is a subquantity of a total stock. The value of the total stock is thus derived from the sum of the partial stocks and their respective stock values.

3.2.3 Special Stock and Special Procurement Forms

Special stock refers to material stock that is maintained separately from other stock and is identified accordingly. This separation is based on special property circumstances and the spatial separation of the location in which they are situated. We differentiate between a company's own and external special stock.

A company's own special stock is material stock that legally belongs to your company but is stored at a vendor or customer. The inventory management of this stock, either unrestricted or in a quality inspection, is done with a neutral storage location, that is, exclusively on a plant level. In particular, consignment stock and stock provided to a subcontractor for subcontracting belong to a company's own special stock. Both types of special stock will be explained in more detail below.

Customer returnable packaging stock is usually returnable transport packaging delivered to a customer as packaging for ordered materials. This transport packaging stock is physically situated at the customer location but is legally the property of the company that shipped it.

Externally owned special stock is stock that legally belongs to an external vendor or customer but is stored at your company. Unlike a company's own special stock, because externally owned special stock is located at your company, it is also maintained at the storage location level. From the viewpoint of material availability, externally owned stock can be unrestricted quality inspection stock as well as blocked stock. A common type of externally owned special stock is vendor consignment goods stock.

Sales order and *project stock* are stock that is available for the fulfillment of a sales order or project. This stock is permanently allocated to the reference object (such as a sales order or PSP element) and serves the production of materials ordered by customers or the execution of a project. Externally owned special stock also includes all transportation packaging that was used in the external procurement from a vendor, and is still the property of that vendor and is to be returned to him by the procuring company.

The following is an overview of consignment and subcontracting, their integration with logistics core processes and their effects on inventory management (Table 3.2).

Consignment stock refers to stock that is initially delivered to the customer without being invoiced and remains the property of the seller until its actual consumption. Only when the stock is withdrawn by the customer is an invoice issued for the amount withdrawn or consumed. Because a purchase commitment generally does not exist, consignment stock can be returned to the vendor until withdrawal by the customer.

Table 3.2 Overview of special stock

	A company's own special stock	Externally owned special stock
Consignment	Customer consignment	Vendor consignment goods
Process stock	Material stock provided to vendor (subcontracting)	Sales order stock Project stock
Packaging	Returnable packaging stock at customer	Returnable transport packaging stock

If, on the other hand, an agreement exists between the customer and vendor stipulating that the customer must keep any remaining consignment stock after a specified deadline, a transfer posting to the customer's stock is performed. Depending on which view is used to examine the consignment, one can differentiate between a customer consignment and a vendor consignment.

For a customer consignment, goods are sent from your company to a customer. Until these goods are withdrawn by the customer, the material is still among the valuated stock of the issuing plant and is the property of your company. The accumulation of special stock at the customer location is done by posting a goods issue (see Fig. 3.9). Supply of consignment goods is done without invoicing the customer. It is only a consignment withdrawal, in which a goods issue reduces customer stock as well as the stock of the issuing plant, that is relevant to billing.

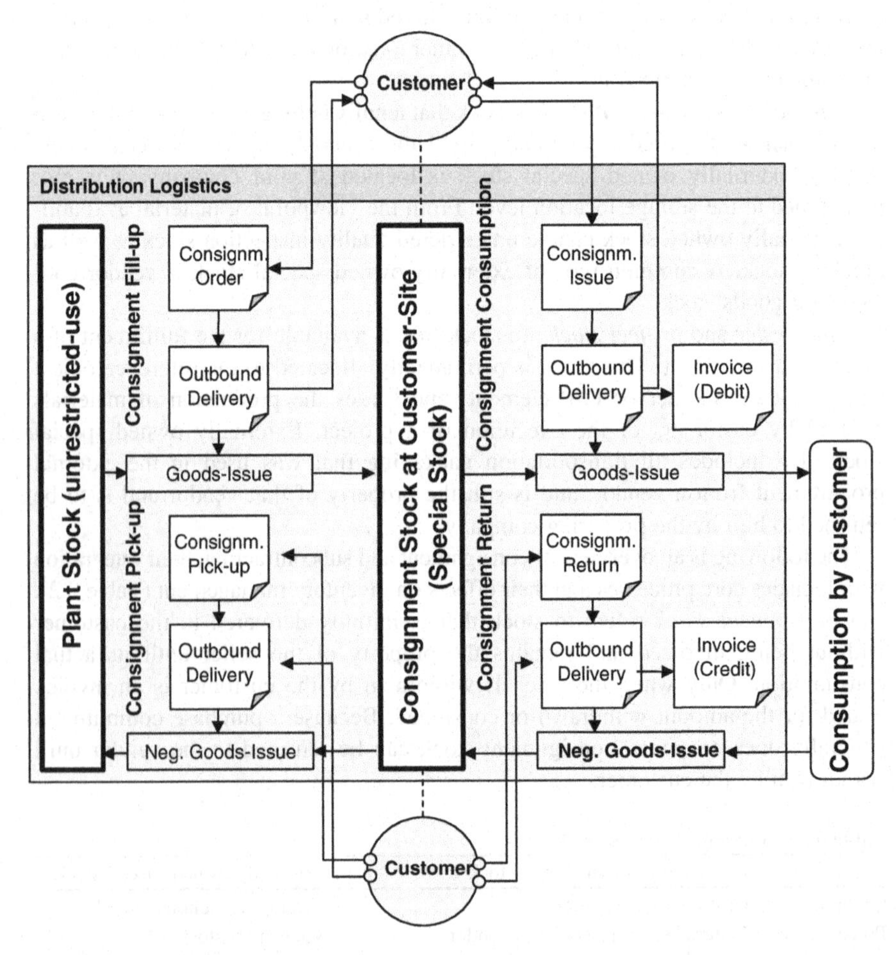

Fig. 3.9 Customer consignment overview

When consignment stock is returned by the customer, known as a *consignment pick-up*, it is removed from the customer special stock and posted back to your company's storage location stock. Reduction of special stock through pick-up is not relevant to billing. Withdrawal of consignment stock can be reversed through returns. Goods receipt for a consignment return replenishes the special stock and leads to a credit for the customer.

From the view of procurement logistics, vendor consignment goods represent a special type of procurement in which a vendor provides material to a company without demanding payment. The vendor remains the owner of these goods until the company withdraws something from the consignment stores. The *consignment stores*, that is, the consignment stock, are also located at the customer site, which in this case is the procuring company. Withdrawal of material leads to a liability toward the customer, who issues an invoice at stipulated periods. From the perspective of inventory management, the consignment stock is maintained under the same material number as the company's own stock. This gives Materials Planning the possibility of considering the non-valuated consignment stock as unrestricted-use stock.

Procuring consignment stock and subsequent invoicing upon withdrawal require a price agreement to be in place with the external vendor. Such a price agreement is stored as a purchasing info record and makes use of the condition technique to save individual rebates and quantity scales in the system (see also the section on purchasing info records in Volume 1, Chap. 4, "Procurement Logistics"). Because the consignment stock of a particular material can come from more than one vendor with varying prices, the stock is managed separately with the price of the respective vendor (see Fig. 3.10).

The actual procurement process for consignment material is done in the same way as for the external procurement of standard materials, via purchasing requisitions, purchase orders and outline agreements. From a purchasing perspective, procurement is completed upon goods receipt. The goods receipt of non-valuated consignment stock can be done with or without reference to unrestricted-use, blocked or quality inspection stock. Invoicing and payment of the material is initiated only after its withdrawal from stock.

The *withdrawal* and consumption of consignment material is done via a goods issue from the unrestricted-use consignment stock of the vendor. Withdrawal leads to a liability toward the external vendor.

Because the vendor cannot directly track the goods withdrawal, the liabilities are settled periodically without an invoice receipt. Payment is done via a settlement run by generating a credit note for the respective vendor.

3.2.3.1 Subcontracting

Another special case in external procurement is *subcontracting*, in which a company orders material from an external vendor and supplies that vendor, the subcontractor, with some or all of the components needed to produce the ordered material. In this section, we will examine the procurement and subcontracting of a finished product from the viewpoint of purchasing and inventory management. The procurement process begins by ordering the finished product from an external vendor.

The finished product is ordered through a purchasing requisition, a purchase order or a scheduling agreement with a subcontracting line. Such an order not only includes information regarding the material to be produced, but also, in one or more subitems, information regarding the components that are provided to the supplier for the subcontracting job. The components are either entered manually or derived with the aid of a BOM (bill of material) explosion for the procured finished product.

The physical provision of components is usually done via a transfer posting from unrestricted-use stock to *vendor consignment stock*. Alternatively, the required components can be provided by a second vendor. In such a case, a purchase order is placed to the second vendor for the required components. In both cases, the components belong to the company and are maintained from an inventory management standpoint as special stock and identified as vendor consignment stock (see Fig. 3.11).

After manufacture or processing, the subcontractor supplies the ordered product. Goods receipt not only leads to an increase in stock of the finished product, but also

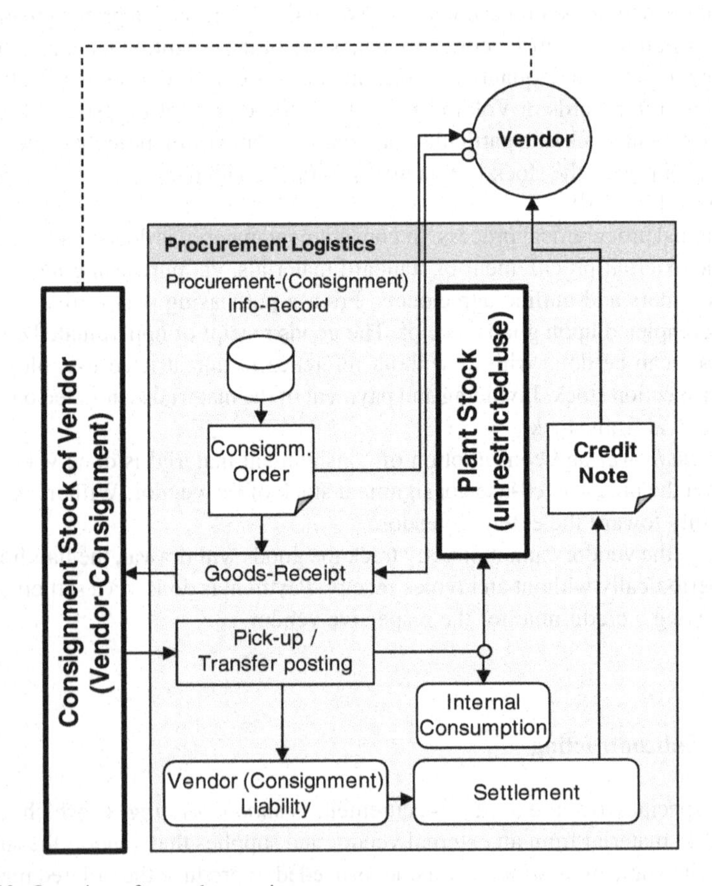

Fig. 3.10 Overview of a vendor consignment

to the consumption of components from the vendor consignment stock. To ensure exact allocation of the receipt of the finished product as well as the component consumption from the vendor consignment stock, goods receipt is performed with reference to the order. The consumption of components for each original goods issue item takes place exclusively from the vendor consignment stock. Consumption of components that deviates from the component quantities indicated in the order is either recorded in the goods receipt or posted as a subsequent adjustment.

3.2.3.2 Third-Party Order Processing

The material itself or the purchasing procedure and the involved system settings determine the way in which a material is procured for a particular sales order. Procurement is carried out from an available storage location stock via internal or external procurement, triggered by a purchasing requisition, an order, a planned or production order, or by a delivery from an external vendor. *Third-party order processing* is an order placed with an external vendor with the stipulation that the

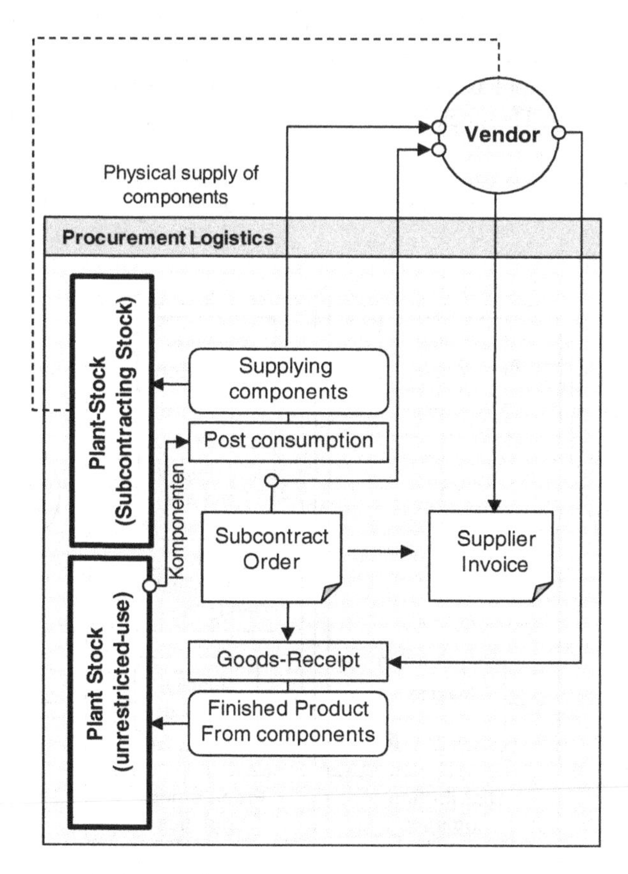

Fig. 3.11 Overview of subcontracting

order is to be delivered to a third party. The customer thus orders from one company, but delivery is performed by an external supplier who sends the goods directly to the customer and invoices the company for them.

The decision to have an item sent by a supplier rather than the respective company is done on the level of the corresponding order item. A sales order can thus consist of several different standard and third-party items. The third-party items can be entered manually or generated automatically, based on parameters in the material master, such as when a sold material is to be exclusively procured externally.

When saving a sales order with third-party items, the items to be procured externally generate a purchase requisition item. The conversion of a purchase requisition to a purchase order is described in Volume 1, Chap. 4, "Procurement Logistics". For the conversion, the system takes on the data of the sales order, including the customer delivery address. The order is transmitted to the supplier, who then delivers the goods to the final customer (see Fig. 3.12).

Third-party processing is goods movement from a supplier to a customer in which the inventory management of the ordering company is not affected. In order to document this step, and to enable a value-based update of stock, a statistical goods receipt can be performed from the view of inventory management. Such a statistical goods receipt has the same effect as a goods receipt for an order with an account assignment: The stock is not updated in its quantity, but only with regard to

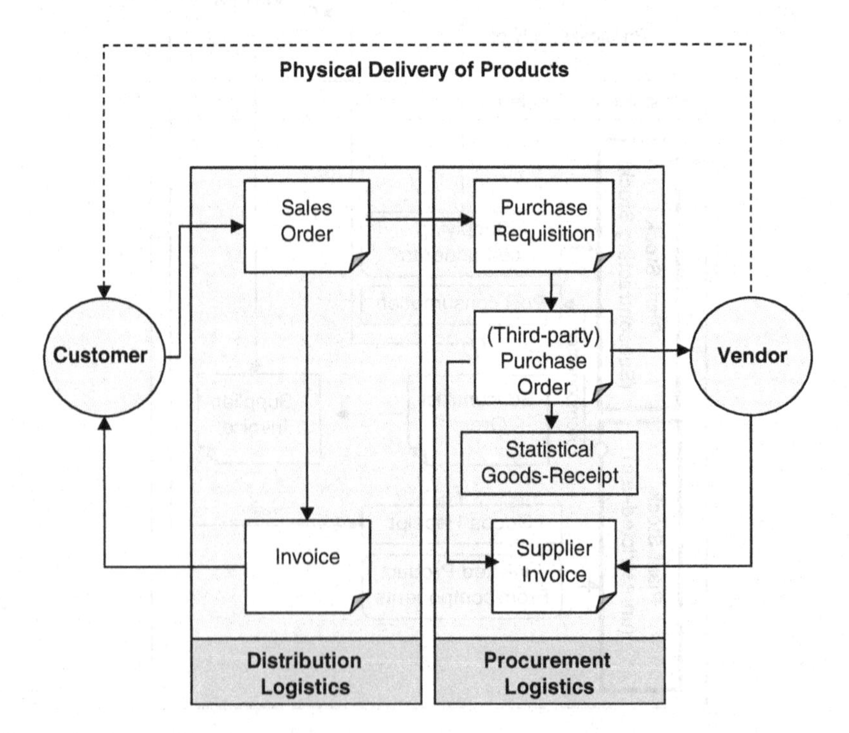

Fig. 3.12 Course of third-party processing

value as consumption. The update of the order value is posted to a clearing account for the subsequent invoice verification.

In the event of an external procurement procedure, the vendor sends a vendor invoice to the ordering company. The purchasing procedure is completed when a billing document is generated for the customer. To prevent the customer from receiving the invoice before receiving the goods from the vendor, the system can be set such that a statistical goods receipt must first have occurred before a vendor invoice can be entered. The goods receipt can be triggered automatically by a shipping notification from a vendor. When a statistical goods receipt has been previously posted, the clearing account is settled through receipt of the vendor invoice. Customer billing based on the calculated and delivered quantities is usually done only after the vendor invoice has been entered in the invoice verification process.

Alternatively, the system can be set such that order-related billing is done with reference to the original order quantity immediately after the sales order is generated.

3.2.4 Handling Units

The so-called *handling unit* (HU) is a physical unit consisting of a material to be packaged and the packaging itself. It can be identified by a specific number. The HU is much more than a simple shipping unit, and in *Handling Unit Management* (HUM), it represents a package management system that includes all logistics processes.

Handling Unit Management (see Fig. 3.13) enables the mapping of purely packaging-controlled logistics by observing not the goods movement of an individual material, but rather the logistics movements of the HUs containing those materials. This type of observation allows a simplified, systematic mapping of goods movements and an optimization of logistics core processes with regard to a purely package-related processing of material movements.

Handling units can be used in all logistics processes and across multiple processes, as well as outside of your own SAP system. That is why every HU has a specific, scannable identification number that is either assigned within an SAP system based on internal number ranges or across several systems. This specific number allows detailed information on the content and packaging of a handling unit

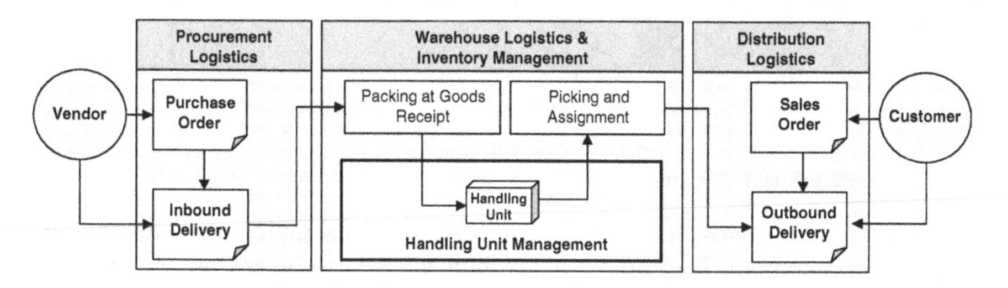

Fig. 3.13 Handling Unit Management

to be accessed at any time, and enables the cross-process, chronological documentation of all goods movements of an HU. To identify the handling unit, a *serial shipper container code* (SSCC) is generally used. This 18-digit code serves to identify logistics units and is part of the coding standard EAN-128.

Handling units are supported in all logistics processes, and in addition to the pure inventory management information (regarding what material in what quantities and which packaging makes up an HU) also include the current status of a handling unit. This includes whether or not a goods receipt or issue has taken place or been announced, and the current location in the warehouse. All business transactions are recorded in the history of the handling unit and enable seamless documentation and evaluation of its life cycle (see also Fig. 3.32).

In SAP applications, handling units consist of an HU header and HU items (see Fig. 3.14). In addition to the specific identification number, the header information also includes the packaging material used, the total weight and dimensions of the handling unit and the history. The items contain the packaged materials and the packaged quantity.

Detailed data of a handling unit

Figure 3.14 shows the detailed data of a "nested" handling unit. On the uppermost level, the image shows a container as transport equipment (Packaging Material TM-080). This container has two packages (Packaging Material PK-101), each of which contains a packaged material, 101-110 and 100-110.

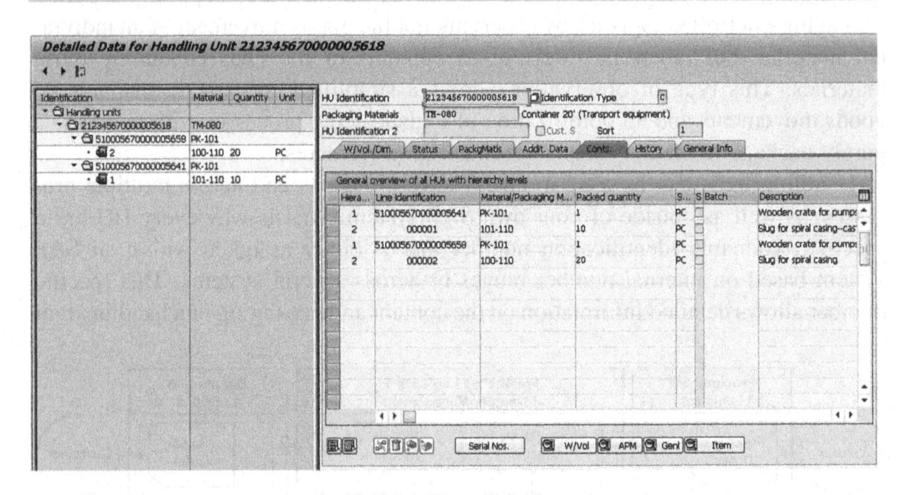

Fig. 3.14 Detailed data of a handling unit

Handling units represent a physical unit consisting of shipping materials and the materials they contain. The purpose of *shipping materials* is to enclose the material or keep it together. We can differentiate between shipping materials serving as loading equipment and "pure" packaging material. Traditional loading equipment includes pallets, pallet boxes and containers. Packaging materials are usually boxes, plastic foils and cardboard.

A handling unit can itself be packaged and then represents a new handling unit. In addition, all packing functions are available to pack, repack or unpack a handling unit. Handling units can be manually created or generated automatically upon goods receipt or in the packing zone of a warehouse. Packing is a separate function within logistics, where a shipping material can be entered in a packaging dialogue at any time and a handling unit can be subsequently created.

Depending on the warehouse management system used (SAP WM or SAP EWM), the automatic generation of handling units is based on the packaging recommendations saved in the system, corresponding regulations and the *packaging specifications*. In SAP ERP, automatic packaging is done according to the settings stored in the system per delivery type and the stored packaging recommendations. The actual shipping materials are maintained in an SAP system as *packing materials*. Packing materials are materials having a special material type that are suitable for use in packaging.

Manual packing in SAP ERP is usually done with a reference document for the inbound or outbound delivery of materials. Manual packing in relation to outbound deliveries with SAP ERP is discussed in Volume 1, Chap. 6, "Distribution Logistics".

In addition to the possibility of creating and editing HUs with a document reference, SAP ERP allows you to use a *packing station* (see Fig. 3.15).

This function is especially designed for employees in the warehouse who physically pack materials and enter data either via a keyboard or scanner. In addition, the exact weight can be determined using a scale connected to the system; the corresponding values are transferred to the dialogue screen. Packing with a packing station can be used for outbound as well as inbound deliveries. The left side of Fig. 3.15 shows the structure of the handling unit with the packaging material, and the right side indicates the material and the partial quantity to be packed.

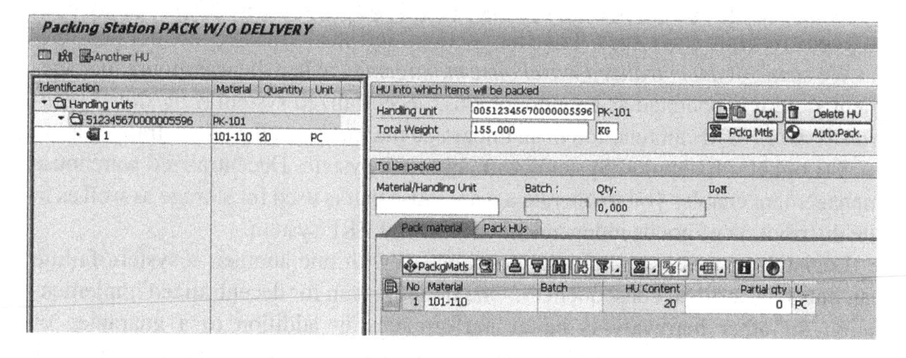

Fig. 3.15 Packing station

3.3 Warehouse Management with WM

SAP ERP is the core of companywide data entry and processing. As described in the previous section, inventory management, from a logistics standpoint, deals with the quantity- and value-based data entry and processing of material stock, and generally serves the documentation, planning and entry of all goods movements. In this section, we turn our attention to the *physical* movement of stock with WM.

Physical goods movement can be processed with the aid of the warehouse management function integrated into ERP. *Warehouse Management* (WM) not only offers an overview of the total quantity of a material in the warehouse, but also provides information on the location of a specific material. Integrating warehouse management with inventory management is of central importance because for the mapping of goods receipt and issue processes, inventory management postings represent either the trigger or the completion of warehouse management activity. One example of this is sales order processing in distribution logistics, in which the connection to delivery processing enables the picking of goods with reference to the delivery. The completion of warehouse management in this case is goods issue with reference to the delivery.

Figure 3.16 shows an overview of warehouse management with SAP ERP and its technical-process connection to inventory management. Integration of warehouse management is primarily based on the fact that goods receipt and issue postings trigger follow-up processes in warehouse management. These transfer requirements, picking and putaway, and operative stock transfer of material are mapped and controlled by a *transfer order*.

A special type of warehouse management with SAP ERP is *Lean WM* – inventory management on the warehouse level that enables the picking of deliveries. In contrast to a WM system in which goods movements and changes in inventory are done on the bin location level, Lean WM performs inventory management exclusively on the storage location level. Stock quantities are thus only visible within the Inventory Management function. Like WM, Lean WM also generates transfer orders. However, a transport order generated for a delivery serves as a picking list containing the articles and quantities to be withdrawn. Because no bin location information is recorded, Lean WM can only be used for the processing of goods receipts and issues. Transfer postings and stock transfers are not possible.

We speak of *decentralized warehouse management* when the warehouse management system is operated as an autonomous, decentralized system (*Logistics Execution System*). WM can function in an integrated manner as well as decentralized as an LES (Logistics Execution System) with any ERP system. Decentralized warehouse management enables you to operate a warehouse that is used for storage as well as in the distribution of goods independent of a central ERP system.

Logistics processes are closely coordinated with one another; a system failure can thus have critical consequences. One major reason for decentralized implementation on other hardware is better performance in addition to a guarantee of continual availability of the decentralized warehouse management system and a minimization of failure risk.

Figure 3.17 shows a diagram of the technical-process connection of a decentralized, ERP-based warehouse management system.

The integration of both systems is primarily done by replicating inbound and outbound deliveries from the ERP system to the decentralized warehouse management system. Actual warehouse management, the physical realization of warehouse activities resulting from deliveries, occurs decentrally with the aid of transfer orders. When the planned goods movement based on the deliveries is carried out, the central system confirms the delivery to the lead ERP system and triggers a change in stock. This decentralized processing causes the physical change in stock to be upstream of the resulting inventory posting, because inventory management takes place in the lead ERP system.

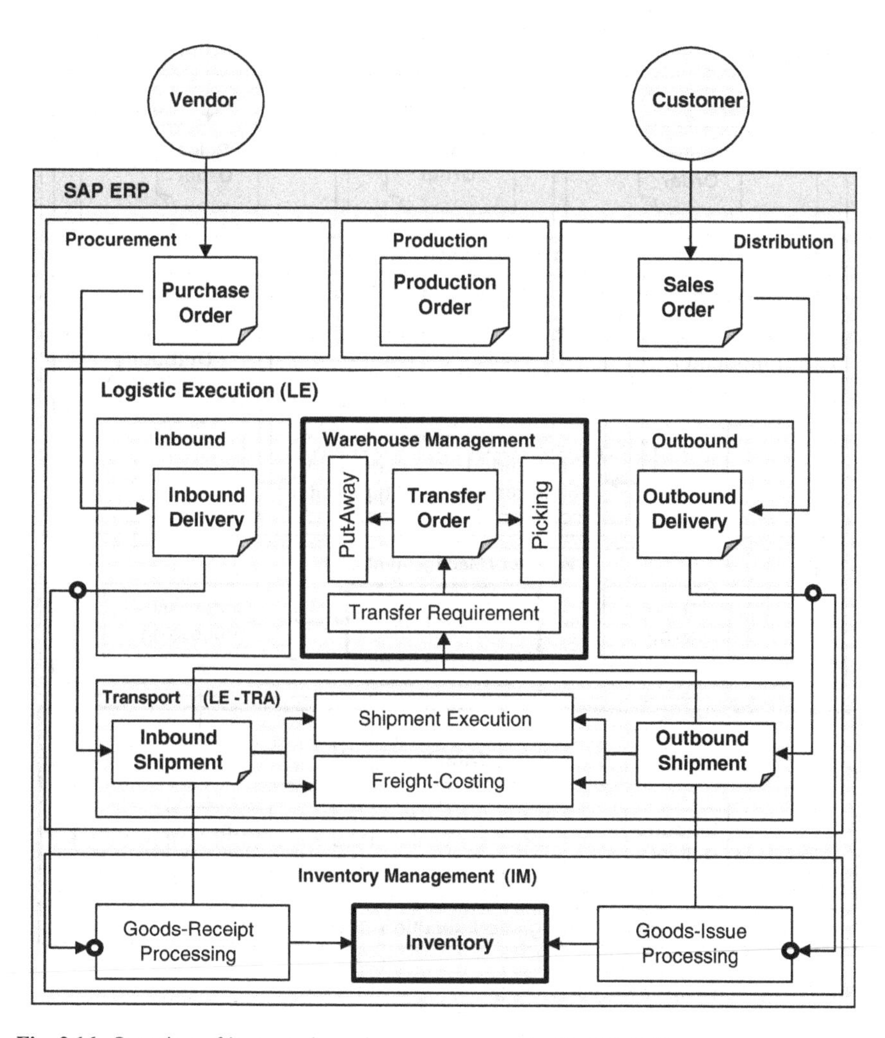

Fig. 3.16 Overview of integrated warehouse management with SAP ERP

As an ERP-based warehouse management system, WM offers the opportunity to map an entire warehouse complex, including its physical and logistics structure, right up to the lowest bin location level. In the following section, we will use a concrete example of goods receipt and issue to provide an overview of warehouse management with WM (SAP ERP) and its core functions.

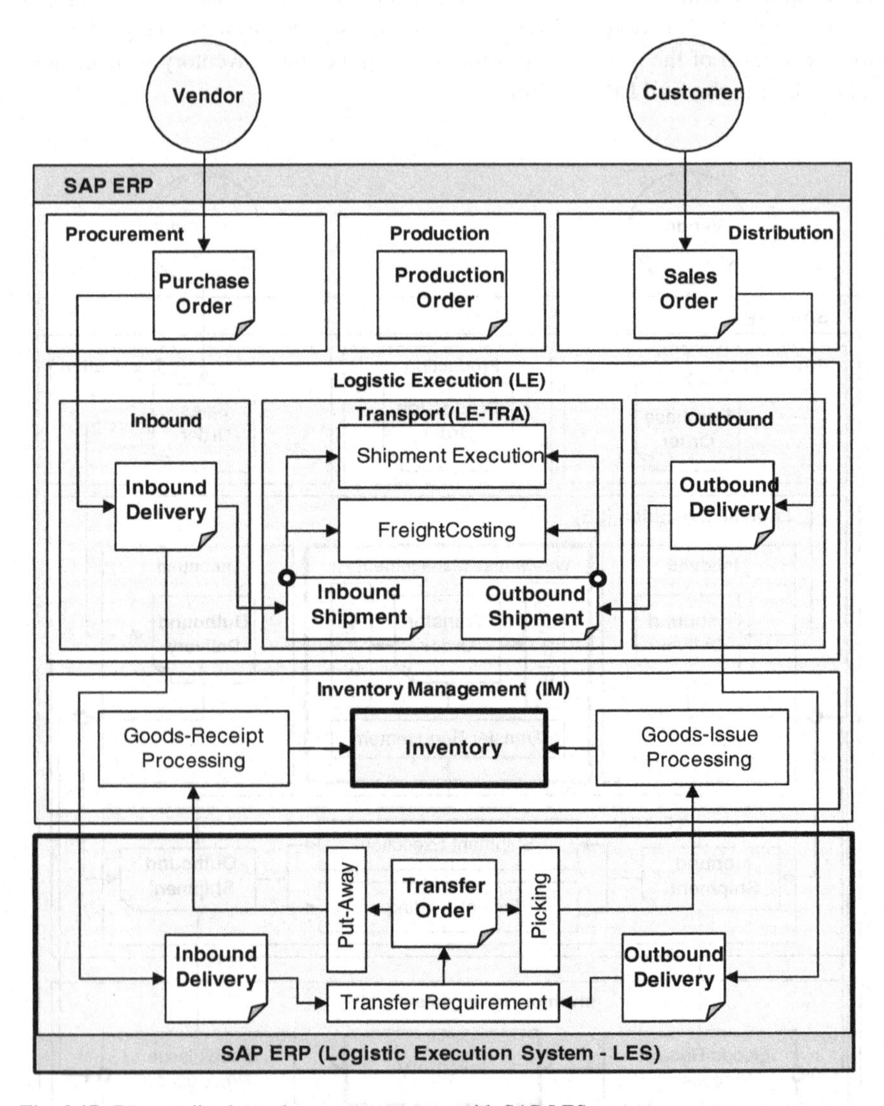

Fig. 3.17 Decentralized warehouse management with SAP LES

3.3.1 Warehouse Structure and Integration

Without the use of a warehouse management system, the storage location would be the lowest level on which stock could be managed. As a rule, the storage location represents the physical location of material stock or reflects the various storage facilities of a warehouse complex. When WM is used as a warehouse management system, quantity-based inventory management is still done on the level of the storage locations. Mapping of the storage facility – the physical structure of the warehouse and the internal structures of the warehouse complex – occurs with the aid of flexible organizational structures.

Every warehouse has its own *structure* according to which goods are stored depending on their composition and space requirements. In automated warehouses, this structure is based on existing automation technology. In the age of *eBusiness* and *just-in-time* practices, the ratio of automated warehouses is steadily increasing. The term automation refers to machines taking on physical warehouse functions as well as warehouse control, or material flow computers performing process control and automatic control tasks.

Automated warehouses are generally either purely pallet warehouses with picking systems and material replenishment functions, or container warehouses in which picking is done on a "goods-to-person" basis. Warehouse automation is often combined with manually operated storage systems in accordance with process requirements. The organizational elements of WM enable individual mapping of the respective warehouse structures.

This book does not delve into warehouse automation technologies and their system integration into SAP Business Suite. The *Material Flow System* (MFS) and the *Radio Frequency Framework* (RF Framework), both components of SAP EWM, are briefly discussed in Sect. 3.4.6, "Cross-Warehouse Functions" below.

The *warehouse number* pools all organizational and physical characteristics of a warehouse and, in practice, typically represents the actual warehouse building or building complex in which the warehouse is located. The warehouse number, as the highest organizational element, thus divides the company from the viewpoint of warehouse logistics and is the basis of various control parameters in WM (see also Fig. 3.19).

A warehouse number is generally divided into several storage types. Each *storage type* defines spatial or organizational circumstances – a storage section, storage equipment or a specific storage zone. In any case, the storage type describes a specific area within a warehouse that is characterized either by an identified space or organizational procedure. Storage types include one or more storage sections and form the basis for storage-type-specific control parameters that especially control putaway and picking as well as the inventory-taking procedure of the corresponding storage type. It is also possible to save storage-type-specific material data. The most common physical storage types are:

- Bulk storage
- Open storage

- High rack storage
- Shelf storage
- Picking areas

A special kind of storage is represented by *interim storage types*. The interim storage type forms the organizational interface between warehouse management and inventory management. Integration is initially based on the type of movement in inventory management, the central instrument for the control of goods movements and a corresponding reference movement type in warehouse management. The *reference movement type* or WM movement type controls the material movements from the viewpoint of warehouse management and decides on the interim storage type to be used. In the case of a goods movement for a purchase order, the system first determines the relevant inventory management movement type and then the allocated WM movement type. Updating of inventory is subsequently done in the goods receipt area, a typical interim storage type – a storage type that is used jointly by inventory management and warehouse management. They especially include:

- Goods receipt areas
- Goods issue zones
- Posting change zones
- Interim storage area for differences

The *picking area* is an organizational element within a storage type or section that logistically pools storage sections for the purpose of removal based on identical picking activities. Unlike the storage section that pools a logistical grouping of storage bins based on their common putaway strategy, the picking area considers the respective picking strategy. For outbound deliveries, the picking area speeds up the shipping procedure by enabling parallel picking by splitting the picking list into corresponding picking areas (see Fig. 3.18).

From the viewpoint of stock placement control, the *storage section* pools those storage bins with the same or similar characteristics. As a physical or logical subunit of a storage type, it represents the counterpart of the picking area, and serves as an organizational aid for material placement. The criteria for pooling depend on the respective characteristics of the storage bins or on the characteristics of the materials stored there. An example would be pooling based on the load transfer frequency of a material (a fast- or slow-moving item) or based on physical characteristics (such as weight or size). These storage-bin-related criteria often have to do with the distance to the load transfer point, the load capacity of the storage bin or the temperature. High rack storage, for instance, mapped as a storage type, consists of several storage bins of varying sizes. The storage bins in the lower sections are often larger to accommodate especially heavy or bulky material. Storage bins in the upper levels are generally smaller, with the front area reserved for fast-moving items and the rear for materials with a lower transfer frequency.

A *storage bin* is the smallest spatial unit in WM and typically represents the physical bin to which a storage type is specifically allocated. Storage bins are specifically assigned using a coordinate system. For example, the coordinates

01-05-05 specify a storage bin in Aisle 1, Column 5 on Level 5. With the exception of a few alphanumeric characters, any letter and number combination may be used for storage bin coordinates, depending on operative needs.

From the view of the warehouse management system, storage bins represent master data to which additional characteristics can be entered. In addition to the maximum weight that a storage bin can carry, such other characteristics can include the total capacity and storage bin type. The storage bin type influences placing strategy, such as when the system searches for a storage bin based on a certain pallet type that is to be placed into a bin.

Physical loading and unloading of trucks is performed at the storage locations. To optimize the picking and placing processes in a warehouse, these doors are situated near so-called *staging areas*. From the system view, the doors and staging areas are organizational units to which warehouse numbers are allocated. The doors are usually allocated to staging areas and configured for subsequent use for goods receipt, goods issue, cross-docking or flow-through.

Unlike the doors, through which materials enter and leave the warehouse, staging areas serve as interim storage sites for materials. Because of their vicinity to the doors, they can be configured for goods receipt or issue. In the case of goods receipt, they serve to temporarily store materials that have been taken in for goods issue and must subsequently be transported into storage. In their function for goods issue, they serve as interim storage for picked materials that must subsequently be loaded for shipment at their assigned door.

In contrast to the physical warehouse structure that is mapped in the system with the organizational elements mentioned above, so-called *quants* are used to manage transactions of quantities on the lowest storage bin level. From a management point

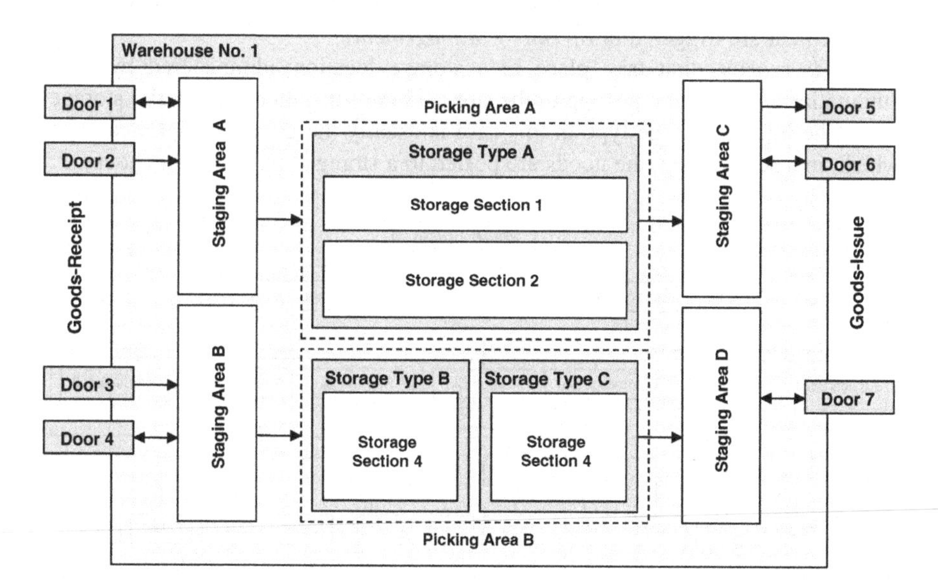

Fig. 3.18 Organizational structure of a WM warehouse

of view, a quant is a certain quantity of materials having the same characteristics that are situated in one specific, clearly identifiable storage bin. A quant is generated by the system when materials are placed in an empty storage bin, updated through picking and placing, and automatically deleted by the system when there is no longer any stock in that particular storage bin. Materials with differing characteristics, such as different batches, are maintained in a single storage bin as two quants (Fig. 3.31 illustrates the inventory data of a quant).

Like a quant, a storage unit aids not only in the structuring of a warehouse, but also in the logistical pooling of physical material quantities within a warehouse. WM employs its *Storage Unit Management* to manage a specific material quantity such as pallets or containers belonging together as a unit. Storage units can be composed of one or more material items, and are always identified by a specific code. Without the use of the activated Storage Unit Management on the warehouse number level, all material stock is managed as quants on the storage bin level (see Fig. 3.19). Activation of Storage Unit Management causes inventory management activity on the pallet and storage unit level, whereby a storage unit may consist of one or more quants. Several storage units can be situated in a single storage bin. The activation of storage units serves to optimize storage capacity and control material flow by allowing the movement of heterogeneous pallets having more than one material as a single unit within the warehouse. One major advantage is that you can determine where each storage unit is located in the warehouse, what material quantity is stored in it, and which activities have already been completed or are planned for that storage unit (see also Fig. 3.30).

WM is fully integrated in the SAP ERP system. Material movements, goods receipts, picking and shipment of materials for sales orders thus lead to physical movement in the warehouse. Most goods movements occurring in Warehouse Management are triggered in *Inventory Management*.

Goods receipts that take place in a storage location administered by WM automatically generate a posting to the storage bin of the allocated interim storage type. The interim storage type in this case is usually the goods receipt area from which, in a second step, the goods are posted to a storage bin in the warehouse.

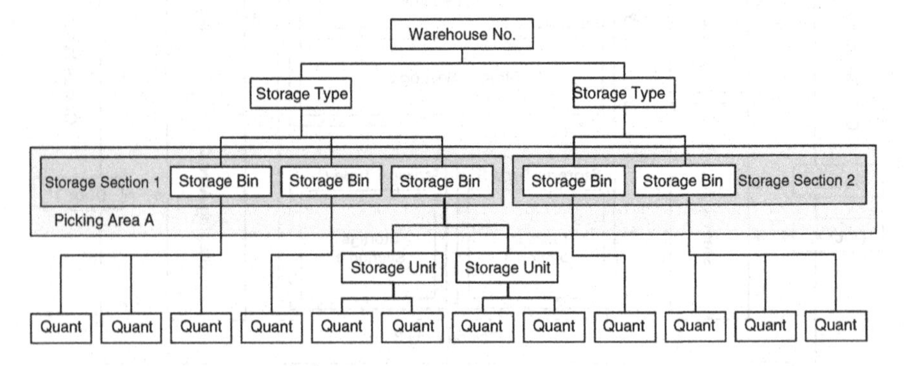

Fig. 3.19 Structure of a WM warehouse

In the case of *goods issues* without an outbound delivery, the material is posted to a goods issue zone after picking. Because the posting causes a corresponding reduction in total stock in Inventory Management, posting to the storage bin of this interim storage type first triggers a quant with a negative quantity. In this case, the accounting goods issue posting is accomplished before actual goods issue in the warehouse. Adjustment of the negative quantity occurs upon physical stock removal to the goods issue zone.

The various possibilities for stock placement and removal and the available strategies will be explored in the following sections.

Storage locations can be set in the system as *HU-managed*. Stock situated in an HU-managed storage location is generally packed and maintained as handling units. Mixed stock composed of packed and unpacked materials is not provided for in such cases. For packing procedures, that is, the packing or unpacking of a handling unit, a second storage location, a so-called *partner storage location*, must also be indicated. The partner storage location is then used for stock transfer from a storage location with HU management to a storage location that is not HU-managed. Stock transfers and goods movements are therefore not possible with the indication of a handling unit. If no handling unit is indicated, the system does not generate a material document for the posting, but rather a delivery.

Figure 3.20 shows integration with Inventory Management from the view of goods movement. Stock movements are done with or without reference to an inbound or outbound delivery. Actual stock movement is controlled with the aid of *transfer orders*.

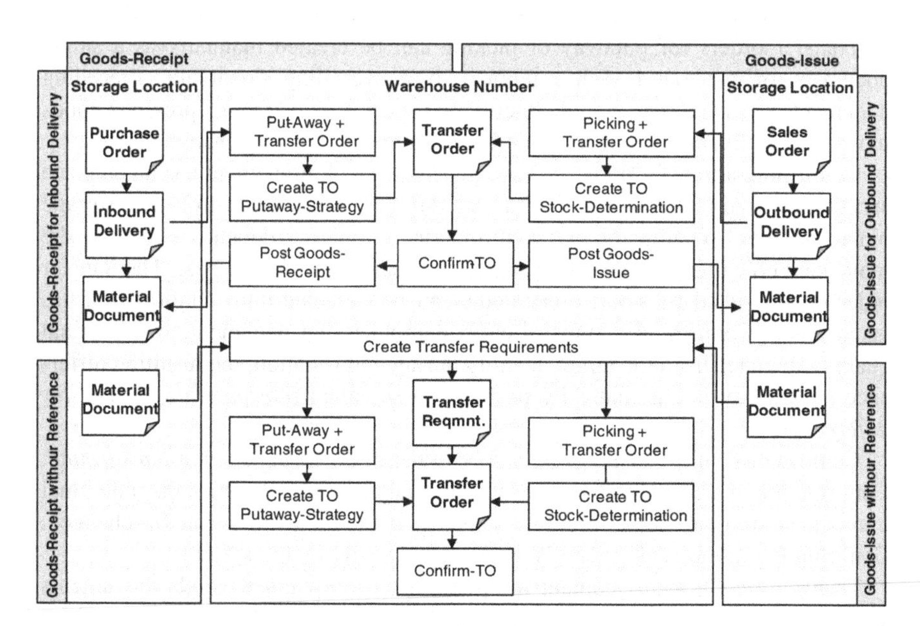

Fig. 3.20 Integration of transfer requirement and transfer order

The *transfer order* is the central document for control of stock movement in WM. Every material movement in a warehouse requires a transfer order that contains all information required to execute the physical transportation into or out of the warehouse or from one storage bin to another. In this regard, there is no differentiation between physical movement and pure transfer postings. Transfer postings can occur when materials are transferred from a quality inspection to unrestricted-use stock.

Transfer orders contain all information needed by the Warehouse Management system to execute stock movement. In addition to the materials and quantities to be moved, such information also includes precise data indicating from where the material is to be moved (source storage bin) and to where (destination storage bin).

- The *source storage bin* indicates the storage bin from where materials are taken or the goods receipt area (interim storage type) from where they are to be taken and placed into storage.
- The *destination storage bin* determines the storage bin in which the materials are to be placed or the goods issue zone (interim storage type) via which they are to be removed.

Return storage bins are used for special situations. When an entire pallet is being removed and picking of a specific partial quantity is to take place, the return storage bin can be used to hold quantities no longer needed if the opened pallet is not to be placed back into storage. Determination of storage bins is done automatically according to the putaway and picking strategies determined by the system. The subsequent sections will focus on the basics of controlling putaway and picking.

Transfer orders for putaway or picking can be created manually as a single document or generated with reference to a preceding document. Preceding documents in such a case would either be transport demand from a previous goods movement, delivery documents, material documents or posting change notices.

Confirming a transfer order acknowledges that stock movement has taken place and the required material quantity has actually been transported from one storage bin to another. Whether or not confirmation is required depends either on the movement type used or on system settings for a particular storage type. The transfer order or individual transport order items can be canceled until confirmation has taken place. Confirming a transfer order completes the document. If the actual quantity deviates from the target quantity during confirmation, the resulting difference in quantity is automatically posted to a so-called *interim storage area for differences*.

Confirmation of a transfer order can be either in the *foreground* or *background*. Confirmation in the foreground refers to a manual procedure in which the individual transaction steps are monitored on screen and default values can be changed. Confirmation in the background is done automatically by the system.

Transfer requirements contain information on the planned goods movements and are used to plan or trigger subsequent stock movements. Unlike transfer orders, which contain detailed information on the stock movement to be executed, transport requirements do not include information on the storage type or bin related to putaway or picking. The header and item data of a transfer requirement includes administrative, material- and quantity-specific information on the planned goods movement, as well as the date and direction (picking or putaway). Transfer requirements are the preceding documents for transfer orders and are usually automatically generated by the system based on stock postings. For goods movements that begin directly in WM, transport requirements can be manually created. The transition from a transfer requirement to a transfer order can either be performed manually or automatically in the background.

3.3.2 *Inbound Delivery*

Inbound deliveries refer to the physical inward movement of materials to a warehouse. Inward movement is generally done on the basis of external procurement, returns or stock transfers. Each goods receipt leads to an increase in warehouse stock and the corresponding stock movement. From the warehouse management view, an inbound delivery can occur with or without reference to a document. When goods receipts are processed with reference, the system refers to an incoming delivery or transfer requirement. An inbound delivery without reference does not refer to a previous goods receipt posting in Inventory Management.

Upon goods receipt, WM records all transactions for the materials to be stored. From reading the barcode, to a warehouse employee identifying the materials, to placing the material in a destination storage bin, the procedures triggered by goods issue in the warehouse are transparent and usually occur automatically. We differentiate between the following goods receipt types:

- Goods receipt with reference to a delivery
- Goods receipt without reference to a delivery
- Goods receipts without prior posting in Inventory Management

We will now turn to an explanation of these types of goods receipt.

3.3.2.1 Goods Receipt with Reference to a Delivery

Goods receipts with reference to a delivery are executed without previous generation of a transfer requirement (see Fig. 3.20). Such a goods receipt with reference can be illustrated with a process example.

Goods receipt based on a purchasing order represents the completion of an external procurement. External procurement, the generation of a purchasing order and the significance of the delivery in the ordering process are covered in Volume 1, Chap. 4, "Procurement Logistics". In a warehouse managed with WM, the receipt of materials delivered by an external vendor is the starting point for the subsequent warehouse activities. In the case of goods receipt with reference to a delivery, the delivery itself represents the starting point. In the selection of the delivery, the operator is supported by the system via its *Inbound Delivery Monitor*. Depending on the selected delivery, the list view shows information such as the deliveries requiring putaway (see Fig. 3.21).

Deliveries to be placed into stock can be selected and form the reference for the generation of the transfer order (see Fig. 3.22). The transfer order takes on the relevant data from the preceding document and, depending on the type of placement saved in the system, determines the destination storage bin in which the material is to be put away.

By activating the storage unit management system, each handling unit to be put away corresponds to a storage unit. When the transfer order is created, the storage unit's destination storage bin has already been determined by the system, and the handling unit and storage unit are known to the system based on a common, specifically assigned number. For subsequent identification via barcode, labels can now be printed for each storage unit (see Fig. 3.23).

Inbound Deliveries for Putaway

Item View | TO in Background | TO in Foreground

Delivery	Deliv.date	Vendor	Name of vendor	Total Weight	WUn	Volume	VUn	OPS	WM
180000136	22.04.2011	20001	TOP Metallverarbeitung GmbH	14.859	LB	0	"3	A	
180000178	18.04.2011	1000	C.E.B. BERLIN					A	
180000179	18.04.2011	1000	C.E.B. BERLIN					A	
180000181	18.04.2011	3000	C.E.B. New York	200	LB	4.800	"3	A	A

Fig. 3.21 Deliveries requiring putaway

Transfer order for an inbound delivery

Figure 3.22 shows the generation of a transfer order for Inbound Delivery 180000181 from External Vendor 3000. The delivery contained several packed material items of 20 units each of the material LEHU-002. Each item was delivered packaged and is identified by a specific handling unit number in the system. According to system settings in this example, a separate transfer order is to be generated for each handling unit.

Create Transfer Order for Inbound Delivery: Overview HU

Warehouse Number: 050
Delivery: 180000181 Delivery Date: 18.04.2011
Vendor: 3000 C.E.B. New York

Active Worklist | Inactive Items | Processed Items

Handling units

Handling Unit	D...	D...	Destination Bin	S...	Pri...	P	C	GR Date	SLED/BBD	Certificate...	T	Item	Material	Plant	St...	Batch	Putaway qty
530005670000005670				P1				18.04.2011				1	LEHU-002	3000	0050		20
130005670000005696				P1				18.04.2011				1	LEHU-002	3000	0050		20

Fig. 3.22 Creating a transfer order

To acknowledge that a handling unit to be placed into stock has actually been put into its destination storage bin, the transfer order is confirmed (see Fig. 3.24). The work list to be confirmed can also be displayed and selected via the inbound delivery screen. Whether or not confirmation is required, that is, if the system must be explicitly informed that the materials have reached their destination, depends on the system settings of the respective movement type.

In addition to a subsequent editing of the destination storage bin suggested by the system, manual confirmation allows you to enter a quantity difference if the actual quantity delivered does not match the planned quantity taken over from the inbound delivery.

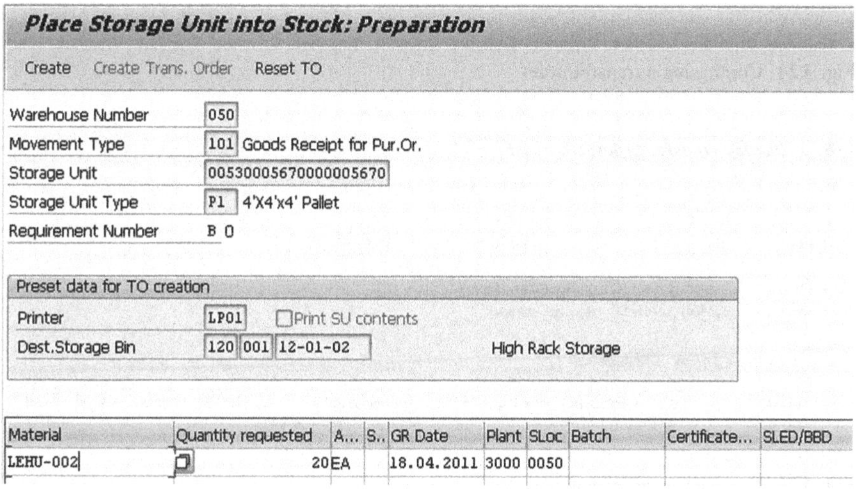

Place Storage Unit into Stock: Preparation

Create Create Trans. Order Reset TO

Warehouse Number 050
Movement Type 101 Goods Receipt for Pur.Or.
Storage Unit 00530005670000005670
Storage Unit Type P1 4'X4'x4' Pallet
Requirement Number B 0

Preset data for TO creation
Printer LP01 Print SU contents
Dest.Storage Bin 120 001 12-01-02 High Rack Storage

Material	Quantity requested	A...	S..	GR Date	Plant	SLoc	Batch	Certificate...	SLED/BBD
LEHU-002	20	EA		18.04.2011	3000	0050			

Fig. 3.23 Putaway with storage unit management

Confirming a transfer order

Figure 3.24 shows manual confirmation of a transfer order in the foreground for the placement of 20 units of Material LEHU-002 to Storage Bin 12-01-03 in Storage Location 0050 of Plant 3000. The Handling Unit number in which this material is situated corresponds to the number of Storage Unit 00130005670000005696.

Figure 3.25 shows the destination data of the transfer order with the confirmed item. The source storage bin data is the goods receipt area for external receipts (Storage Type 902) with Storage section 001 and reference to Delivery 0180000181. The system automatically determined the destination storage bin. In this example, the material was placed in Storage Type 120, high rack storage, in Area 001 and in Storage Bin 12-01-03.

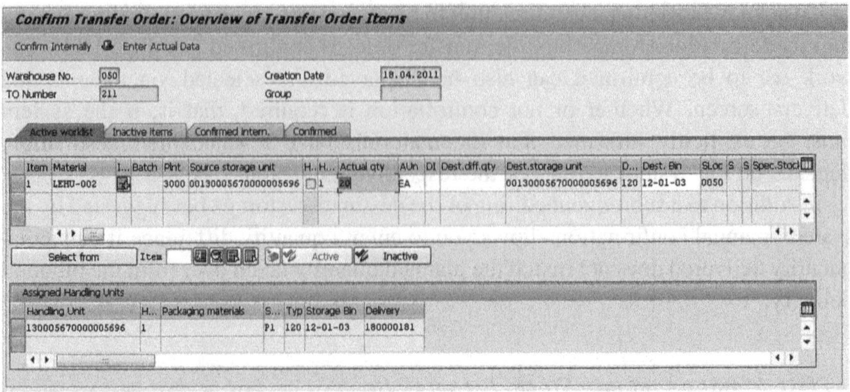

Confirm Transfer Order: Overview of Transfer Order Items

Confirm Internally | Enter Actual Data

Warehouse No. 050 | Creation Date 18.04.2011
TO Number 211 | Group

Active worklist | Inactive items | Confirmed intern. | Confirmed

Item	Material	I...	Batch	Plnt	Source storage unit	H..	H...	Actual qty	AUn	DI	Dest.diff.qty	Dest.storage unit	D...	Dest. Bin	SLoc	S	S	Spec.Stock
1	LEHU-002			3000	00130005670000005696		1	20	EA			00130005670000005696	120	12-01-03	0050			

Select from | Item | Active | Inactive

Assigned Handling Units

Handling Unit	H...	Packaging materials	S...	Typ	Storage Bin	Delivery
130005670000005696	1		P1	120	12-01-03	180000181

Fig. 3.24 Confirming a transfer order

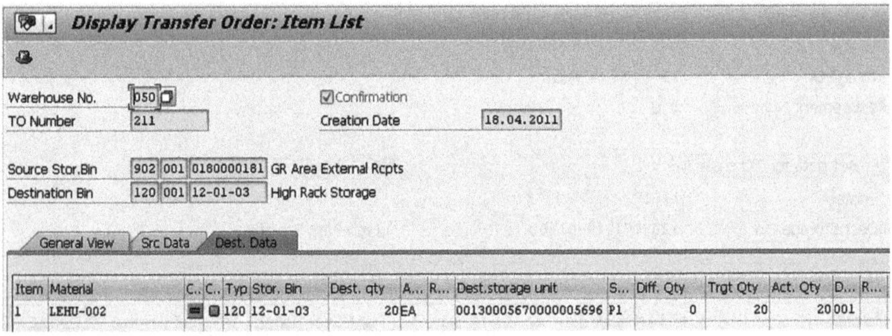

Display Transfer Order: Item List

Warehouse No. 050 | Confirmation
TO Number 211 | Creation Date 18.04.2011

Source Stor.Bin 902 001 0180000181 GR Area External Rcpts
Destination Bin 120 001 12-01-03 High Rack Storage

General View | Src Data | Dest. Data

Item	Material	C..	C..	Typ	Stor. Bin	Dest. qty	A...	R...	Dest.storage unit	S...	Diff. Qty	Trgt Qty	Act. Qty	D...	R...
1	LEHU-002			120	12-01-03	20	EA		00130005670000005696	P1	0	20	20	001	

Fig. 3.25 Transfer order for putaway

From the *warehouse management* view, stock movement is completed: The materials moved with reference to a delivery are stored in the storage bin. To update stock data, the goods receipt is posted in Inventory Management. This posting, the quantity- and value-based update of stock, is documented in the material document. The posting of the goods movement in Inventory Management is either done automatically in the background or manually, for instance with the aid of the Inbound Delivery Monitor.

3.3.2.1 Goods Receipt Without Reference to a Delivery

Stock movements in WM can also be triggered without a direct document reference to a delivery. Putaway in the warehouse management system in such a case is preceded by goods movement in Inventory Management that has already generated a quant in the storage bin of the goods receipt interim storage area (interim storage type) and a transfer requirement (see Fig. 3.20). A transfer order for the transfer requirement can be created manually or automatically by the system.

3.3.3 Transfer Order and Quantity Deviations

The transfer order serves the internal transfer of materials from the goods receipt area to one or more storage bins that have been determined automatically by WM (see Fig. 3.25). After the stock movement, a warehouse employee confirms the transfer order by entering the moved materials manually or with a scanner. Any quantity deviations between goods receipt posts already entered in Inventory Management and the quantities actually confirmed are subsequently adjusted in Inventory Management.

3.3.3.1 Goods Receipts Without Prior Posting in Inventory Management

Goods receipts without a prior posting in Inventory Management and without reference to an inbound delivery are executed directly in Warehouse Management through the generation of a transfer order. The material to be stored is generally already in the goods receipt interim storage area of the warehouse. The transfer order is generated without reference and creates a *negative quant* (negative stock quantity) at the storage bin of the goods receipt interim storage area. The materials are placed from the goods receipt interim storage into the determined storage bin and the transfer order is confirmed. Once the procedure is confirmed, the material is available in the system. Adjustment for the negative quant in the goods receipt interim storage area is done when the goods receipt is posted in Inventory Management. The result is a positive quant on the destination storage bin of the material.

3.3.3.2 Stock Placement Control

The purpose of *Stock Placement Control* is to efficiently find an optimal storage bin while utilizing available stock capacity and considering operational requirements. Automatic determination of a storage bin upon creation of a transfer order is performed in a flexible order stored in the system.

Determination procedures are primarily based on search strategies that first attempt to find a suitable storage type. Storage type determination is controlled by a search sequence in which a *storage type indicator* from the material master as well as the indicator of the stock type and special stock are factors.

After the system has determined the storage type, *storage section determination* is performed for the appropriate storage type if the storage section check is activated. Actual determination in this case is done in a search sequence defined in the system. The storage section indicator and storage class influence this search. The *storage section indicator* controls the search, for instance with regard to fast- or slow-moving materials. The storage class classifies dangerous goods with regard to the storage conditions to be determined. Depending on the storage class (such as whether a material is an explosive substance or flammable liquid), the system determines the storage type and storage section in which materials may be placed.

Determination of the destination storage bin is performed according to the *putaway strategy* stored in the system: WM selects the storage bin (according to operational requirements) using a series of preconfigured strategies, bin capacity restrictions and master data parameters for the materials to be stored. WM has putaway strategies that you can alter with your own extensions and logic. The most important strategies are:

- **Fixed storage bin**
  With the fixed storage bin strategy, the system selects a storage bin based on the bin to which a material has been directly assigned in the material master.
- **Open storage**
  Open storage refers to storage sections of a storage type for which one single storage been has been defined.
- **Addition to existing stock**
  This putaway strategy selects storage bins in which stock of the material to be stored is already situated. The prerequisite for adding additional quantities is the presence of sufficient residual capacity, which is automatically verified by the system. If no capacity is available, the system continues storage bin determination with the next strategy, attempting to find the next free bin.
- **Next empty storage bin**
  This search strategy, in which the system suggests the next empty storage bin for putaway, is frequently used for high rack and shelf warehouses. The search for the next empty storage bin is performed using an alterable sorting sequence that prevents a one-sided utilization of the warehouse and can be controlled via storage bin coordinates.
- **Putaway according to pallets**
  The goal of this strategy is to store only the same storage unit types in a single storage bin. The storage unit type refers to a certain combination of loading equipment and packed materials. Depending on the system settings, it can be defined for a specific storage bin type, such as only pallets having certain dimensions, or so that mixed storage with a variety of storage unit types is permitted.
- **Bulk storage**
  Like the storage unit strategy, the "bulk storage" strategy defines how many columns and what stack height per storage bin are allowed. The stack height often depends on the material and can be controlled via the bulk storage indicator. Such material usually has a large space requirement that needs to be accessed quickly and requires clear structuring.
- **Near the picking bin**
  Depending on system settings, this strategy first tries to determine the fixed bin location of a material. If putaway is not possible for capacity reasons, the

material is placed in a reserve storage space close to the fixed bin and is picked during from-bin transfer. Alternatively, the material can be immediately placed into the reserve storage space without a previous fixed bin check.

3.3.4 Goods Issue

Goods issue refers to the physical departure of materials from the warehouse. This departure is usually due to internal material consumption, material issue or goods issue to customers within the context of sales and distribution logistics. Every goods issue results in a corresponding stock movement and reduces the material stock. Just as in the case of goods receipt, goods issue can occur with or without reference to a reference document. For goods issue processing with reference, the system refers to a delivery or transfer request (see Fig. 3.20).

In the goods issue process, WM records all transactions for materials to be removed. Processes triggered by goods issue in the warehouse (such as picking, packaging and staging of the materials) are transparent and generally executed automatically. We differentiate between the following types of goods issue:

- Goods issue with reference to a delivery
- Goods issue without reference to a delivery
- Goods issue with manual picking

All three will be explained in more detail below.

3.3.4.1 Goods Issue with Reference to a Delivery

For a goods issue with reference to a delivery, the delivery replaces the actual transfer requirement, and the transfer order is generated with reference to a delivery note. We provide a process example of a goods issue with reference to a delivery below.

Goods issue for a delivery represents the conclusion of the sales process in distribution logistics. Sales and sales order processing and the creation of deliveries were discussed in Volume 1, Chap. 6, “Distribution Logistics”. In a warehouse managed using WM, goods issue to a customer is the culmination of a series of warehouse activities beginning with the selection of deliveries to be processed. The system supports the operator with its *Delivery Monitor*, which allows you to choose the day’s workload to be picked. For manual processing, the delivery to be picked is selected and a transfer order is subsequently generated for the issue.

The delivery to be issued is selected, representing the reference for the generation of the transfer order (see Fig. 3.28). The transfer order assumes the relevant data from the preceding document and, depending on the issue data stored in the system, determines the source storage bin from which the material is to be picked and the destination storage bin to which the material is to be moved. Several transfer order items can result from a single goods issue item, depending on whether or not the quantity to be moved is to be picked from various source storage bins.

Outbound delivery and the day's workload

Figure 3.26 shows Outbound Delivery 80016882 for one unit of Material LEHU-002. Delivery is executed from the central warehouse with the Warehouse Number 050, which is managed with WM. The goods issue requires a transfer order (WM-TO), and picking has not yet occurred.

Figure 3.27 shows the Delivery Monitor with the deliveries to be picked. Delivery 80016485 is selected, and the transfer order is generated in the foreground.

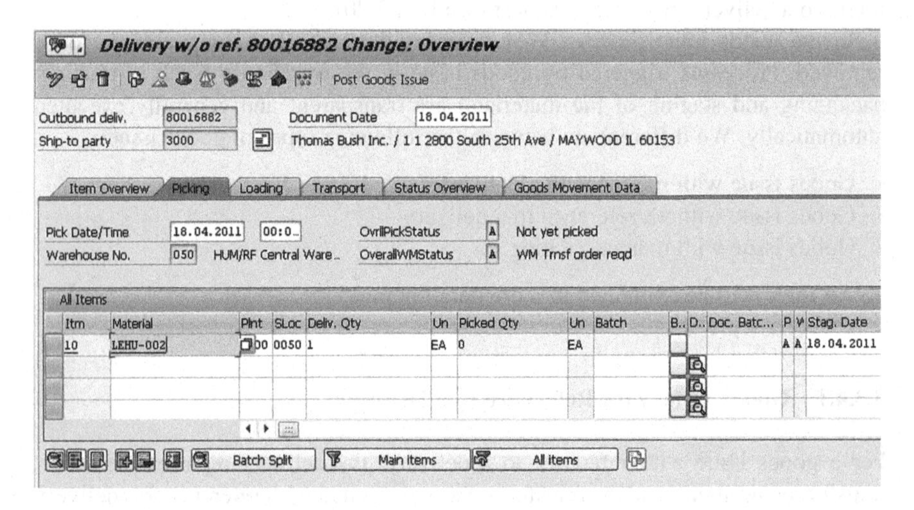

Fig. 3.26 Outbound delivery from a WM-managed warehouse

Day's Workload for Picking

Item View | TO in Background | TO in Foreground | TO for Group

ShPt	Pick Date	Total Weight	WUn	Volume	VUn	ProcTime	Nr Items			
Delivery	**GI Date**	**DPrio Route**	**Total Weight**	**WUn**	**Volume**	**VUn**	**OPS**	**WM**	**Nr Items**	
3000	18.04.2011	142,268	KG	120	"3	0,00	2			
80016866	20.04.2011	000001	140	KG	0	"3	A	A	1	
80016882	10.04.2011	2 000044	2,268	KG	120	"3	A	A	1	
3000	21.04.2011	330	KG	183.072	"3	0,00	1			
80016845	25.04.2011		330	KG	183.072	"3	A		1	

Fig. 3.27 The day's workload to be picked in the Delivery Monitor

Manual transfer order before saving

Figure 3.28 shows the transfer order generated in the foreground for Delivery 80016882 before it is saved by the operator. The system has already determined the quantity to be picked of Material LEHU-002, as well as the source storage bin and storage unit. The material is taken from Storage Type 120 (high rack storage), Area 001 (fast-moving), the storage bin with the coordinates 12-01-01 and moved to Shipping Area 916. A log of this storage bin determination is shown in Fig. 3.34.

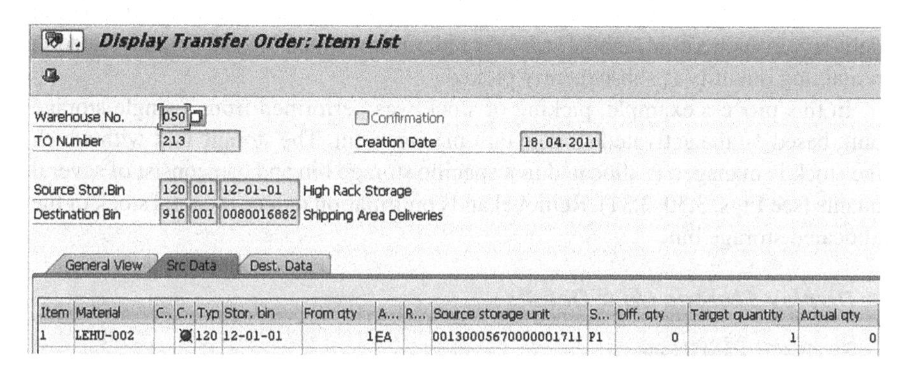

Fig. 3.28 Generating a transfer order for picking

Picking refers to actual removal of material from the storage bins and moving it to a destination storage bin. Depending on operational needs, shipping materials can also be determined, and a *pick HU* (picking handling unit) can be automatically generated for the operator to pack the picked quantity. The pick HU is then directly assumed in the delivery; further packaging is usually not necessary. For an HU-managed storage location or if expressly desired, picking is done by removing the respective handling unit, which can then be repacked in a pick HU.

To confirm that the picking quantity has in fact been removed from the source storage location and a goods movement has taken place, the transfer order is confirmed. The work list to be confirmed can also be displayed and selected via the Delivery Monitor (see Fig. 3.29).

Two-step picking is a process in which the materials to be removed for several deliveries or transfer orders are first taken from the storage bins. In a second step,

Outbound Deliveries for Confirmation

Item View For Group

Delivery	Pick.reqst	GI Date	Ship-to	Name of the ship-to party	C	W
80016882	213	10.04.2011	3000	Thomas Bush Inc.		B

Fig. 3.29 Deliveries to be confirmed in the Delivery Monitor

they are allocated to their corresponding reference documents. The advantage of this two-step process is the ability to pick a large amount of business transactions, thus minimizing the total number of picking actions. Because removal and subsequent allocation to deliveries or transfer orders represent two separate warehouse procedures, a transfer order is generated for each step.

In the event that the quantity actually picked differs from that indicated in the transfer order, the difference quantity is determined by the system and recorded in the transfer order item. The partially picked quantity is updated in the delivery document and the respective status is set for *partial picking*. If partial picking has been stipulated with the customer, the delivery is generated as a partial delivery and another delivery is created for the quantity not yet processed. It is also possible to perform further picking for the delivery, especially if you want to avoid a partial delivery. In such a case, another transfer order is generated for that delivery, and the remaining quantity is subsequently picked.

In this process example, picking of stock was performed from a single storage unit, based on the activated storage unit management. The storage unit with which the stock is managed is allocated to a specific storage bin and can consist of several quants (see Figs. 3.30, 3.31). Removal and confirmation reduce the total stock of the allocated storage bin.

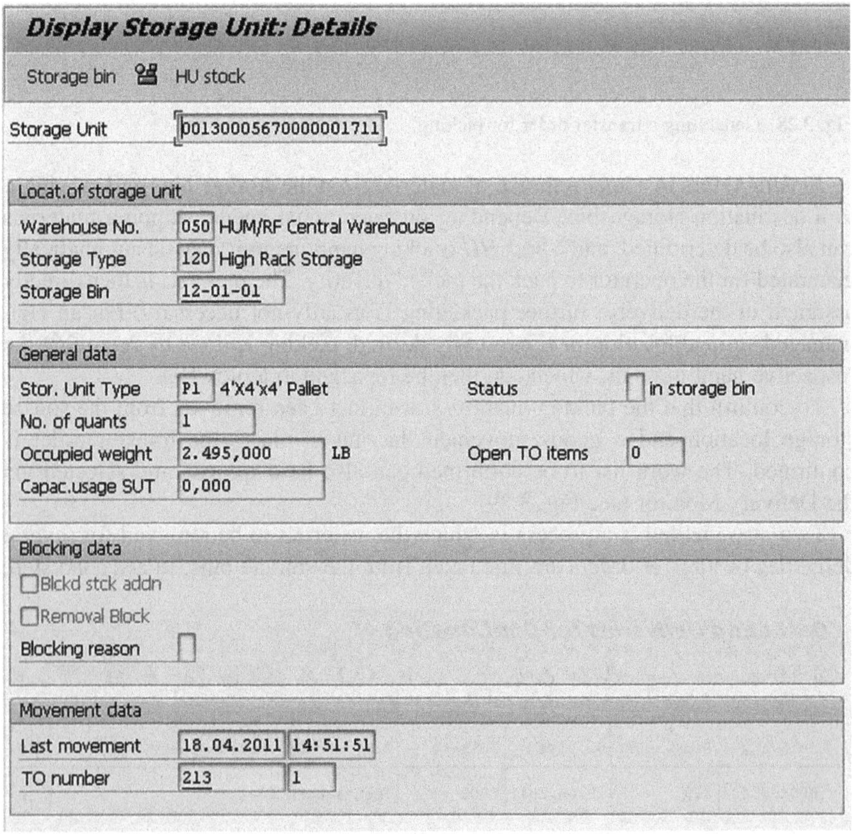

Display Storage Unit: Details

Storage bin | HU stock

Storage Unit: 00130005670000001711

Locat.of storage unit

Warehouse No. 050 HUM/RF Central Warehouse
Storage Type 120 High Rack Storage
Storage Bin 12-01-01

General data

Stor. Unit Type P1 4'X4'x4' Pallet
Status in storage bin
No. of quants 1
Occupied weight 2.495,000 LB
Open TO items 0
Capac.usage SUT 0,000

Blocking data

Blckd stck addn
Removal Block
Blocking reason

Movement data

Last movement 18.04.2011 14:51:51
TO number 213 1

Fig. 3.30 Storage unit in a source storage bin

Storage unit

Figure 3.30 shows Storage Unit 130005670000001711, which is located at Storage Bin 12-01-01 in the high rack storage of the central warehouse. In addition to general data such as weight and status, the storage unit includes information regarding any blocks or the recent movement data. In this example, movement was in the form of Transfer Order 213.

From the viewpoint of warehouse management, confirming the transfer order generally represents the conclusion of stock movement in a distribution process. To update inventory data, a goods issue takes place for a delivery. As in the case of a goods receipt, this posting serves to update quantity and value of stock, and is also documented by a material document. Posting of goods movement to inventory management is either automatically done in the background or manually performed, such as with the aid of the Delivery Monitor.

Display Quant

Physical inventory Additional data...

Material	LEHU-002	LES Handling Unit Part	
Plant/Stor.loc.	3000 0050	Warehouse No.	050
Batch		Storage Type	120
Stock cat.		Storage Bin	12-01-01
Special stk		Quant	2
Picking Area		Stor. Unit Type	P1
		Storage Unit	00130005670000001711

Stock data

Total stock	499 EA	GR Date	31.05.2002
Avail.stock	499	GR Number	1 1
Stock for put.	0	Last movement	18.04.2011 14:51:51
Pick quantity	0	Document number	213 1
Weight	2.495,000 LB	Certificate No.	
Cap.consumpt.	0,000		
☑HU quant		Delivery	180000069 1

Blocking indicators

☐Putaway block	☐Curr.stk plcmnt
☐Removal Block	☐Current Removal
Blk.reason	☐Inventory Act.

Fig. 3.31 Quant in a source storage bin

Quants were explained above in Sect. 3.3.1, "Warehouse Structure and Integration". Based on the goods movement presented in our example, we can now examine the effects on stock. The quant not only includes organizational allocation to a storage bin and storage unit, but also precise information on the total stock and available stock of the stored materials. The available stock is based on the quantities yet to be picked or placed in the transfer orders not yet confirmed. In addition to the most recent goods movement and the related transfer order, the stock data also includes data on any special stock or blocks.

As in the case with the quant, which contains exact information regarding the last goods movement, object references are also updated in moved handling units. The handling unit history (see Fig. 3.32) ensures the seamless transparency of handling unit movements with the indication of the *packing objects*. These are the respective documents of procurement, distribution or warehouse logistics to which a handling unit has been allocated in the process.

As is the case with the handling unit history, the *document flow* of a sales process contains all documents generated with reference to a preceding document in a business transaction. In a traditional shipping process, this document chain generally consists of a source sales order and the delivery generated with reference to that order.

Handling unit history

Figure 3.32 shows the history of the handling unit with the external number 130005670000005702. This handling unit, a pallet, was used in the previous examples for the shipping process. Figure 3.30 shows the storage unit for the "to" storage location in Transfer Order 213. This transfer order, Outbound Delivery 0080016882, and Goods Issue Document 4900004792011, generated with reference to this handling unit, are updated as a packing object in the handling unit history.

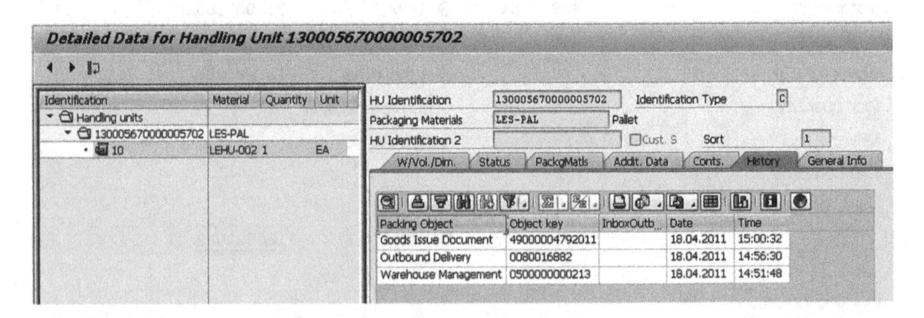

Fig. 3.32 History of a generated and removed handling unit

In a warehouse managed with WM where goods issue is executed with reference to an outbound delivery, a transfer order and handling unit are subsequently generated. The handling unit is noted in the document flow with its internal number, which differs from its external one. The delivery culminates in a goods issue, documented in the system with the material document, which, in the document flow, is called a *GD goods issue* (see Fig. 3.33).

3.3.4.1 Goods Issue Without Reference to a Delivery

For a goods issue without reference to a delivery, the physical stock movement is generally preceded by an accounting inventory posting of a goods issue (see Fig. 3.20). In Inventory Management, the goods issue posting creates a negative quant in the goods issue interim storage (interim storage type), and the system automatically generates a transfer requirement.

Based on this transfer requirement, a transfer order is created for issue and the storage bins from which the material is to be picked are determined. After the materials have been moved in the warehouse and brought to interim storage, the transfer order is confirmed. By confirming the stock movement, the negative stock of the goods issue interim storage is cleared.

Internal stock movements for which no goods issues are performed can be conducted directly by manually creating a transfer order. These types of movements are generally stock transfers within a warehouse for which a transfer order is created without a preceding document.

Typically, goods issue is the result of a previous stock removal. The stock removal is controlled via a transfer order that contains the source as well as the destination storage bin. Thus, the first step in the stock removal procedure is the search for a suitable source storage bin. Such a search is conducted via *stock removal control.*

As in the case of stock placement control, each stock removal procedure involves a search according to a predefined sequence and strategy. This search, triggered by a transfer order, is also influenced by the material to be removed and its

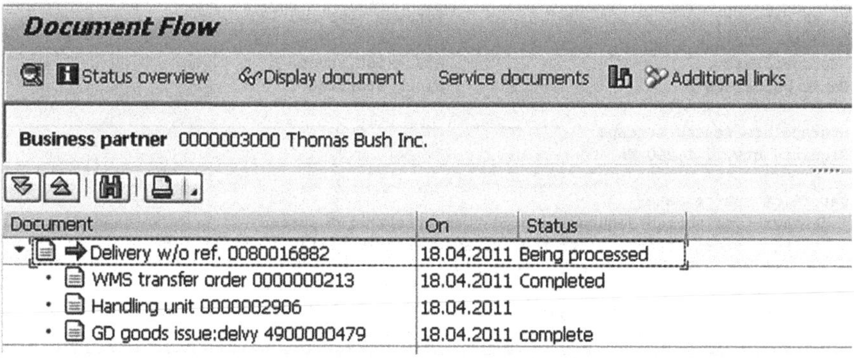

Document Flow

Status overview   Display document   Service documents   Additional links

Business partner 0000003000 Thomas Bush Inc.

Document	On	Status
Delivery w/o ref. 0080016882	18.04.2011	Being processed
WMS transfer order 0000000213	18.04.2011	Completed
Handling unit 0000002906	18.04.2011	
GD goods issue:delvy 4900000479	18.04.2011	complete

Fig. 3.33 Document flow of the sales process

characteristics. Especially important is whether or not a material is batch-managed (for more information on batch management, see Volume 1, Chap. 3, “Organizational Structures and Master Data”).

The search for a storage type and area for removal is done according to a search sequence stored in the system much like that of storage type and area determination for stock placement control. Determination of the storage type is particularly based on the storage type indicator, the special stock indicator and the stock category. The *stock category* indicates whether the stock is unrestricted-use, return stock, blocked stock or stock in quality inspection. The storage section is determined using a preference list that has a certain sequence with which the system determines the storage section. This search is performed for the determined storage type using the storage type indicator for stock placement control and using the storage class.

Storage bin search

Figure 3.34 shows the log of a storage bin search for the transfer order used in the previous example. The system first determined the storage type and area for Material LEHU-002 with the aid of Picking Type Indicator 120 and Storage Section Indicator 001. Storage Bin 12-01-01 in Storage Type 120 was determined using the stock removal strategy FIFO.

The actual determination of a storage bin is done using a stock removal, or picking, strategy stored in the system. WM selects the most suitable quant within a

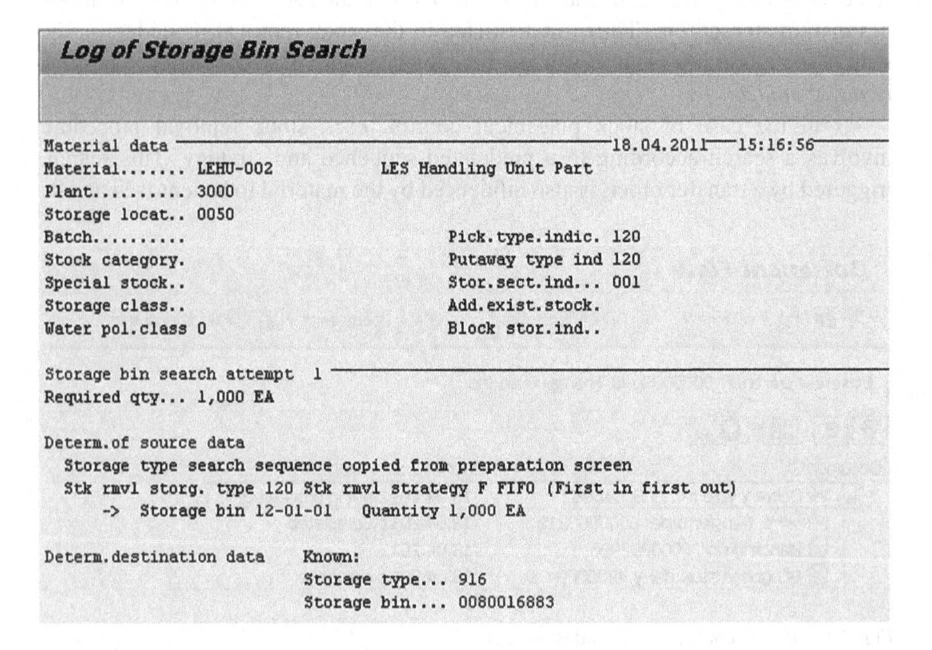

Fig. 3.34 Log of storage bin search

storage type from a series of preconfigured strategies, depending on operational needs. The following stock removal strategies are offered, which can be altered through a user's own extensions and logic:

- **FIFO and stringent FIFO**: **"First In, First Out"**
  With this stock removal strategy, WM suggests removal of the oldest quant for each storage type determined in the storage type search. The age of the quant and thus that of the materials stored in it are determined using the date of the inward stock movement posting. For stringent FIFO, the system searches an entire warehouse number, taking into account all storage types.
- **LIFO**: **"Last In, First Out"**
  This picking strategy suggests the most recently added quant for each storage type.
- **Partial quantities first**
  The partial quantities strategy attempts to achieve an optimization of the warehouse situation by trying to take the quantities required for a transfer order from a complete storage unit, with the goal of producing as few partial units, or *partial pallet quantities*, as possible.
- **According to quantity**
  For this removal strategy, selection of the storage type is influenced by the required quantity. The *control quantity* is stored in the material master, and causes a material to be taken from fixed bin storage for small quantities and from high rack storage for large quantities.
- **Shelf life expiration date**
  The expiration date or shelf life of a material is stored in the material master record. WM tries to find the material with the oldest expiration date. If the expiration date has not been stored, the system uses the FIFO strategy.
- **Fixed bin**
  Using the fixed bin strategy, the system determines the storage bin using the fixed bin stored in the material master. Since the storage bin found using this picking strategy is also suggested even if negative stock exists, a replenishment control should be set for fixed bins.

The following section provides an overview of selected warehouse-internal and cross-warehouse processes.

3.3.5 Warehouse-Internal Processes

Warehouse-internal processes not only include warehouse-wide procedures such as picking, placing and internal goods movements, but also WM functions that are available in all storage processes. The processes of hazardous material management and warehouse automation will not be explored further in this book.

Every company is legally obligated to conduct an accounting balance of stock, or the taking of *inventory*. By determining the physical inventory, the book inventory balance stored in the system can be compared with the actual inventory. Measuring, counting, weighing or estimating inventory can be done by Inventory or Warehouse Management. Because of its higher degree of precision when WM is used, inventory is performed by Warehouse Management, by taking the inventory of each quant in their respective storage bins. We will not detail the taking of inventory within Inventory Management, as it would extend beyond the scope of this book. If the inventory analysis results in a difference between the target and actual values, this is noted in the physical inventory document and subsequently posted in Inventory Management.

WM supports legally-accepted *inventory processes*, which can be set for each storage type:

- **Periodic or annual inventory**
  For the periodic inventory method, the physical inventory of a company is recorded on a key date (usually the end of a fiscal year).
- **Continuous inventory**
  This type of inventory is conducted throughout the entire fiscal year by recording partial stocks at desired times. All materials are counted at least once per year, and the book inventory balance is regularly compared with the actual stock. For continuous inventory during putaway, inventory is conducted when stock is initially placed into a storage bin. No further inventory is counted during the fiscal year, because all goods movements at a particular storage bin have to be verified by transfer orders, which, by law, must be archived.
- **Inventory sampling**
  In the current fiscal year, inventory is taken based on a random sample method and recorded in the system. This sampling is done on a random part of the material stock used to make projections.
- **Cycle counting**
  This type of inventory enables you to count materials in warehouse stock repeatedly at specific intervals during the course of the fiscal year. The procedure is based on a consumption- or requirement-based ABC analysis for the selection of material stock that is to be considered. Materials whose value-based share of total consumption or requirement amounts to a particular percentage stored in the system are inventoried as low-value materials.

To take inventory in the system, the storage type to be inventoried is first blocked and a physical inventory document generated (see Fig. 3.35). The physical inventory document includes the inventory process to be employed and the storage bins to be checked. If no open transfer order items exist for those storage bins, the physical inventory document is activated and the respective storage bins are locked for goods movement.

In order to perform physical inventory on storage bins, the physical inventory document is printed and forwarded to the responsible employees. Inventory is then determined and manually recorded on the printout of the physical inventory document.

After physical inventory is taken, the counting results are entered in WM. This can be done manually, with the aid of scanners or, if an external system is employed, through the automatic import of large amounts of data. Any discrepancies in the inventory items and differences between actual, counted quantities and the book inventory balance are entered in WM as inventory differences and serve as the basis for an inventory adjustment. Adjustment in Warehouse Management is done with the aid of the *interim storage area* (interim storage type), by generating a negative quant for insufficient quantities and a positive quant for positive differences. These differences are subsequently posted from Warehouse Management to Inventory Management. In the case of considerable deviations, a recount can be requested by the having the system issue a new physical inventory document.

From the viewpoint of Inventory Management, a posting change is a change in bookkeeping information of such data as stock type, special stock categorization or the material or batch number. Stock transfers, on the other hand, are related to physical goods movements in the warehouse. A posting change can lead to a stock transfer.

Warehouse Management considers a posting change with regard to the storage bins in which the respective material quantities are situated. The posting change, representing a change in stock data, is usually done in Inventory Management.

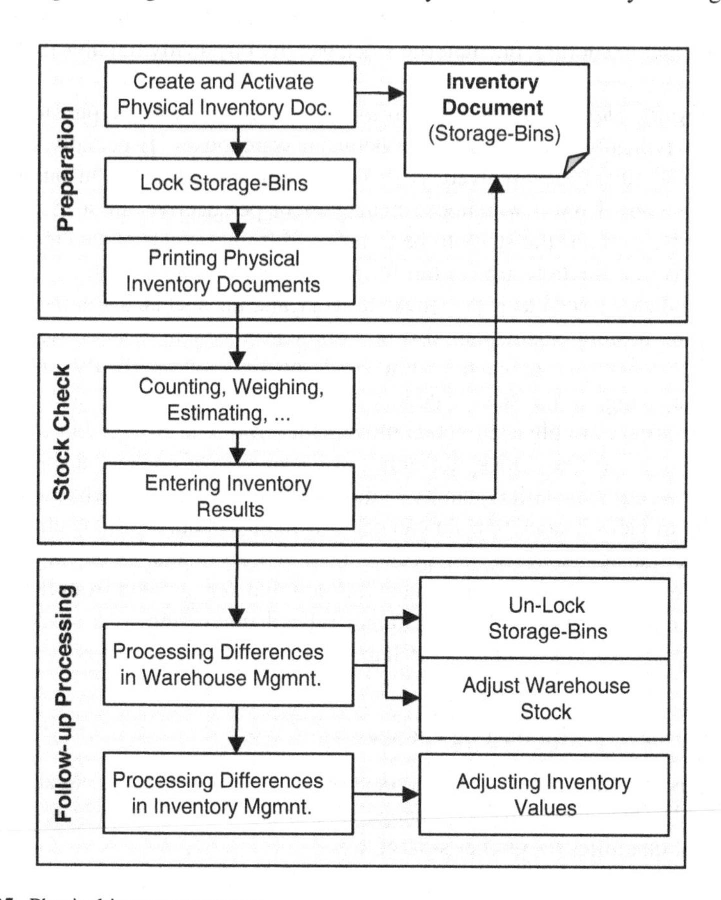

Fig. 3.35 Physical inventory process

If a posting change affects a certain plant/warehouse combination managed by WM, it also affects Warehouse Management. In such a case, the system must be informed as to where the altered material quantities are located and which storage bin stock is affected.

Posting changes are normally not connected to a physical goods movement in Warehouse Management. Because only stock attributes are changing, the material stays in its storage bin. The posting change in Inventory Management triggers a *posting change notice*, comparable to a transfer requirement, with a posting change interim storage area (interim storage type). As is the case for an outbound delivery, a negative quant is generated at the interim storage area representing the material quantity to be reposted before the posting change process. At the same time, a positive quant is created for the material quantity with the stock attributes it will have after the posting change. The actual stock transfer is done via a transfer order with reference to the posting change notice. There are two possibilities for this: You can perform a posting change from plant to plant or between storage locations.

In the case of a *posting change from plant to plant*, stock is generally stored within the same warehouse complex. When the ownership of stock is transferred from one plant to another, the material stock usually physically remains in the same storage bin.

For *posting changes from one storage location to another*, physical goods movement typically occurs within or between warehouses. Especially if several warehouse numbers are involved, such movement is considered an outbound or inbound delivery from a warehouse management perspective. In such cases, the transfer order contains the stock to be transferred from a single stock category and the respective destination storage bin for physical goods movement.

In special cases, such as if the stock category and the special stock indicator are changed, the posting change can also be initiated directly in Warehouse Management. The system is set such that a stock posting automatically takes place following a change in the stock category.

Stock transfers are physical goods movements from one storage bin to another. Unlike posting changes, stock transfers always involve a change in the physical location of a certain material quantity in the warehouse. From a warehouse management point of view, a stock transfer can occur with or without a posting change.

A stock transfer from the perspective of inventory management is explained above in Sect. 3.2.1, "Goods Movement". Stock transfers as seen from the process view, especially with regard to document-related processing as a stock transfer order, were discussed in Volume 1, Chap. 4, "Procurement Logistics".

Stock transfers with posting changes

In a posting change, the batch allocation of a certain material quantity is changed. This stock attribute change makes it necessary to transfer those material quantities to another storage bin.

For stock transfers within a single warehouse number, SAP ERP supports users by providing detailed information on the stock movements to be performed, as well as managing and displaying these movements. In everyday business, such stock movements often result from a consolidation of partial quantities from various storage bins into one storage bin, such as through the provision of materials from high rack storage to a picking area or from the transfer of material for technical reasons, such as because a certain capacity limit has been reached. From the perspective of inventory management, the total stock does not change based on these internal stock movements. Stock movements are thus performed without the use of Inventory Management. As in all stock movements, a stock transfer between storage bins within a single warehouse number is done using a transfer order.

Stock transfers between storage locations generally begin in Stock Management and are completed in Warehouse Management. When stock transfer is required from one WM-managed storage location to a non-WM-managed location, it is processed as a goods issue with stock removal by Warehouse Management. In the opposite case, if a transfer is done from a non-WM-managed storage location to a storage location with Warehouse Management, transfer is performed as a goods issue with putaway. Stock transfers between two WM-managed storage locations are initially conducted in Stock Management. From the perspective of Warehouse Management, stock transfer is conducted in the usual way, with a transfer order issued with reference to a material document or via a list with open transfer requirements.

A special form of stock transfer is replenishment to refill stock in fixed storage bins. In accordance with settings in the material master and the current stock situation, the system calculates the stock to be maintained in the storage bins in the function *replenishment for fixed storage bins*. For replenishment planning for fixed storage bins, the system calculates the current stock situation as well as planned removals due to deliveries that are to be picked from fixed storage bins. Actual filling of the storage bins can be performed based on a previous replenishment plan by generating transfer requirements for the required replenishment quantities. As usual, the transfer requirements are subsequently carried out with transfer orders.

A further possibility to generate stock transfers required for replenishment is by confirming a transfer order for removal. In such a case, a transfer order can be directly generated for replenishment without the prior need to generate a transfer requirement.

3.4 Warehouse Management with SAP EWM

SAP Extended Warehouse Management is a decentralized warehouse management system. Decoupled from SAP ERP, as an autonomous application of *SAP Supply Chain Management* (SAP SCM), integration with an ERP system is generally required for master and transactional data. The EWM release SAP SCM 2007 also provided the opportunity to centrally operate EWM from within SAP ECC 6.0 as an add-on and thus use it as an integrated warehouse management system with an SAP ERP system (Fig. 3.36).

3.4.1 System Integration

From a purely technical standpoint, SAP EWM can also be operated as an autonomous system, without direct connection to SAP ERP or a non-SAP system. However, experience has shown that – depending on the project base of the interfaces to be established and the processes to be implemented – standard integration with an ERP system is preferred. In this chapter, for our illustrations and processes, we assume that the EWM system is linked to an ERP system and close integration of the existing standard interfaces is established. Figure 3.37 illustrates the technical process connection of the SCM-based EWM systems to SAP ERP.

We have already discussed Inventory Management in SAP ERP in this chapter. Warehouse and stock management are done in SAP EWM, and are the focus of this section.

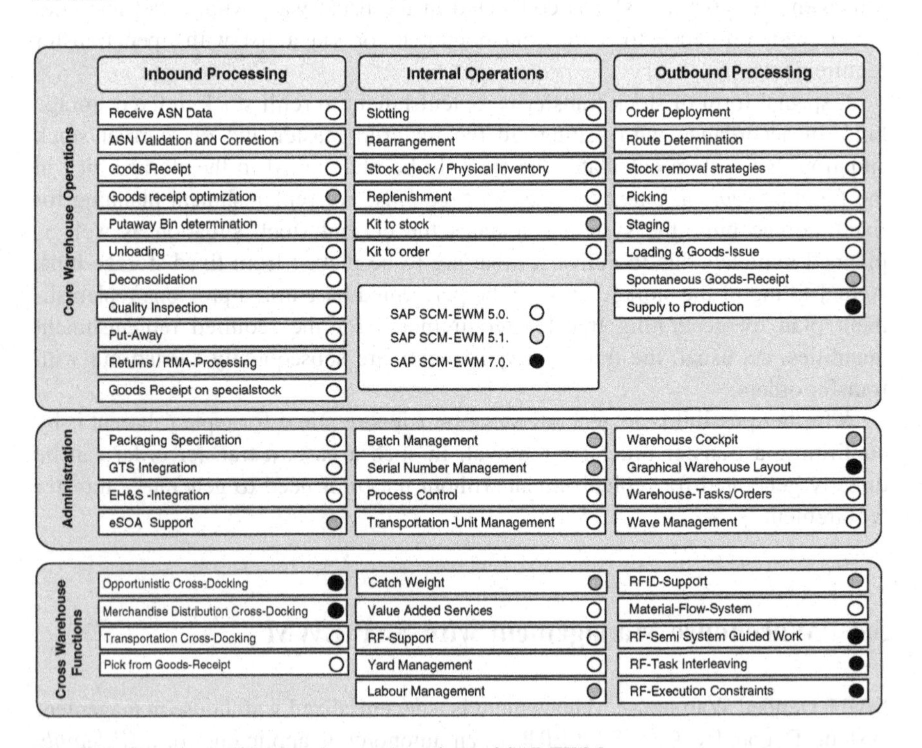

Fig. 3.36 Overview of the functions and versions of SAP EWM

3.4.2 *Warehouse Organization and Stock Movement*

The acceptance of a warehouse management system is, among other things, based on its seamless integration in an ERP system and on its flexibility in mapping logistics processes. SAP EWM was developed for complex warehouse and distribution centers with a variety of products and a high document volume. That is why the design of this new, decentralized warehouse management system has special emphasis on the flexible mapping of warehouse-internal processes. The extent of

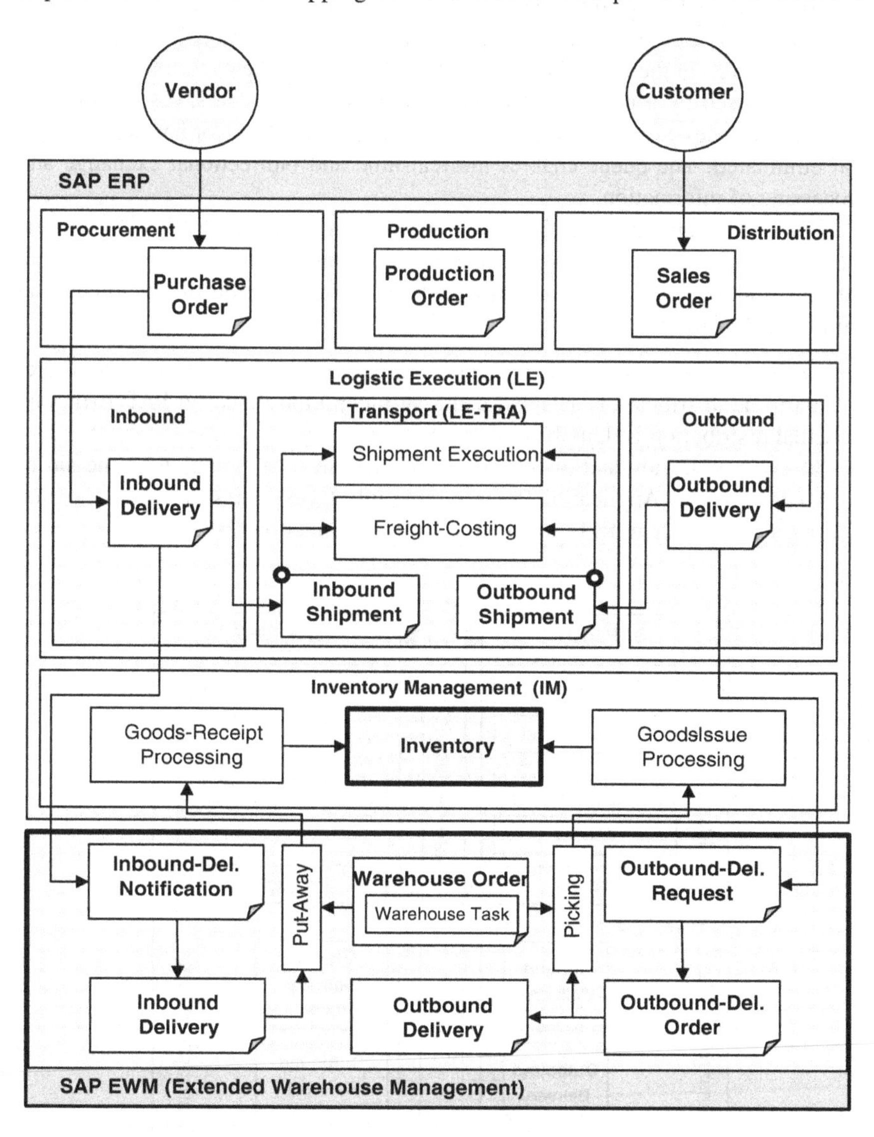

Fig. 3.37 Decentralized warehouse management with SAP EWM

functions has been expanded considerably in comparison to the existing WM and supplemented by a number of warehouse structure elements.

In this section, we will provide a short introduction to ERP integration, followed by an overview of warehouse organization and the special features connected to stock movements in SAP EWM.

The technical linkage of EWM to ERP and such functions as the transfer of inbound and outbound deliveries between the systems take place in real time via defined interfaces. These interfaces enable the seamless integration of both systems by distributing, altering and returning data relevant to delivery.

Inbound and outbound processing is performed asynchronously, based on the sequence stored in the *inbound* and *outbound queues*. In the event of an error, for example caused by a missing network connection, this queue saves all transfers and allows processing to continue seamlessly as soon as the error has been located and eliminated. The queue enables the real-time and bidirectional exchange and processing of information.

Figure 3.38 shows the technical integration of SAP ERP and SAP SCM. For the transfer of master and transactional data, two different procedures are used: master data distribution via CIF and communication and distribution of document data with the aid of BAPIs:

- The distribution of master data is done via the *Core Interface* (CIF). The master data to be distributed is selected using an integration model in SAP ERP, and actual distribution is done from ERP to SCM.
- The communication and distribution of document data is done with the aid of *BAPIs* (Business Application Programming Interfaces). These are interfaces that are accessed with an RFC connection (Remote Function Call).

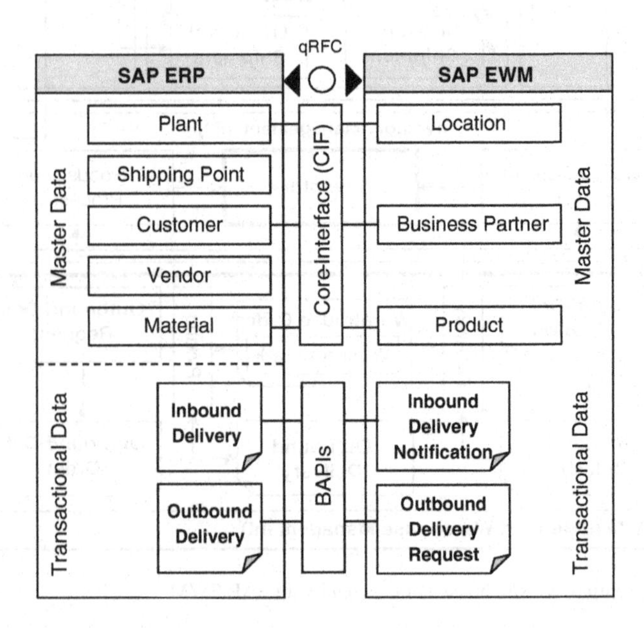

Fig. 3.38 Overview of technical integration

You can find more information on the integration and distribution of master data and the interface technology used in that process in Volume 1, Chap. 2, "Organizational structures and Master Data".

From a technical standpoint, close integration of EWM and ERP is achieved via *interfaces*, while process integration is primarily done via *organizational data* (see Fig. 3.1). As in the case of integration of the WM system, in SAP EWM, organizational allocation of warehouse numbers is initially achieved by their allocation to specific plant/storage location combinations. Stock management is always executed in ERP, and all quantities of materials stored in an EWM warehouse are allocated on the plant/storage location level. However, the assigned WM warehouse number is identified in the system settings as a decentralized warehouse, and informs the system that warehouse-logistics-related processing is to be conducted in a decentralized warehouse management system, and the respective document is replicated to the integrated EWM system. This ERP warehouse number is assigned an EWM warehouse number in the EWM system settings.

Allocation of a decentralized warehouse

From the viewpoint of SAP ERP, decentralized warehouse management, specifically the integration of an EWM system, is executed through the allocation of a WM warehouse number to a specific plant/storage location combination. The WM warehouse number in this case is identified in the ERP system settings as a decentralized warehouse number and represents a type of "interim storage number" for the technical system integration of SAP EWM. Based on this procedure, both a decentralized connection to an EWM system as well as central warehouse management with WM can be achieved simultaneously for a single plant, depending on the warehouse locations determined for a specific material. This flexibility allows the separation and selection of the warehouse management system on the material level. The interim storage number is used exclusively for integration purposes and has no further logical significance in ERP.

The structure of a warehouse is typically based on the goods and their required space, and according to any existing automation technology and requirements. As in the case with WM, SAP EWM also offers several possibilities to map an individual warehouse structure according to its operational requirements. This organization data also plays an important role in the control of processes (see Fig. 3.40).

Comparable to WM, the *warehouse number* in EWM also represents a physical warehouse or complete warehouse complex in which stock is stored. Even though the two systems are not related, the warehouse number in EWM also summarizes the individual physical areas of a warehouse complex in a logical manner, and with its four digits, it represents the highest organizational level.

The *storage types* generally serve to categorize the physical sections in a warehouse according to technical, spatial and organizational aspects. Based on spatial circumstances and physical characteristics of the storage equipment used, block storage, open storage and high rack storage are defined by a company's own physical storage types. In addition, the use of a certain storage type can be changed in the system settings of SAP EWM with the aid of a *role code*. This code indicates whether a storage type is a "regular storage type" based on the previous definition or a spatial area for the identification, provision, removal or inspection of materials. A *yard*, meaning the parking area assigned to a warehouse, is also a storage type to which the respective "yard role" has been allocated.

Within a certain storage type, *storage sections* combine all storage bins having certain homogenous characteristics. Such characteristics can relate to materials stored in the bins. The respective criteria are user-defined, and serve as an organizational aid and control and optimization parameters for storage. Examples of storage sections include certain zones within a storage type (racks) with regard to inventory turnover and the composition of a material (see Fig. 3.40).

Storage section

A storage section exists in a warehouse for high rack storage. Based on the inventory turnover of the materials stored there, two storage sections are defined. Materials with a low turnover frequency, or *slow-moving items*, are stored in a rear storage section. To minimize travel times during material removal, the *fast-moving items* are stored near the aisle in the front section of the rack.

Storage bins represent master data and signify the physical locations, that is, the bins with their specific coordinates (see Fig. 3.39). In addition to the specific allocation of a storage type and section, storage bin master data includes precise information on the storage bin type, volume and maximum weight that can be stored in that bin. The storage bin type is particularly used in the determination of the putaway strategy and the search for a suitable storage bin, and determines the type of pallet that can be stored in that storage bin.

Storage bin

Figure 3.39 shows the storage bin master data for Storage Bin 01-01-A in Storage Type 0050 (fixed bin storage). Based on the allocation of Storage Bin Type P002, the storage bin is set up for pallets having a height up of to 2 m. In addition to general master data, the storage bin data also includes information on the current stock situated in that storage bin, as well as on the most recent stock movements.

The current stock of a certain material quantity having the same characteristics is mapped using *quants*. We have already discussed quants in Sect. 3.3.1, "Warehouse Structure and Integration", with regard to WM. They aid in the transactional management of stock on the lowest storage bin level. Quants are created in an EWM system in the same way as in WM, through the putaway of material in a storage bin. Picking and stock transfers cause the stock, and thus the quant, to be updated and automatically deleted by the system if stock is no longer situated in that particular storage bin.

Activity areas are a special feature in SAP EWM and represent a logical grouping of storage bins with regard to the warehouse activities to be performed, such as putaway, inventory-taking and picking. Activity areas serve the optimization of warehouse activities. To do so, the storage bins for which an activity is to be executed are sorted according to the criteria stored in the system before the system generates warehouse tasks along the lines of the *warehouse task creation rules*. Creating warehouse tasks to perform warehouse activities – a further special feature of SAP EWM – is explained in the following section.

Doors are allocated to a specific warehouse number and describe the physical location where transport units and their vehicles are loaded and unloaded and materials arrive at or leave the warehouse. From a system viewpoint, a door is a storage bin to which a certain storage type with the storage role "door" is assigned and for which a specific loading direction has been determined. The loading direction determines whether a door is only to be used for goods receipts or goods issues or both movement directions. In operative practice, these doors are in the vicinity of the staging areas in which putaway and picking processes are controlled.

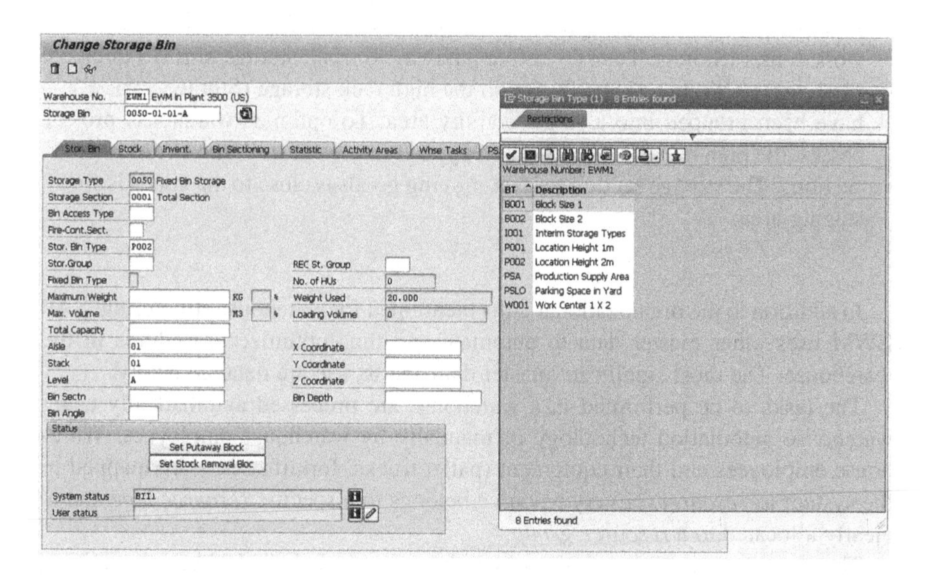

Fig. 3.39 Storage bin master data

Staging areas, in their function as interim storage areas, control inbound and outbound delivery processes. From the system view, staging areas represent a special storage type (storage type role) that has been assigned a warehouse number and at least one door. From the viewpoint of goods receipt, staging areas are used to temporarily store unloaded materials to be put away. For goods issue, they serve as interim storage of material quantities already picked and waiting for loading.

A *work center* is a physical location in a warehouse where certain warehouse activities can be carried out and to which a material quantity can be moved for that purpose. Such activities include:

- Deconsolidation
- Packing
- Counting
- Quality inspection

Depending on the activity to be performed, the *work center layout* determines which tab is shown in the respective system work center (see Figs. 3.54, 3.56, 3.68). From a technical standpoint, a work center is also a storage type with an allocated storage bin and corresponding settings for the required storage sections. This flexibility enables you to customize a work center, especially for deconsolidation or quality inspection, and assign inbound and outbound sections.

Organizational structures in EWM

Figure 3.40 shows Warehouse Number 1000. Goods receipt takes place at Door A, and goods issues via Door B. Storage Type A refers to the goods receipt staging area in which materials are kept after unloading until their final putaway in Storage Type D or E. Mixed pallets are deconsolidated at a work center (Storage Type C) before putaway. Certain storage bins (P) in the rack storage (Storage Type D) and in the high rack storage (Storage Type E) have been grouped into a single activity area. To optimize warehouse processes, the high rack storage (Storage Type D) has been split into two storage sections. The storage section for fast-moving goods is close to the goods issue staging area.

In addition to the organizational units mentioned for warehouse structuring, SAP EWM uses other master data to automate and thus optimize procedures in the warehouse. The most significant master data and its use are detailed below.

The tasks to be performed in a warehouse are processed automatically using warehouse automation technology or manually by warehouse employees. Warehouse employees and their equipment (pallet trucks, forklifts, etc.) are mapped in the system as *resources*. Every resource belongs to a specific *resource type* and is clearly allocated to a *resource group*.

The resource type groups resources with similar technical or physical qualities, such as their horizontal speed, qualification and preferences. The resource type allocated to an employee with its assigned *bin access type* determines such issues as whether that particular resource (in this case, a person) is only allowed access to certain stock situated in a particular section of the warehouse, or that only handling units of a certain size (handling unit type) may be moved.

The resource group determines the sequence in which warehouse tasks are selected for the corresponding resource. To execute and accept warehouse tasks, the resource can log in to a *radio frequency environment* (RF environment).

Because SAP EWM is an SCM component, the material master data is distributed from ERP to SCM via the APO Core Interface (described in Volume 1, Chap. 2, "Organizational Structures and Master Data"). A *warehouse product* is the warehouse-number view of the material master, and contains EWM-specific parameters only applicable to a specific warehouse and a specifically authorized employee. Such data includes information on the control of picking and putaway, parameters governing the determination of a storage section and the characteristics of storage bins in which a product can or must be placed (see Fig. 3.41). The relevance of these parameters is described in more detail in reference to storage control.

The general procedures for serialization and batch management are detailed in Volume 1, Chap. 3, "Organizational Structures and Master Data". Both functions are also available in SAP EWM.

Serial numbers are a series of characters used in addition to a material number to allow differentiation between an individual item and other materials with the same material number. In particular, companies that supply customers with high-priced

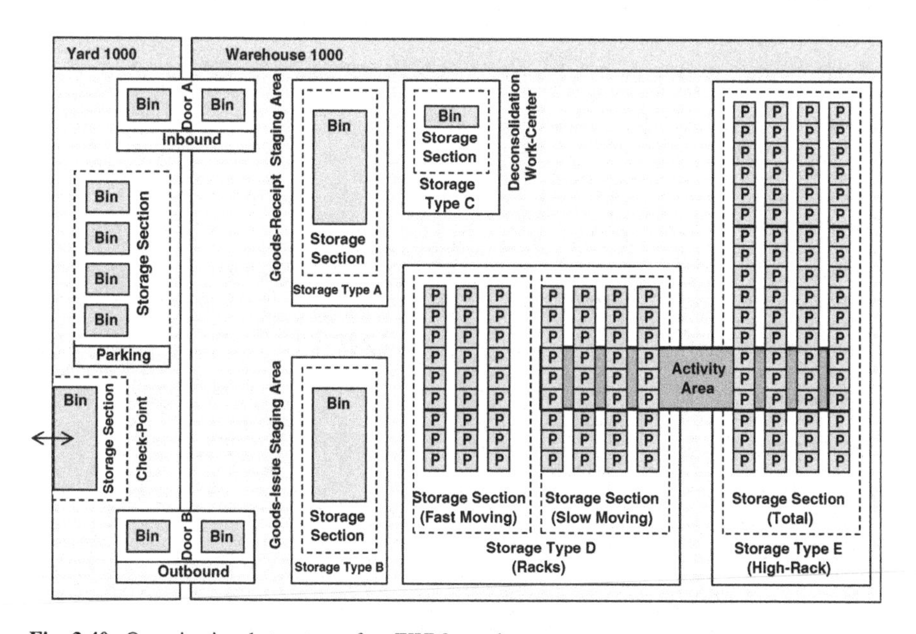

Fig. 3.40 Organizational structure of an EWM warehouse

items often want to give each product a specific number for easy identification, such as for warranty and guarantee reasons. EWM offers the opportunity to give each item a 30-digit serial number in addition to the product number. These serial numbers are managed by profiles, which are assigned to products and can control such questions as whether entering a serial number at goods receipt is obligatory or optional. In addition, the assignment of serial numbers can also be location-dependent: It is possible to issue serial numbers only to a product located at a specific site, or to make the assignment of a serial number obligatory at only one site.

Batches have become very important to many industries, and regulations sometimes require their use. Companies in the food and animal feed industries, as well as chemical and pharmaceutical companies, are especially subject to laws requiring the traceability of critical components or ingredients in all production, processing and distribution stages and to establish a suitable crisis management program. Based on the complexity of batch characteristics or shelf-life issues – such as in the food industry – many companies need comprehensive batch management. SAP EWM includes the complete integration of batch management. In all processes, such as goods receipt or picking, batches or their characteristics can be taken into

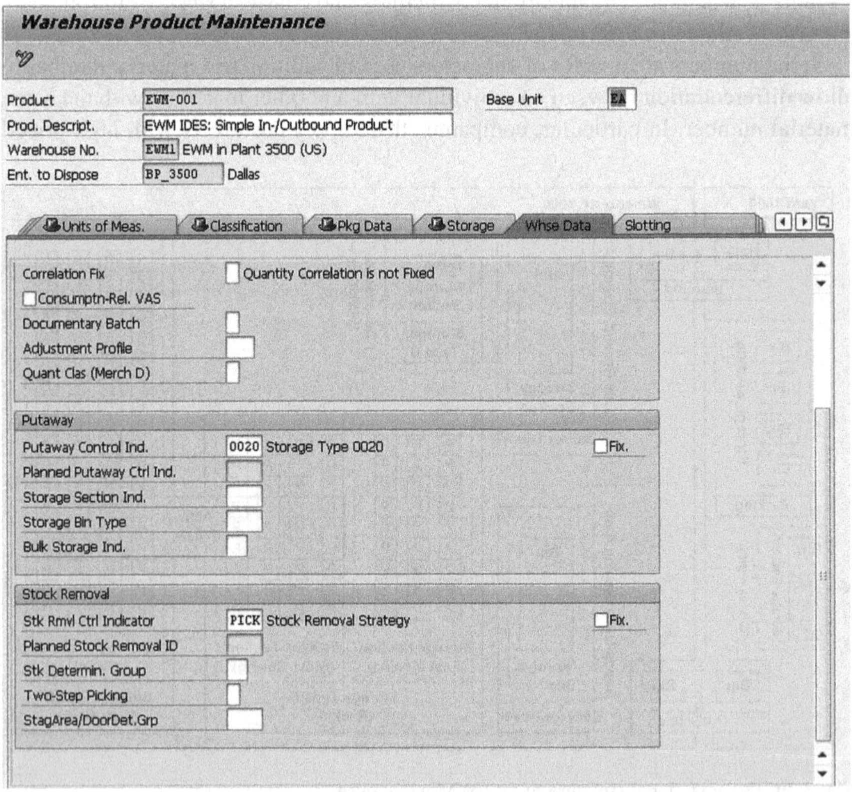

Fig. 3.41 Warehouse product with EWM-specific parameters

account. EWM can process batch specifications from the ERP system, but also can determine batches itself, based on the corresponding batch search strategy.

Dangerous goods in logistics require special storage circumstances for legal reasons alone. It is especially important for a warehouse management system to monitor processes surrounding the safe handling of hazardous material and support employees in observing legal and operational specifications. With its component *SAP Environment, Health, and Safety Management* (SAP EHS Management), SAP EWM can enable safe handling, storage, and the secure transport of dangerous goods.

The EHS functions supplied by EWM include phrase management, with which text modules in various languages can be managed to save information on dangerous goods and issue it on dangerous goods documents. EWM allows you to save information pertaining to a hazardous substance in the dangerous goods master record, such as data required for a dangerous goods check. In addition, EWM provides a function with which you can print a list of dangerous goods for the fire department, including all pertinent information.

A *packaging specification* provides information on all packaging requirements for materials to be put away or transported, and contains information on each material regarding the packaging materials required, as well as what steps are involved in the packaging process. The packaged content refers to the material to be packaged or another packaging specification. Generally, this information is used for the automated palleting of materials, and controls which storage bins are possible for putaway at the time of goods receipt. Packaging specifications are also supported by the following processes and work steps:

- Packaging at work centers
- Packaging during confirmation of warehouse tasks
- Packaging and deconsolidation using the RF Framework

Determining what packaging specification to use is done automatically on the basis of SAP consolidation technology. The option of defining your own conditions, characteristics and circumstances and using them as a condition record in the system lets you determine packaging specifications and tailor them to operational needs.

Packaging specifications can be structured hierarchically and enable you to map complex packaging procedures: The packages are organized in a hierarchy and must be packaged according to instructions saved for the respective packaging level (see Fig. 3.42). Packaging specifications can be printed out to serve as instructions for employees in the warehouse for such tasks as how to affix labels or stack products on pallets.

Means of transport such as trucks are mapped in EWM as *vehicles*. Vehicles are used to group deliveries and are loaded and unloaded at warehouse doors. These procedures lead to a change in status of the allocated delivery document, and enable integration with transportation management, for instance, to generate bills of lading. Means of transport can be further specified with transportation units, the actual load carriers. A transportation unit is the smallest loadable unit of a vehicle, and is either fixed to that vehicle or represents a unified structure with it. Thus, a truck with a cargo area and an additional trailer consists of the vehicle itself with its own loading capacity (the vehicle and the transportation unit) and a trailer that is mapped in the system as a further transportation unit.

Transportation units are units consisting of a load carrier and packaged goods, and are mapped in the EWM as handling units. The actual transportation unit is stored in the system as a packaging material, as is the case with handling units. This allocation to packaging materials allows for flexible construction of a vehicle in which you can determine how many transportation units a vehicle should have and in what order they are to be allocated. Movements of vehicles and transportation units outside the warehouse are done via *Yard Management*. Transportation units that are situated in the yard aid in the inventory management of the materials contained in them. Yard Management as a cross-warehouse function is explained in more detail below.

In WM, stock movements are triggered by transfer orders that are either generated with reference to sales and distribution logistics delivery documents, or based on transfer requirements (see Fig. 3.20). In EWM, warehouse tasks (picking, removal and placement, transfer postings and stock transfers, as well as the scrapping of material) are triggered by *warehouse requests*. A warehouse request is a work list for purposes of placing, moving or picking the "requested" materials.

In WM, the transfer order contains information on the material to be moved – specifically, from what location (source storage bin) to what new location (destination storage bin). Then, that stock movement is confirmed by the warehouse employee. In EWM, a warehouse request is generated from a preceding document.

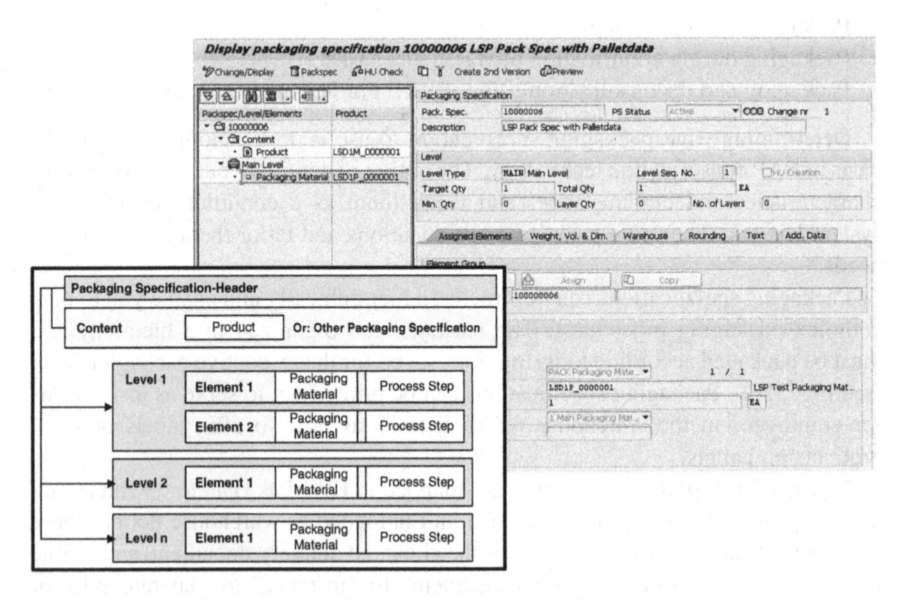

Fig. 3.42 Structure of a packaging specification

Such preceding documents are usually inbound delivery notifications (see Fig. 3.48), transfer posting notifications and outbound delivery notifications (see Fig. 3.61) that have been replicated from WM to EWM. The generation of warehouse requests for these preceding documents is done automatically, and the generated warehouse requests are enhanced with EWM data.

Warehouse requests are the basis for the creation of *warehouse tasks*, which in turn fulfill those requests.

A warehouse task contains all necessary information to move a certain material quantity or handling unit in the warehouse, and simultaneously serves to document the actual stock movements achieved. All necessary data is stored in the warehouse task for physical internal transport to, from or within a warehouse, from one storage bin to another. These stock movements are based on logical or physical goods movements. Depending on the type of unit moved, we differentiate between product and HU warehouse tasks (see Fig. 3.43).

A *product warehouse task* is a warehouse task based on a physical goods movement or stock change. The number of items in a product warehouse task depends on the stock movement to be executed. In addition to an indication of the material to be moved, the product warehouse tasks include the quantity to be moved and the source and destination storage bins. With regard to the runtime, the corresponding material quantity is reserved in the system.

When a product warehouse task is confirmed, the EWM is informed that the material quantities in question have reached their destination. For manual confirmation, the destination storage bin, or *target location*, can be changed. For a goods receipt or issue posting, the product warehouse tasks consists of an item whose confirmation increases or decreases stock. Transfer postings are made up of two items: one with the source storage bin whose stock is reduced, and one with the destination storage bin whose stock is increased. In the case of a partial

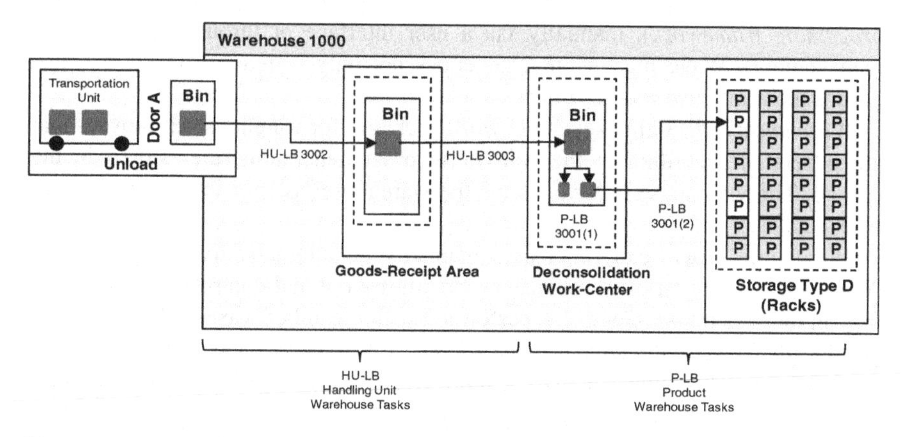

Fig. 3.43 Product and handling unit warehouse tasks

confirmation, an item is generated for every partial quantity. In this way, difference quantities can be recorded and justified with an *exception code*.

Unlike product warehouse tasks, in which material quantities are moved, HU warehouse tasks are for the internal transport of packages. These packages, HUs, are generally pallets that are moved within a warehouse from one storage bin to another or for loading and unloading of a transportation unit. The warehouse task contains the source and destination storage bins, as well as precise information on the HU to be moved. In contrast to product warehouse tasks, this precise information makes a quantity reservation unnecessary, because the HU is specifically known to the system.

Handling unit and product warehouse tasks

Figure 3.43 shows a simple goods receipt process. A transportation unit parks at Door A and contains two pallets that are unloaded via an HU warehouse task. The source storage bin corresponds to the warehouse door, while the destination storage bin represents the goods receipt area. One pallet contains two different products with varying storage requirements. This HU is moved from the goods receipt area to a work center and subsequently opened. The actual deconsolidation step is also executed via a warehouse task. As the result of the deconsolidation, the pallet is divided into two product quantities that are then placed into rack storage. The placing of these products is done using product warehouse tasks.

Both product and handling unit warehouse tasks can be created with reference to a warehouse request, and for internal stock movements, without a reference document as well, and they let you achieve multi-level, internal warehouse processes. The actual generation of these tasks is done automatically via a PPF (*post-processing framework*), manually via a user interface or through *waves* (wave management). Wave management generates not only warehouse tasks, but also groups them into *warehouse orders*.

Warehouse orders are executable work packages for warehouse employees that are generated according to the warehouse order generation rules stored in the system. The warehouse order contains the warehouse tasks or inventory items to be executed.

Warehouse orders can be created manually or automatically by the system. Automatic generation of warehouse orders allows optimal control of the material flow in the warehouse. Control is performed with the aid of waves. Making waves has the advantage that warehouse tasks can first be collected and released together at a later time. This allows them to be more effectively grouped into work packages for warehouse employees, which has a positive effect on travel times, and potentially the picking and base times as well. In addition, the material flow can be optimized, for instance by initially not undertaking replenishments after picking.

Wave Management groups a certain kind of warehouse request into a wave. Regardless of the point in time in which a wave is to be released, that is, the point in which warehouse tasks can be processed, wave management generates the necessary work packages in the form of warehouse orders and their corresponding warehouse tasks. The criteria for wave management can be kept in wave templates. At the time of the warehouse task generation, EWM uses putaway strategies to determine the storage bin for putaway or, using a picking strategy, the stock for material removal. You can also determine the time that is needed to execute a stock movement. This time span depends on the material, the quantity to be moved and the geographical distance of a particular stock movement. During the warehouse task generation process, these target times are added together and, if necessary, supplemented by a defined setup time. If EWM Resource Management is used, travel time is also taken into account.

Wave generation and release to create warehouse tasks are typically done automatically. In order to automatically allocate warehouse tasks to waves, criteria from the delivery can be utilized. Due to the automated processing, this flexibility enables optimized grouping of the outgoing deliveries and an optimization of picking by drastically reducing picking times.

The processes in a warehouse, stock movements and the process steps to be performed depend on the individual operational needs and on spatial circumstances in the warehouse. In practical situations, it is thus rare that the material flow is the same within a warehouse for all products and in all areas. Several people and resources are involved in the warehouse processes. Pallets might need to be deconsolidated in the goods receipt area or consolidated in the goods issue area. For flexible, tailor-made control of the material flow across several stations, and in order to enable cross-resource stock movement through various stations, EWM offers a storage control.

The purpose of *storage control* is to map complex, multiple-level stock movements for putaway and removal and for internal warehouse transfer. Storage control is done with relation to the spatial circumstances along the lines of the predominant warehouse processes and the stock to be moved. Using storage control, EQM can specify the putaway or removal route across several stations in a process- or layout-oriented manner. This allows processes such as counting or deconsolidation in the goods receipt area or packing in the goods issue area to be performed in an automated fashion. Storage control can be executed on multiple levels, enabling material flow via several interim storage bins.

Stock movements are controlled via *storage processes*. Each process and all process steps in goods receipt and issue are allocated to a storage process. The possible goods movement types and the direction of movement are assigned to a storage process via a warehouse process type and an activity. In EWM, the warehouse process types include the following:

- Putaway
- Removal
- Goods receipt posting

- Goods issue posting
- Inventory
- Posting change
- Cross-line placement

The determination of the storage process depends especially on product and document information (see Figs. 3.53, 3.65). The storage process is automatically selected by the system when the warehouse request is created, based on the document type, the product and the delivery priority. By using document characteristics and control indicators in the product master, the storage process can be controlled with great flexibility. For simple goods movements, the storage process itself can contain the storage type and bin from or to which a material is to be moved. For complex movements, it can contain the storage process or process-oriented storage control, or a storage creation rule.

Process-oriented storage control allows you to map complex picking and putaway processes. The individual steps, such as unloading, quality inspection, the execution of supplementary logistics services and subsequent putaway, can be adjusted as desired and are allocated to a storage process in the system. The determined procedures and their activities are assumed by the handling unit to be picked or placed. The HU thus possesses the information regarding which process steps are necessary for putaway, picking or warehouse-internal stock movements. That is why process-oriented storage control only works with HUs.

Figure 3.44 shows the multi-level picking and putaway processes. Depending on which storage process has been selected by the system, materials can be brought

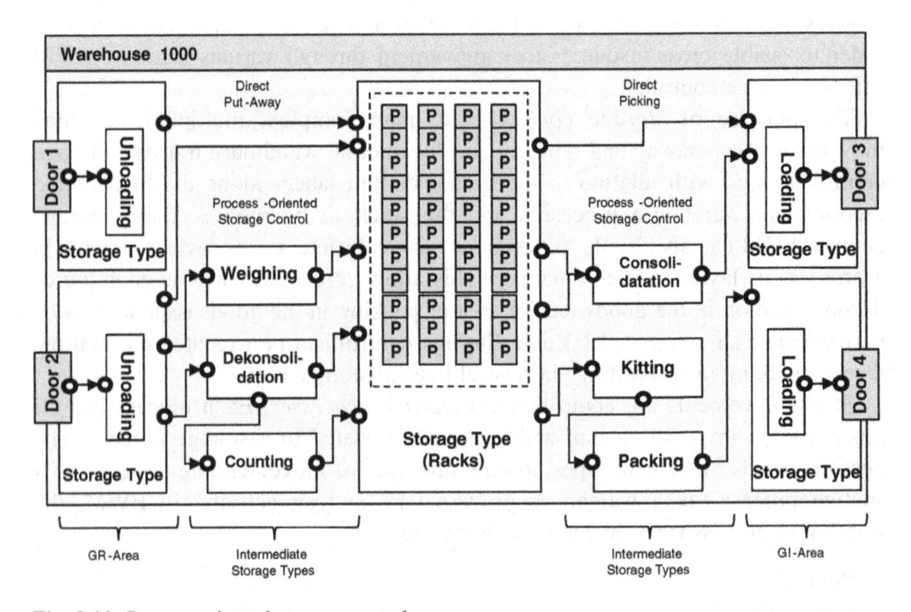

Fig. 3.44 Process-oriented storage control

directly from the goods receipt area to their storage bin in a single step. The destination location can be altered manually.

Process-oriented control also allows multi-step processes in which, for example, pallets with certain materials are weighed or isolated before putaway. To do this, the system generates warehouse tasks from the source location to an interim storage type. These interim locations are either work centers in which specific activities are to be performed or areas in the warehouse in which stock must be temporarily stored. Depending on the system settings, a specific destination storage location can be determined at the beginning of the putaway process or at a subsequent time.

In goods issue, the multi-step process can include the creation of kits, a process known as *kitting*, as well as a variety of packaging or consolidation steps before the goods are staged in the goods issue area for loading.

Layout-oriented storage control is used when stock movements cannot take place directly from a source to a specific destination or interim location because of spatial circumstances. Such circumstances are commonly related to the layout of the warehouse or the conveyor technology used. In the case of layout-oriented control, SAP EWM automatically deactivates the original warehouse task and generates the necessary intermediate steps.

Layout-oriented storage control

A pallet is to be brought from the goods receipt area to a work center. The system generates a warehouse task for this purpose. The warehouse is equipped with a conveyor system that moves pallets from a transfer point directly to a work center. This spatial circumstance is taken into account in storage control. The system deactivates the original warehouse task and automatically generates an intermediate step in the form of a warehouse task from the goods receipt area to the transfer point.

Process-oriented storage control can be combined with layout-oriented control. For this, the system first executes process-oriented control and then checks to see if the determined process steps are also possible from the layout view. The necessary intermediate steps are automatically generated by the system.

3.4.2.1 Warehouse Monitoring

Process control in SAP EWM is done with the aid of warehouse tasks. With the warehouse management monitor, EWM offers warehouse employees a tool with which they can be constantly updated on the current situation in the warehouse and react accordingly (see Fig. 3.45). All activities and documents are displayed on the monitor, from deliveries to the stock situation and warehouse tasks, to the efficiency of individual employees.

Warehouse management monitor

Figure 3.45 shows the warehouse management monitor in EWM. The screen is divided into three sections. On the left side, there is a hierarchy list with predefined nodes depending on desired work area. The work area, documents to be selected, processes or messages can be found using the selection criteria. The upper-right area contains the superordinate data, while the lower area has the subordinate detail information on the selection. In this example, an inbound delivery has been selected. The screen shows a number of inbound deliveries. The subordinate detail information for the selected delivery document indicates putaway as the corresponding warehouse activity status. A further example of the warehouse management monitor is shown in Fig. 3.72.

The warehouse management monitor, or *Warehouse Monitor*, is not only a display instrument, but also offers the possibility of actively influencing processes. For instance, you can quickly react to unplanned events, such as by blocking storage bins or allocating warehouse tasks to resources. The warehouse management monitor also possesses functions for alert monitoring that highlight current and potential problematic situations for the warehouse manager and provide tools to handle exceptions. The monitor can be tailored to specific needs and expanded.

With the graphical warehouse layout GWL, available in EWM releases from Version 7.0 on, you can depict the structure of your warehouse in two-dimensional form (see Fig. 3.46).

Any master data errors within the warehouse structure can be rapidly identified and corrected using this function. Graphical depiction is based on the storage bin coordinates in the storage bin master data. In addition to storage bins, the GWL can display other structures, such as walls or offices, as well as process-relevant objects like handling units, forklifts, conveyor lines and rack feeders. Thus, the GWL

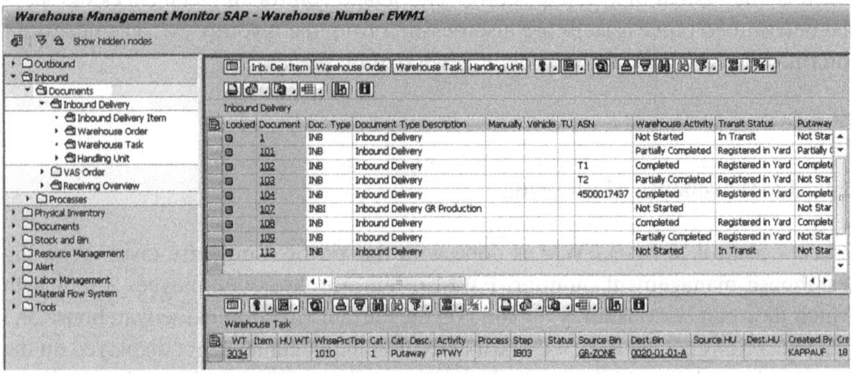

Fig. 3.45 Warehouse management monitor with warehouse task for putaway

enables you to visually monitor processes as well as influence them. In step with the framework concept of EWM, the design and functions of the GWL can be tailored and expanded according to your specific needs.

As an enhancement to the primarily text-based warehouse management monitors, the *Easy Graphics Framework* (EGF) offers the opportunity to display data in a Warehouse Cockpit (see Fig. 3.47).

With the aid of various chart types, you can monitor such things as resource capacity or the status of certain system messages. You can adjust the layout and displayed information to fit your needs.

3.4.3 Goods Receipt

Goods receipt is one of the core processes in the warehouse, and it represents a significant interface between external and internal logistics. It is used for the controlled establishment of stock and can be developed in a variety of ways, based on spatial circumstances, the materials and quantities to be stored, volumes and weights, as well as on the related information flow from warehouse to warehouse.

In this regard, EWM offers various design possibilities and enables a flexible integration of front- and back-end process steps into the overall process (see Fig. 3.48). For example, you can integrate front-end transportation, yard or unloading processes as well as back-end deconsolidation, packaging, counting and checking processes into the process chain in physical goods receipt. The

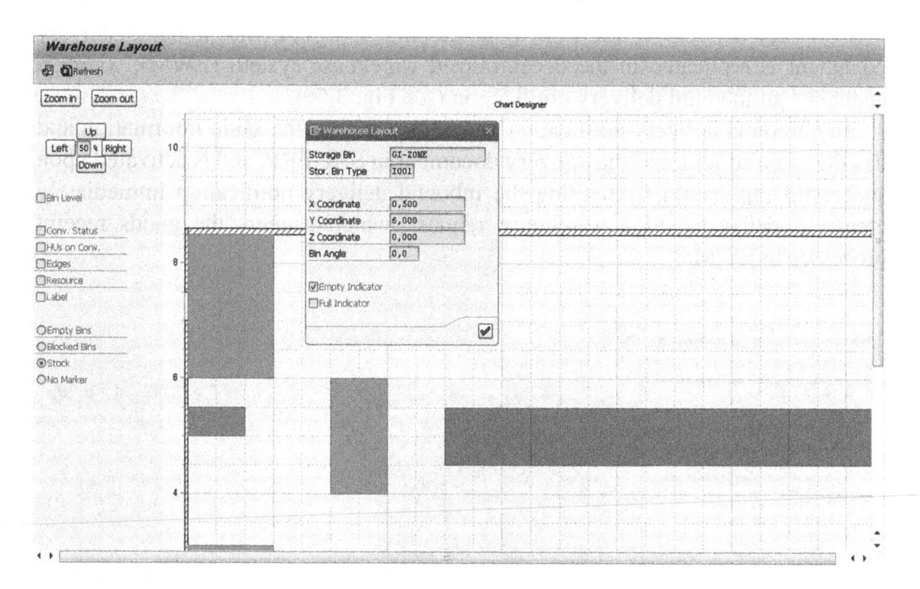

Fig. 3.46 Graphical warehouse layout

following section uses a process example to illustrate the basic functions and steps in inbound goods deliveries.

3.4.3.1 Goods Receipts Documents

The inbound delivery is the central document in goods receipt. It contains all data relevant to the delivery and corresponds closely with the ERP system.

An inbound delivery is generated in the ERP system with reference to a purchase order or the receipt of a shipping notification from the vendor (see Fig. 3.49).

In addition, an inbound delivery can be created based on a customer return or as the result of a production process in which finished products must be stored.

Shipping notification in ERP

Figure 3.49 shows Inbound Delivery 180000183, generated in the ERP system with reference to an order. The warehouse number EW1 is set in the system as a decentralized warehouse, and is linked to EWM warehouse number EWM1 in the EWM system. The goods receipt for this shipping notification is thus done in a decentralized EWM system. The document has been distributed accordingly and generates an *inbound delivery notification.*

Based on a certain combination of plant and storage bin and the allocated WM warehouse number, the system determines for every item of the delivery whether the goods receipt warehouse is decentralized or not (see Fig. 3.1). If at least one item of that delivery is relevant to processing in EWM, the inbound delivery document is replicated in the decentralized warehouse system (EWM), where it generates an inbound delivery notification (see Fig. 3.50).

The inbound delivery notification generally includes the same information and has the same structure as the delivery document in SAP ERP, and is activated upon successful replication. Activating the inbound delivery notification immediately creates a delivery and a warehouse request, which triggers the goods receipt processes in EWM.

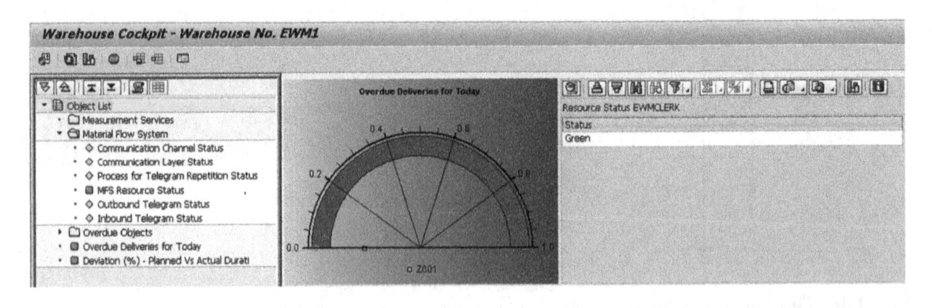

Fig. 3.47 Graphical Warehouse Cockpit

Inbound delivery notification

Figure 3.50 shows Inbound Delivery Notification 180000183 with regard to ERP Delivery 180000183 (see Fig. 3.49). The document has been successfully replicated and activated, and has the same document number as the document in ERP. Upon activation, SAP EWM automatically generates an inbound delivery.

To initiate the goods receipt process, the ERP system generally creates an inbound delivery from a shipping notification. In practice, however, it is possible that a goods receipt may have to be performed spontaneously without a shipping notification having been received. This process can be mapped in EWM as an *expected goods receipt* (see Fig. 3.51).

The purchasing document from WM or a production order generates an expected goods receipt in EWM. First, a notification regarding the expected goods receipt is created, consisting of a copy of all relevant logistics data from the replicated preceding document. After it is activated, EWM automatically generates the

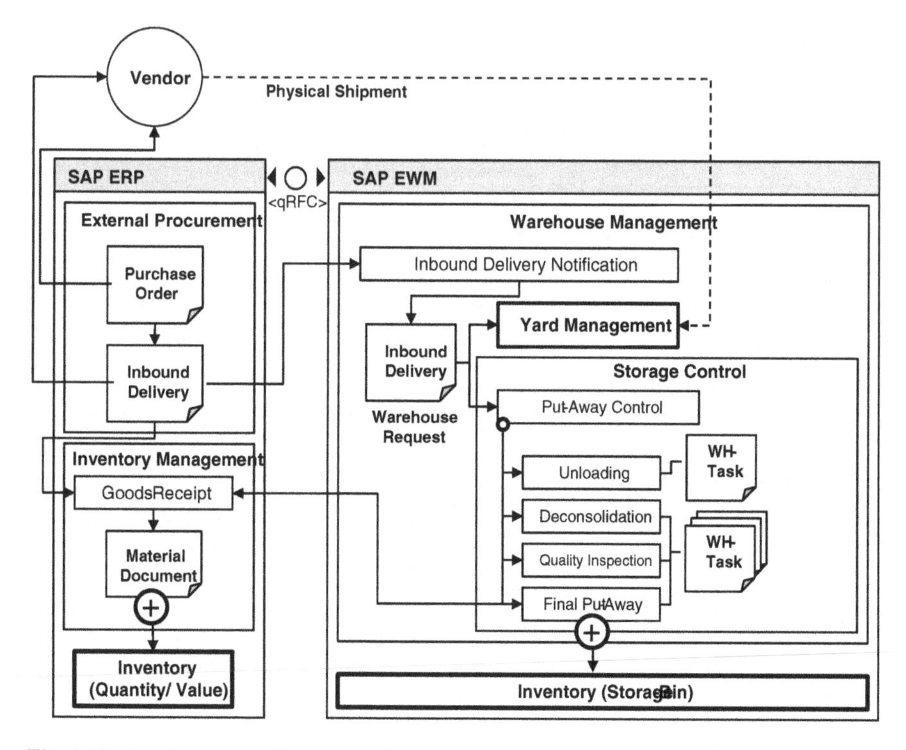

Fig. 3.48 Goods receipt with an inbound delivery

expected goods receipt. The subsequently arriving deliveries are generated without a previous shipping notification upon arrival of the delivery truck and verified by comparing them to the expected goods receipt. Tolerance checks are performed on the delivered materials. The inbound delivery also serves to trigger goods receipt processing for expected goods receipts. The preceding documents are no longer required and are regularly deleted.

From an operational standpoint, expected goods receipts have the advantage that the goods receipt can take place in the warehouse without a prior shipping

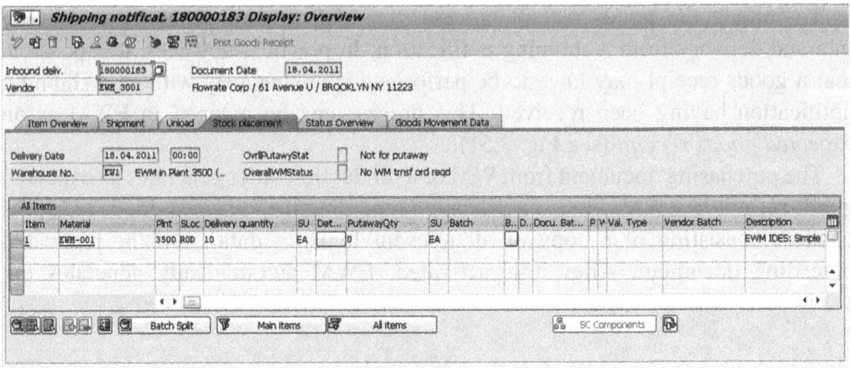

Fig. 3.49 Inbound delivery for an order in ERP

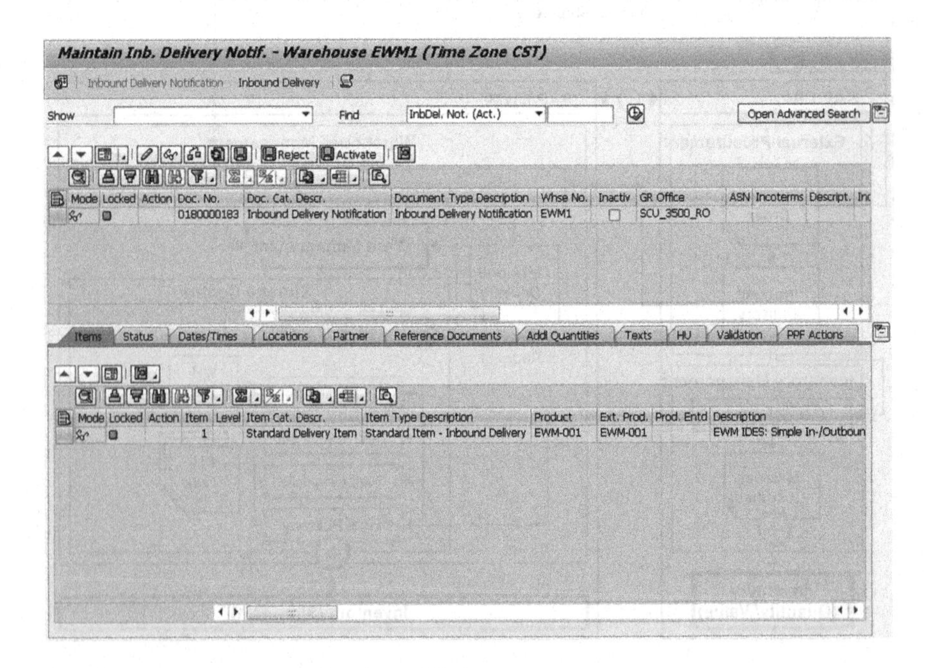

Fig. 3.50 Inbound delivery notification in EWM

notification. In addition, from a warehouse management perspective, expected goods receipts are used to assess the workload, and future goods receipts can be planned based on the number of earmarked shipping notifications and order items. To process expected goods receipts and forecast workload, SAP EWM offers comprehensive transaction and evaluation options.

3.4.3.1 Controlling Putaway

The *inbound delivery* represents a warehouse request and the starting point for subsequent activities in EWM (see Fig. 3.52). In addition to the data taken from the preceding document, the inbound delivery contains all necessary information to trigger and monitor the goods delivery process in EWM. This process typically begins with processing a transport unit and unloading a delivery, and ends with putaway of the materials in the warehouse. After the warehouse process type and storage bin have been determined for putaway, EWM generates the warehouse tasks necessary for the putaway process based on the process- or layout-oriented storage control.

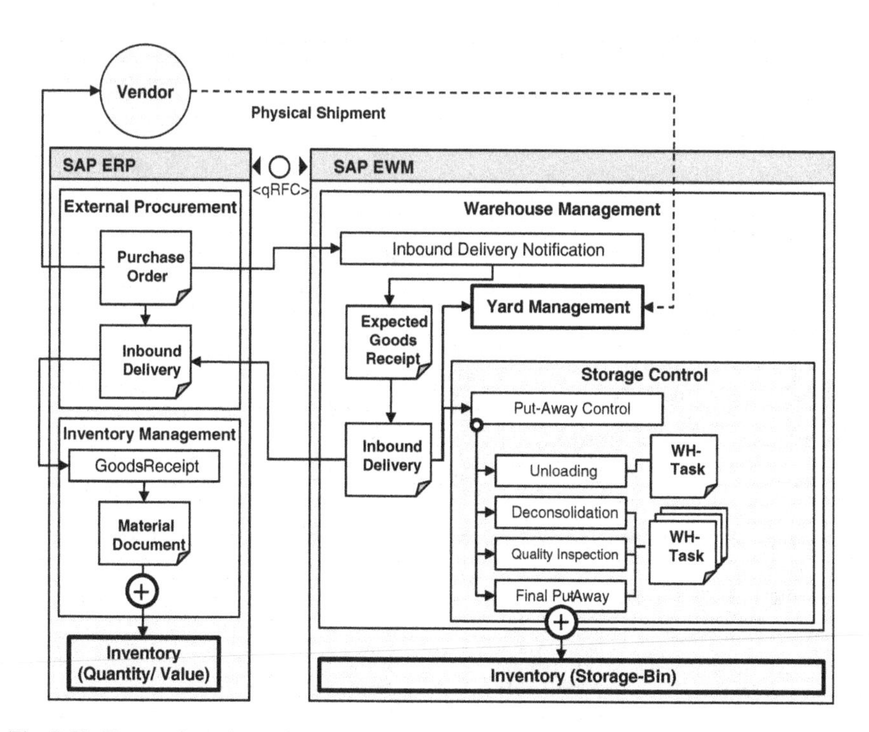

Fig. 3.51 Expected goods receipt

Inbound delivery in EWM

Figure 3.52 shows Inbound Delivery 112. The inbound delivery is the follow-on document of Inbound Delivery Notification 180000183. It represents a warehouse request for the putaway of externally procured materials and on the item level receives a seamless reference to the preceding documents. For this delivery, Warehouse Process Type 1010 (putaway) was found. On the item level, the source and destination storage bins are known. The inbound delivery item is to be moved from the goods receipt area (GR-Zone) into Storage Bin 01-01-A in Area 0001 of Storage Type 0020.

The warehouse process type is determined by EWM when the inbound delivery is generated. The source storage bin and storage type are established using the determined warehouse process type. The destination storage bin for putaway is found according to the chart shown in Fig. 3.53.

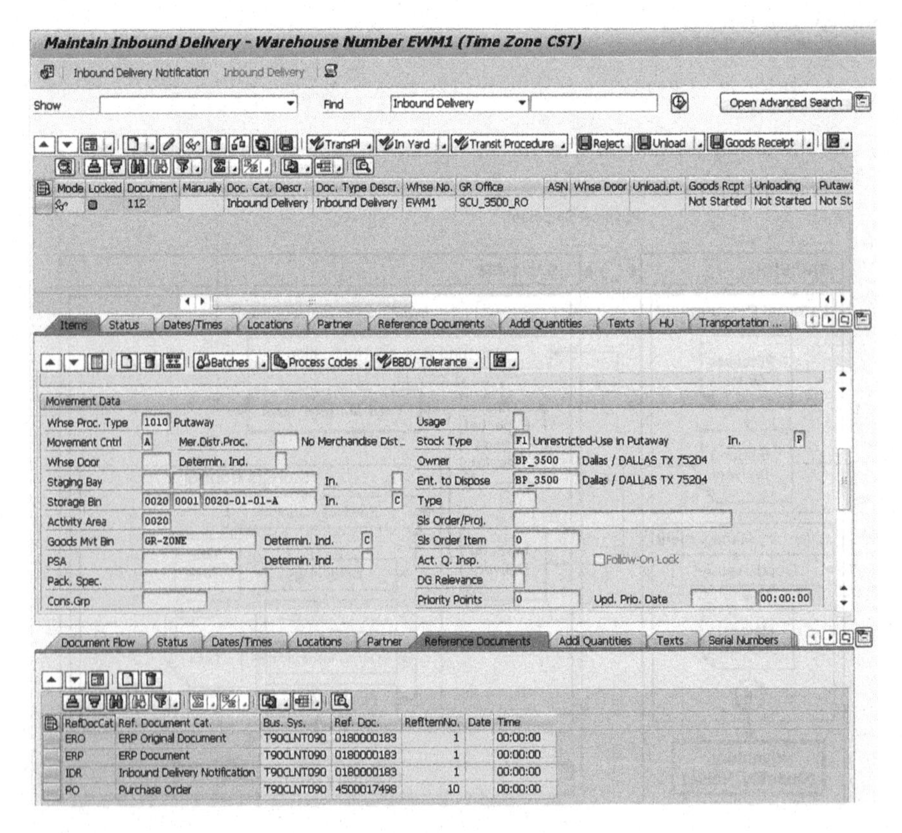

Fig. 3.52 Inbound delivery as a warehouse request in EWM

In the first step, EWM uses the storage type search sequence to determine the storage type of the destination storage bin for every delivered item. The search is done using document and product data. The sequence of a storage type search contains all storage types that can be searched for storage bins. As soon as a storage section is found for a specific combination of storage type and product attributes, such as a hazardous material designation, EWM attempts to determine the actual storage bin. The system follows the sequence and tries to determine the storage sections for the first storage type found. The search can also be influenced by so-called *weighting factors*. These factors change the sequence of storage types and storage sections in the putaway search sequence and enable such actions as the preference of certain storage sections.

Actual determination of the storage bin is based on the storage section indicator of the storage bin type in the product master and with reference to the putaway rule or strategy and the capacity of the storage bin. If during the search of all storage bins no free bin is found in the currently determined storage section, the system checks the other storage sections. If bins are also not found in these sections, the check is done for the next storage type in the storage type search sequence.

The *putaway strategy* serves to find suitable storage bins for putaway. For this, it accesses parameters in the product master, attempting to utilize storage capacity as advantageously as possible. In addition to automatic determination, EWM also allows you to manually edit certain goods movements and change source and destination storage bins, normally automatically determined. When products are put away, you can generally choose between various strategies, which are configured in the system or supplemented with internal program logistics with the aid of a business add-in (BadI). The following standard strategies, in addition to manual entry, are available:

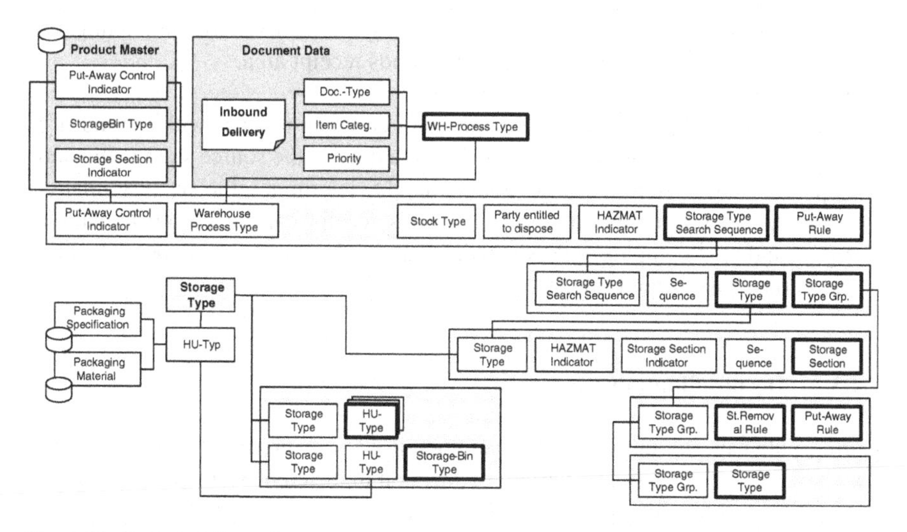

Fig. 3.53 Technical flow of putaway control

- Fixed storage bin
- General storage
- Addition to existing stock
- Empty storage bin
- Near fixed picking bin
- Pallet storage
- Bulk storage

Additional information on putaway strategies

These putaway strategies exist in similar form for WM and were explained briefly above with reference to WM putaway strategies. Thus, we refer to that explanation, as it also illuminates the operational view of putaway strategies in EWM.

3.4.3.1 Unloading

After the inbound delivery has been generated and the storage process type determined, the vehicles and transport units are unloaded. During *unloading*, the goods are moved out of the transport unit from the door to a staging zone, consolidation zone or a work center for quality inspection, depending on operational needs. When EWM Yard Management is used, the unloading process begins with the recording of the vehicle or transport unit at the control point. Actual unloading can be done in a simple or complex manner.

- **Simple unloading**
  For unloading of this type of delivery, only one status is set. The delivery to be unloaded can be determined using an unloading transaction. After unloading, the goods are ready for final putaway in the goods receipt area.
- **Complex unloading**
  The putaway of HUs can also be done in a complex manner with the aid of a warehouse task. The warehouse task cites the door as the source storage bin and, depending on system settings, automatically determines the deconsolidation zone as the destination storage bin. Goods receipt can be posted at various times and, for complex unloading, is done at the latest when the unloading warehouse task is confirmed.

3.4.3.2 Deconsolidation

During putaway, the unloaded HU can contain various products. These products may have to be put away in a variety of ways on in different sections. Distribution of the delivered handling units in several putaway HUs is called deconsolidation.

From the putaway control view, deconsolidation – the intermediate step in the distribution of the delivered HU – is done using an interim storage type. This is typically a physical work center, or *consolidation work center*, in which the HU is deconsolidated (see Fig. 3.54). The display of the consolidation work center shows the employee the work list of the HUs to be processed. The warehouse employee can allocate the HU contents to be put away via Drag & Drop to putaway HUs and generate the subsequent warehouse tasks.

Whether or not deconsolidation is necessary – or whether a deconsolidation HU is to be distributed among several putaway HUs (Fig. 3.55) – depends on the *consolidation group*, which controls whether deconsolidation must take place for an HU. Within the context of putaway control, the system determines the putaway strategy and then the putaway bin for each material. The storage bin can be allocated to an activity area. The consolidation group is determined from the activity area, which is allocated to a particular storage bin. Based on this logic, deconsolidation is always required, for example, when storage bins are determined for products within that HU which are allocated to different activity areas.

Deconsolidation

Figure 3.55 shows the basic deconsolidation process. In this example, storage bins for the putaway of the content of HU1 are determined. Both storage bins are allocated to two different activity areas. That is why deconsolidation is necessary. HU1 is moved to a deconsolidation work center for deconsolidation. During deconsolidation, EWM generates a warehouse task per HU item that takes consolidation into account. HU2 and HU3 are then put away or processed further in accordance with a process-oriented control.

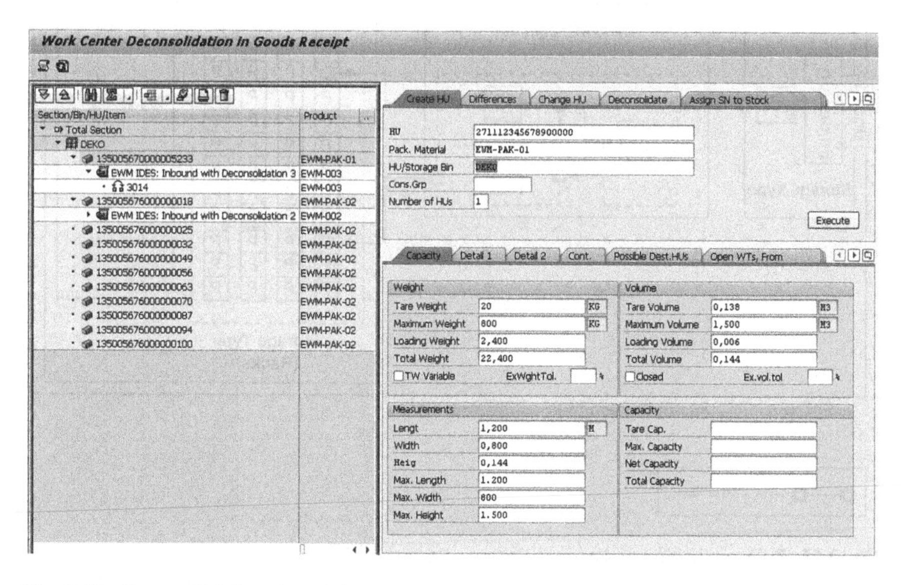

Fig. 3.54 Deconsolidation at a work center

After deconsolidation, it may be necessary to perform a quality inspection or count of the materials to be put away. Generally, warehouse employees in goods receipt perform the inspection and, depending on the result, decide on the further placement of the materials.

With its *Quality Inspection Engine* (QIE), SAP EWM gives you the option of integrating inspection processes into the process chain and control subsequent activities. QIE supports quality management processes in a heterogeneous system environment and is a further development of the functions of the previous Quality Management (QM) in ERP. In this development, performance flexibility of the quality management function has been increased, as has the number of processes that can be integrated into a quality inspection.

The quality inspection can be done in EWM at the time of goods receipt on various levels, such as on the level of the complete delivery, the delivered HU or product-based. The goods receipt process also completes the inbound delivery of customer returns. Inspections can be performed several times for a goods receipt. Depending on process requirements, a general sight check can be conducted upon arrival of a vehicle, for example within the context of a preliminary inspection. An additional inspection is then conducted after unloading on the level of the delivered HUs (see Fig. 3.56).

In addition to the inspection at goods receipt, a quality inspection in the warehouse can also be triggered manually at any time, such as if goods have been damaged.

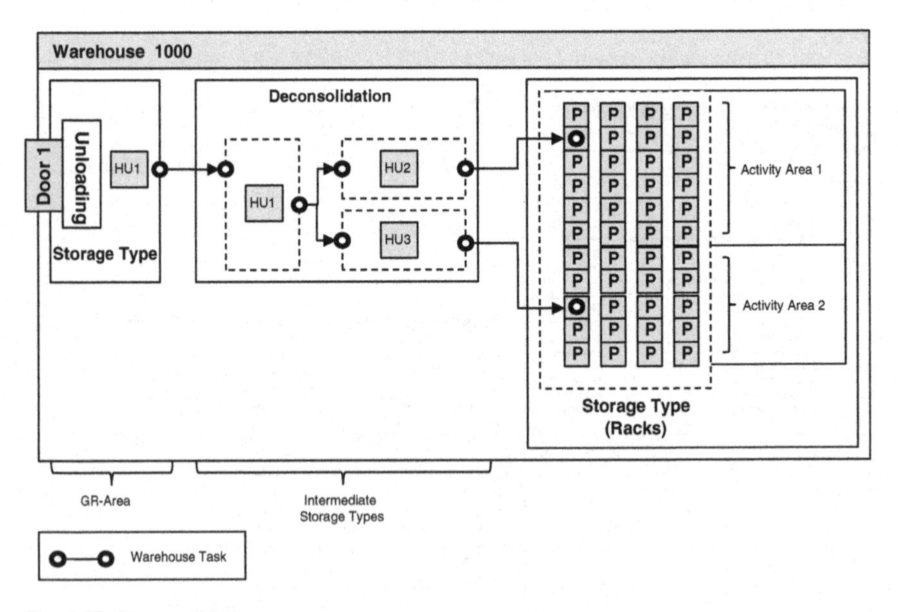

Fig. 3.55 Deconsolidation process

The actual inspection is initiated through the creation of an *inspection document*. Automatic generation of this document is based on the objects to be checked, or the inspection object types, and the applicable inspection rules.

The *inspection object type* defines the software component, process and object for which an inspection document is to be generated. Examples include the counting inspection of a complete inbound delivery, the quality inspection of individual received products or batches, or the inspection of products already situated in the warehouse.

The *inspection rules* are assigned to the inspection object types and basically contain the parameters specifying the extent and process of an inspection. They include the inspection procedure, guidelines for generating the inspection document, the method of determining the extent of an inspection, whether it is a random sample and the code for the actual inspection decision. Also, inspection instructions can be attached to inspection rules in the form of a document.

The purpose of an *inspection document* is to collect data for the inspection, and following the inspection, to enable the findings and indications to be entered. Such error sets serve as a valuable information source, particularly in the vendor selection process within the context of procurement logistics. They also enable quality improvements and process optimization by allowing you to set measures required for the elimination of error sources.

After a quality inspection is completed, a decision regarding the use of the tested objects must be made. This decision is based on the inspection findings and, depending on the application decision, can trigger follow-up actions. They include:

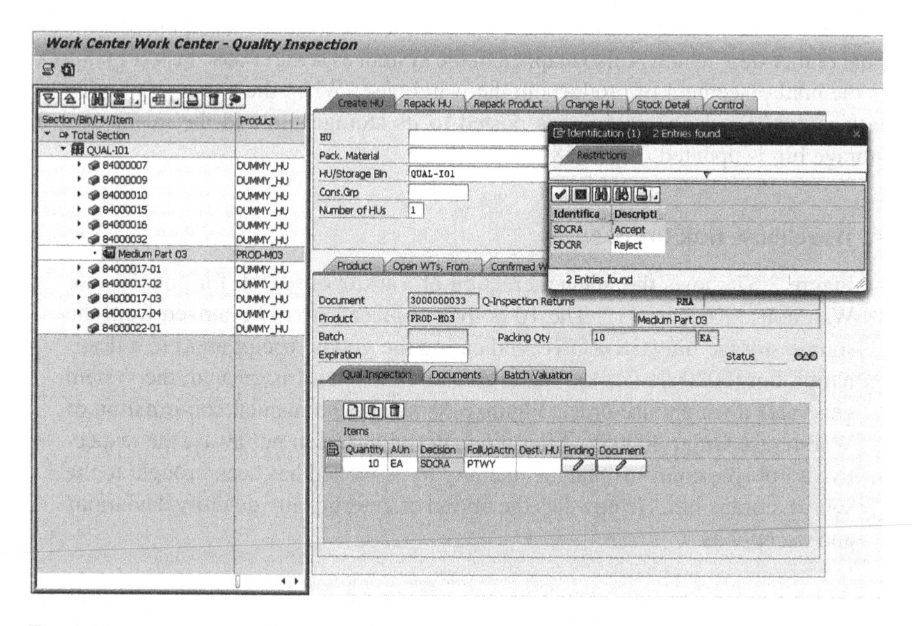

Fig. 3.56 Work center quality inspection

- Putaway
- Scrapping
- Stock transfer
- Return delivery

Quality inspection at the work center

Figure 3.56 shows a work center for quality inspections. The left portion of the screen includes the storage section, the HUs to be inspected and the number of the generated follow-up activity.

The right side contains information on the product and reference to the inspection document (3000000033). In this example, the contents of HU 84000032 are inspected. This HU is a customer return. The results are recorded in the inspection document, and the system automatically determines a follow-up activity. In this case, that action is putaway (indicated by the abbreviation PTWY).

Through integration with *SAP Customer Relationship Management* (SAP CRM), inspection findings can be evaluated and logistics follow-up actions triggered, such as a stock transfer or scrapping. Based on the findings, in the case of customer returns, CRM can dynamically determine the credit amount depending on the condition of the returned goods. In addition to SAP CRM, quality inspections can be integrated in various applications of SAP Business Suite as well as non-SAP applications with the aid of the Quality Inspection Engine.

Putaway in the destination storage bin completes the goods receipt process. The goal of this step, which is also mapped in the system as a warehouse task (Fig. 3.57), is the final placement of products in the warehouse. When the putaway warehouse task is confirmed, the material is posted to its storage bin and the quant of that storage bin is updated (Fig. 3.58).

Warehouse task for putaway

Figure 3.57 shows the manual creation of a warehouse task for putaway for Warehouse Request 112. The 10 delivered pieces are to be moved from the source storage bin (GR-ZONE, indicating the goods receipt area) to a destination bin (0020-01-01-A). When the warehouse task is created, the system generates a warehouse order. Figure 3.58 shows the manual confirmation of Warehouse Order 200055. After the material has been put away, the respective employee confirms that the quantity to be moved has been brought to the source storage bin. He now has the option of entering any quantity deviations and exceptions.

3.4.4 Warehouse-Internal Processes

Several warehouse-internal processes that cannot be specifically allocated to goods receipts or issues exist in the same or similar form in WM (SAP ERP) and have previously been mentioned. In this section, we will provide a brief explanation of the functional characteristics and distinctive features of warehouse-internal processes from the viewpoint of SAP EWM.

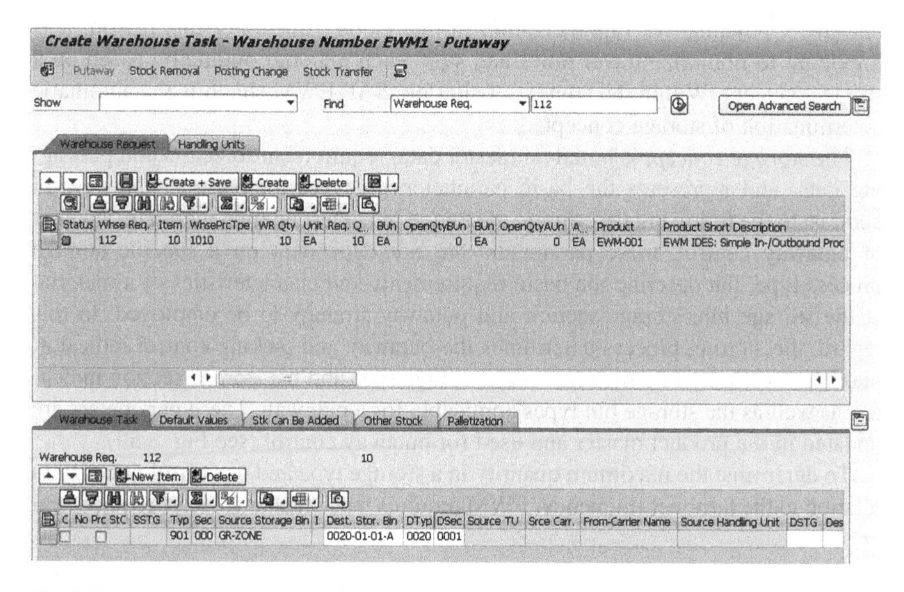

Fig. 3.57 Creating a warehouse task

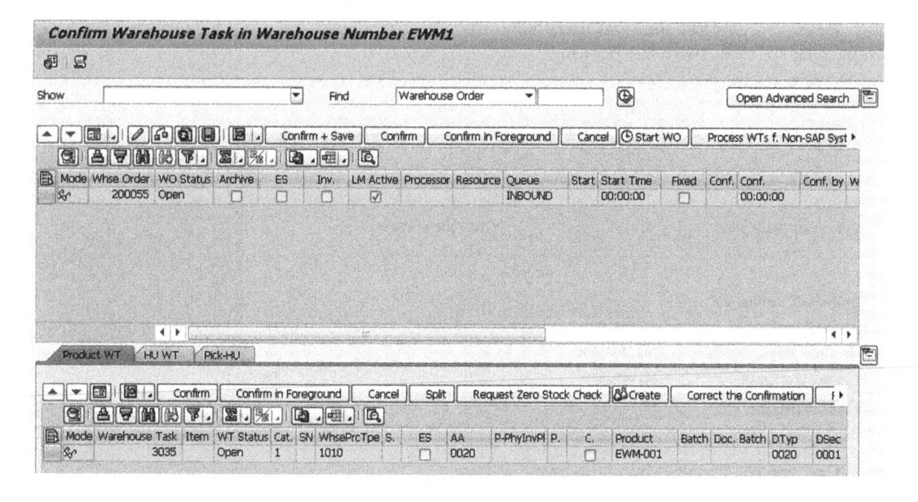

Fig. 3.58 Confirming a warehouse task for putaway

3.4.4.1 Slotting

Travel times are decisive factors in the evaluation of logistics costs, especially for picking. The variety of products in a warehouse and an ever-changing product portfolio often make optimal storage of each individual product even more challenging. With EWM's integrated function of *slotting*, a storage bin can be automatically determined for the putaway of all products from which picking and placing is the most efficient. The putaway of materials in stock in an economically efficient manner not only represents an optimization of warehouse-internal processes but also a potential for rationalization in many companies. For this reason, it is important to minimize travel times and store each material optimally, based on a storage concept. Within the context of slotting, SAP EWM supports the automatic determination of storage concepts.

The *storage concept* is based on master data, requirement forecasts and packaging data, and represents the basic parameters relevant for optimal putaway. In contrast to the following description of storage bin determination within the context of putaway control, these parameters are not dependent on a specific storage process type, but describe the basic requirements and characteristics of a material at the storage bin, storage section and putaway strategy to be employed. In this regard, the slotting process determines the putaway and picking control indicator, maximum quantity of a material in the storage type and the storage section indicator, as well as the storage bin types applicable for a putaway. These parameters are updated in the product master and used for putaway control (see Fig. 3.59).

To determine the maximum quantity in a storage type and to control putaway in relation to the turnover frequency, EWM requires the actual or forecast requirement

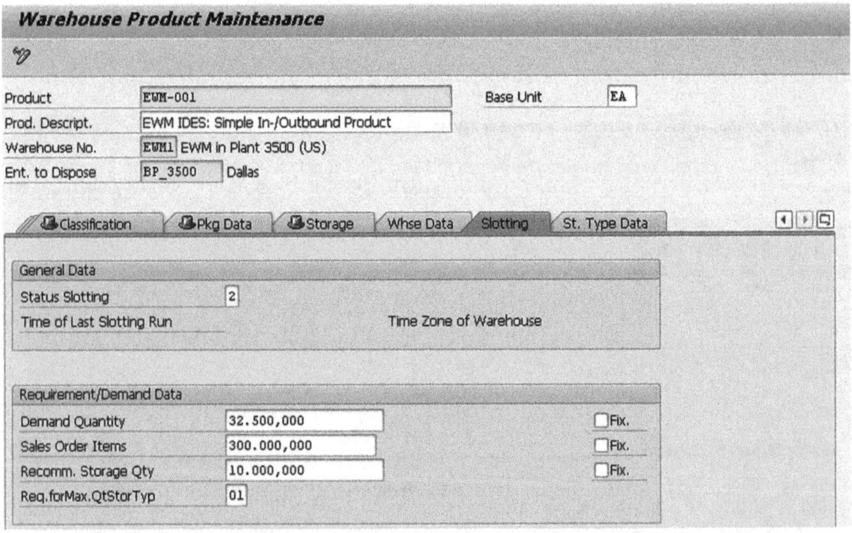

Warehouse Product Maintenance

Product EWM-001 Base Unit EA
Prod. Descript. EWM IDES: Simple In-/Outbound Product
Warehouse No. EWM1 EWM in Plant 3500 (US)
Ent. to Dispose BP_3500 Dallas

Classification | Pkg Data | Storage | Whse Data | Slotting | St. Type Data

General Data
Status Slotting 2
Time of Last Slotting Run Time Zone of Warehouse

Requirement/Demand Data
Demand Quantity 32.500,000 Fix.
Sales Order Items 300.000,000 Fix.
Recomm. Storage Qty 10.000,000 Fix.
Req.forMax.QtStorTyp 01

Fig. 3.59 Slotting in the storage product master

of a material. This requirement data is provided to EWM by SAP APO and updated in the *slotting data* of a storage material.

For the determination of the optimal storage bins, EWM accesses the slotting data. To optimize the organization of materials in a warehouse, for instance with regard to their turnover frequency, EWM offers a warehouse organization option. This option, called *rearrangement*, enables the reorganization of products in the warehouse based on optimal parameters of the slotting function, thus contributing considerably to the reduction of travel times. For rearrangement, the current storage bin is compared with the optimal storage bin. If the optimal storage bin is not already in use, EWM suggests the optimal bin. The warehouse tasks for stock transfer to the optimal bins can be consolidated into waves. They can then be executed together and do not have a disruptive effect on other processes.

Parallel to the functions in WM, Inventory in EWM is also executed on the warehouse level and guarantees that the right product is situated in the right place and amount. Not least for legal reasons, inventory is performed as a physical counting of products and handling units. The system can be controlled in a flexible manner with regard to the frequency and method with which certain items should or must be inventoried. In addition, EWM enables the product- or storage-bin-related taking of inventory.

The storage-bin-related inventory refers to taking inventory of a specific storage bin and all materials or HUs situated in that bin. On the other hand, product-related inventory relates to a certain material that can be located in several storage bins or various HUs. The inventory processes used correspond to legally permitted possibilities and, from a functional standpoint, to the inventory processes in WM that have already been explained.

As in WM, EWM also supports periodic inventory, continuous inventory and cycle counting. For continuous inventory, the inspections discussed above at storage bins or product-related inspections are supported for ad-hoc inventory-taking.

A special feature of storage-bin-related inventory is the *low stock check*. For this inventory simplification process, which is nothing more than an accompanying inventory during putaway or picking, inspection takes place when the warehouse tasks are confirmed. Using a limit value, remaining low stock in a storage bin is counted during picking without explicitly causing a physical inventory taking. During physical picking, the system often checks to determine whether the low stock captured in a sight check corresponds to the actual stock situation. Upon confirmation of the warehouse task, EWM automatically generates the inventory document. In addition to the inventory procedures, EWM offers the possibility of combining various procedures for low stock check.

The inventory-taking process is comparable with that described for WM (see Fig. 3.35). In EWM, an inventory document is also generated and lists the storage bins and materials that need to be counted. The option of a recount is based on the determined quantity or value differences with the aid of a difference analysis.

To analyze differences and subsequently post stock in ERP's inventory management, in EWM you use the *Difference Analyzer*. It accumulates any detected differences and reconciles them with configurable tolerance groups. Before a difference is ultimately deleted from the accounts of the ERP system, it can be blocked from further analysis. An employee can enter notes or make corrections on the detail level. As soon as the status of a blocked difference permits, the stock is offset against the ERP system. In ERP stock management, a corresponding goods movement is generated for the difference quantity. If and to what extent certain users can post differences can be set in EWM and influenced by maximum amounts or percentage-based limits.

3.4.4.2 Scrapping

Some of the daily processes in a warehouse are those that cause a reduction in stock in the warehouse management system. These processes, known as *scrapping*, are, functionally speaking, goods issues in which the material to be scrapped is taken from its storage bin and removed from stock. Whether or not scrapping is necessary depends on the condition of the material, that is, if it is either damaged or has become unusable in some other way. In EWM, we differentiate between the following cases, depending on where the decision to scrap is made:

- Warehouse-internal scrapping
- Scrapping in a quality inspection

In both cases, actual scrapping represents a goods issue that is mapped in the system through a posting change notice, which forms the basis for generating a warehouse task for scrapping. The material to be scrapped can be picked and moved to a scrapping work center. Once there, and after any necessary packaging of the material, an outbound delivery and goods issue are generated.

For *warehouse-internal scrapping*, the stock is allocated to a storage bin or a specific partial quantity of another stock category. After this block has taken place, a goods issue is posted for the respective stock. This process can be triggered by a transfer posting from the ERP system or from the warehouse (EWM). The decision to scrap a material that is made internally generally occurs during picking or while taking inventory.

Another option is scrapping within the course of a *quality inspection* when you have an inspection lot and results. Here, too, the process can be triggered warehouse-internally or externally. Thus, an employee from the quality department can make the decision to scrap based on the results of an inspection procedure. From a quality inspection point of view, which we have described in Sect. 3.4.3, "Goods Receipt", scrapping represents a follow-up process (see also Fig. 3.56).

3.4.4.3 Kitting

Kitting is a term from the realm of procurement and production logistics. It refers to the combination of individual parts according to customer specifications and their delivery as a kit. Kits are either delivered unassembled or assembled. However, delivery to customers (as opposed to stock transfers, for instance) is always done in kit form, that is, unassembled. The advantage of kitting to the customer is that it can speed up production. Kitting becomes especially attractive when procurement logistics must be cut.

The kitting process in EWM allows you to combine kits within a warehouse. For this, EWM uses orders for logistics *value-added services*. The combination procedure is typically done in work centers within the warehouse. EWM supports two processes for kitting:

- Kit-to-stock
- Kit-to-order

The *kit-to-stock* process involves producing kits in advance and storing them in stock. Kitting begins either in ERP, manually on the basis of a production order, or directly in EWM with the aid of a *direct outbound delivery order*. In both cases, kitting requires an outbound and inbound delivery document for stock. When kitting is generated based on an ERP production order, the inbound delivery is automatically generated and already contains the components it requires. Both delivery documents are replicated by the ERP system to SAP EWM. The outbound delivery order for kitting represents a warehouse request and generates the order for logistics value-added service. After the warehouse tasks for picking of kitting components have been generated, the order is moved to the kitting work center. Once a kit has been put together, the processed components are deleted from stock with a goods issue posting. The components have been turned into a kit whose addition to stock is posted to an inbound delivery with a goods issue.

Unlike kitting to stock, *kit-to-order* processes involve compiling the required components based on an order, similar to make-to-order production. If the required kit is not in stock, it is either generated as a part of picking or as a logistics value-added service at a work center. When a kit is composed at the time of picking, the required components are taken from stock and directly assumed in a picking handling unit without having to take place at a work center. When kitting takes place as a logistics value-added service, the order for the logistics value-added service is either automatically or manually generated, depending on system settings. At the time of the order generation, EWM selects the work center in which kitting is to take place and generates the warehouse tasks for picking kit components. The destination storage bin of these warehouse tasks corresponds to the work center for kitting. The finished kits are then processed further, depending on the storage control stored in the system, packaged if needed, and subsequently moved to the goods issue area.

The *kit* consists of the kit header and kit components (see Fig. 3.60). The kit header shows the material ordered by a customer, especially for a kit-to-order procedure. In our example, this material is “Spare parts components”. This material, the so-called *header material*, is divided into its components, which together form the *kit structure*. Our example lists two components to our spare parts kit.

iPPE structure in APO

Figure 3.60 shows the assembly structure in APO. The *integrated Product and Process Engineering* (*iPPE*) contains the kit structure with header and components. In this example, the kit header "SPM_KIT_01" consists of two components. Both the header and the allocated components are stored in the system as product masters. You can also indicate the quantity of the required products for each component.

Kits are not saved as master data in EWM. For a kit-to-order process, EWM receives the kit structure from the ERP system. Depending on whether or not APO is used for the availability check, and whether the order was generated in the CRM system, the kit structure can also be determined using *integrated Product and Process Engineering* in APO. Within the context of this determination, the sales order of the CRM system performs a rule-based availability check in APO. The kit header material is dissolved according to the structure stored in APO, and the required components are determined. The actual availability check is then performed for the required kit components.

If neither SAP APO nor SAP CRM is used, the structure can also be determined in ERP with the aid of a bill of materials (BOM). Along with the packing specifications, this data serves as the basis for the compilation of the kit in EWM. When kitting is performed to stock, EWM always consults a BOM stored in ERP.

To break down an existing kit back into its components, EWM can generate an order for a logistics value-added service. Like kitting-to-stock, *reverse kitting* begins with an outbound delivery order and an inbound delivery. A goods issue is posted for the disassembled kit, and a goods receipt is posted for the former kit components. By confirming the respective warehouse tasks, stock is adjusted accordingly in ERP, and a goods receipt and issue are posted in stock management.

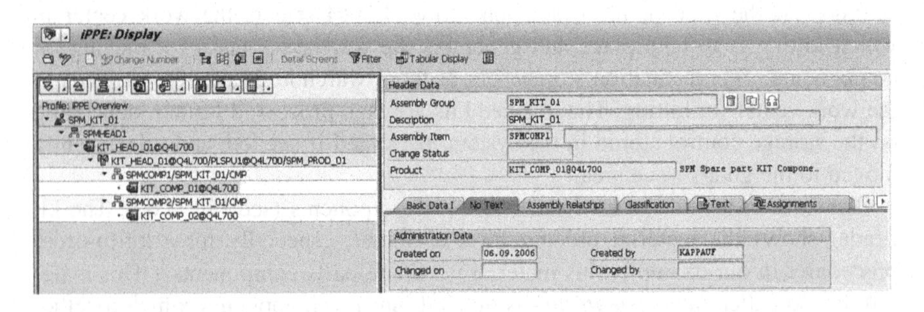

Fig. 3.60 Kits in the iPPE workbench

3.4.4.1 Replenishment

Within a warehouse, picking is often only done in one specific subarea. Thus, efficient and economical picking requires the necessary materials to be immediately available at the time of picking. EWM supports efficient picking with various replenishment strategies that ensure a needs-based control of stock in the respective picking areas.

Replenishment is flexibly integrated in existing logistics processes and can be executed either at set or dynamic times using minimum and maximum quantities. The warehouse product master contains the required information on minimum and maximum stock levels, as well as on requested replenishment quantities of a material. If a defined minimum stock level for replenishment is reached, a replenishment order is automatically generated and the storage bin is filled again with pallet units until the maximum level is reached. Demand-based replenishment plans are based on open warehouse requirements. Direct or exception-based replenishment processes react to incorrect quantities that are detected during warehouse task confirmation.

With regard to this function, we differentiate between the following types of replenishment control:

- **Planned replenishment**
  Planned replenishment is either started manually or automatically. The replenishment quantity is planned by the system on the basis of defined maximum and minimum quantities and is triggered precisely when the current stock is lower than the defined minimum quantity.
- **Order-related replenishment**
  For order-related replenishment, the replenishment quantity is determined based on open warehouse requests. Replenishment is triggered when the actual stock is lower than the determined required quantity. This type of replenishment can also be executed manually or in the background.
- **Automatic replenishment**
  For automatic replenishment, the system automatically determines replenishment upon confirmation of a warehouse task based on the actual stock and the stored minimum quantity.
- **Direct replenishment**
  Direct replenishment control can be initiated directly through an exception code during picking. Direct replenishment can then be performed by a picking employee.

Transfer posting and stock transfer are also warehouse-internal processes.

3.4.4.2 Transfer Posting and Stock Transfer

The major characteristics and differences between transfer postings and stock transfers were discussed in this chapter with regard to inventory management and illustrated with the example of warehouse management in WM. In this section, we will explain the special technical process features of transfer postings and stock transfers in EWM.

The *transfer posting* of a material, in which no change in the physical location of stock generally takes place, can be triggered by ERP as well as directly in EWM. For a transfer posting initiated in ERP, an outbound delivery is generated in ERP and replicated in EWM. These outbound deliveries represent a warehouse request in EWM, which first determines the storage process type. Depending on system settings, a transfer posting can be posted at that point or, if the material is also to be moved to another bin in the warehouse, a warehouse task can be created. When the transfer posting is booked in EWM, the goods issue for outbound delivery is posted in ERP and the stock attributes are adjusted accordingly.

A transfer posting can also be exclusively generated in EWM. This is either done as a direct transfer posting or a posting via a warehouse request document that serves as the basis for the generation of warehouse tasks if the material must also be moved. When the transfer posting is booked in EWM, a goods movement is generated in ERP.

The physical movement of a material within a company from one organizational unit to another, and typically between one warehouse to another, is known as a *stock transfer*. In ERP, a stock transfer and outbound delivery are generated for this purpose. The outbound delivery is distributed to the issuing EWM system, where it triggers an outbound delivery order. Goods issue in EWM leads to a goods issue in ERP. If the receiving storage location is also managed with EWM, an inbound delivery is generated in ERP and replicated to the receiving EWM system (see Fig. 3.61). If the receiving storage location is not managed with EWM, the goods receipt in the receiving location can also be posted to ERP together with the goods issue posting of the outbound delivery. In such a case, if WM is used for the warehouse management of this storage location, this goods movement represents a WM transport requirement (see Fig. 3.20).

3.4.5 Goods Issue

From a logistics perspective, goods issues serve the controlled reduction of stock. Such stock reduction is executed based on sales and distribution processes. From the viewpoint of distribution logistics, a goods issue thus represents the completion of a shipping procedure to the customer and serves as an interface between internal and external logistics. Like a goods receipt, the information flow in the goods issue process is primarily based on spatial circumstances, the materials to be picked and individual process requirements in the warehouse. These requirements can vary from warehouse to warehouse, and may even be dependent on the respective goods recipient. For this core process, SAP EWM offers a variety of design options and enables the integration of individual process steps. In this section, we will use an example to illustrate the basic goods issue process in EWM.

3.4.5.1 Documents in the Goods Issue Process

The central document in goods issue is the outbound delivery. It typically represents a follow-on document to a sales order, but can also be created directly, without reference to a preceding document in ERP. The physical shipment, which forms the completion of a goods issue procedure, thus begins with the generation of an outbound delivery document in ERP (Fig. 3.62).

Whether it is necessary to distribute this document to a decentralized EWM depends on the WM warehouse number that is allocated to a delivery item. As in the goods receipt process, the system checks to determine whether or not the warehouse number indicates a decentralized warehouse (see Fig. 3.1).

If an outbound delivery is relevant to processing in EWM, the document is replicated in the decentralized warehouse system (EWM), where it generates an *outbound delivery request*. This document basically contains the same information and has the same structure as an outbound delivery in ERP, and is activated upon successful replication (see Fig. 3.63).

Outbound delivery in SAP ERP

Figure 3.63 shows Outbound Delivery 80016886, which has been generated in ERP without reference to a sales order. The respective plant/storage

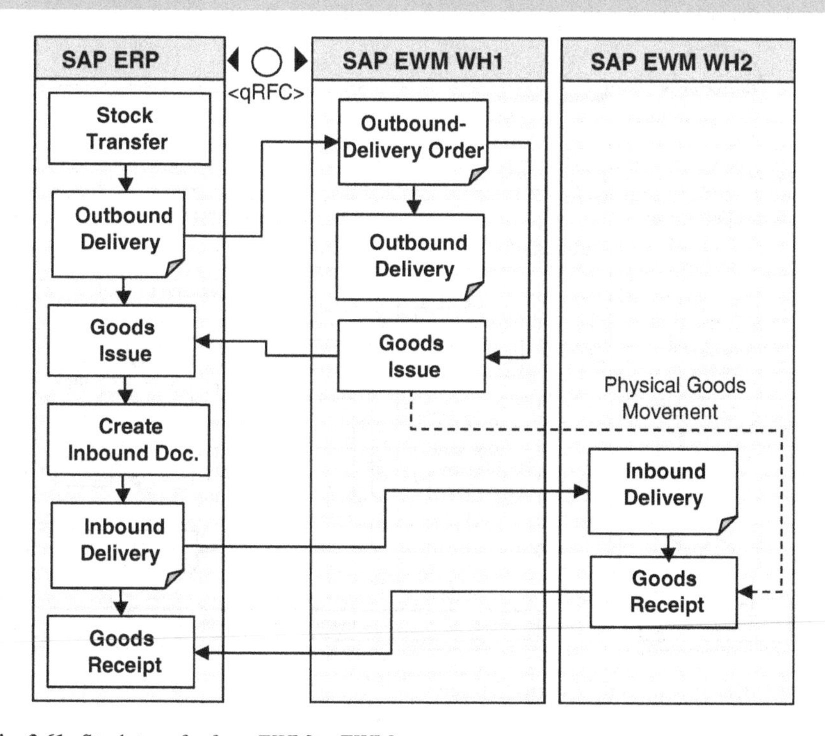

Fig. 3.61 Stock transfer from EWM to EWM

location combination is assigned the warehouse number EW1. This indicates a decentralized, EWM-managed warehouse. Thus, the document is replicated as an outbound delivery request to EWM.

The activation of an outbound delivery request immediately triggers an outbound delivery order, the actual *warehouse request*, which initiates the goods issue process in EWM (see Fig. 3.64).

The *outbound delivery order* (see also Fig. 3.66) contains data assumed from the preceding document as well as all necessary information to trigger the goods issue process and monitor it accordingly. From the perspective of warehouse management, the outbound delivery order represents a work list that is only completed when the picked materials have been loaded and shipped. When the outbound delivery request is activated and the outbound delivery order is automatically generated, EWM begins all procedures required to supply the document with the necessary information and map the process in accordance with the settings in the system based on process- or layout-oriented storage control.

A special kind of goods issue processing is a manual *direct outbound delivery order*, which is created directly in EWM and initially without reference to an ERP document. It has a structure corresponding to an outbound delivery order. After it is

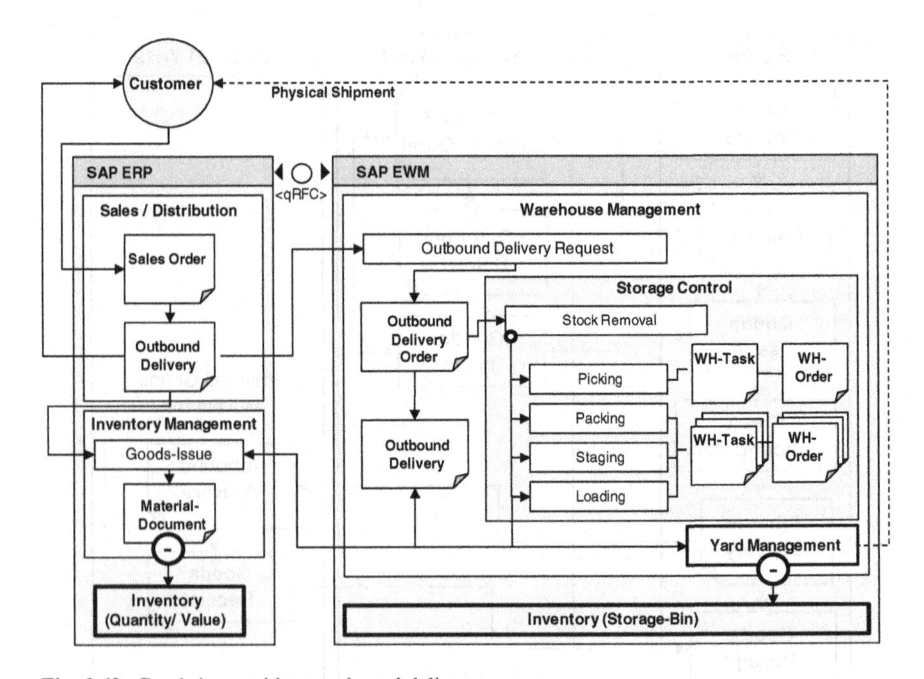

Fig. 3.62 Goods issue with an outbound delivery

created, the document is transmitted to ERP and subsequently contains the ERP document number.

We briefly mentioned direct outbound delivery orders for the kit-to-stock process. In this regard, they reserve the components necessary for the kit. They serve as the final step in the scrapping process by deleting the material to be scrapped from stock upon goods issue. In addition, this document can be used when customers pick up materials or for material removals posted to an account. For direct sales, such an order is a pick-up from the warehouse without having to previously create a sales order. In order for this process to be invoiced, the direct outbound delivery order generates an outbound delivery in SAP ERP.

When the outbound delivery order is generated, the system first determines the warehouse process type and route, and then conducts a rough estimate of the picking storage bin (source storage bin) based on a picking strategy. At the same time, the staging area (destination storage bin) is determined using the warehouse process type.

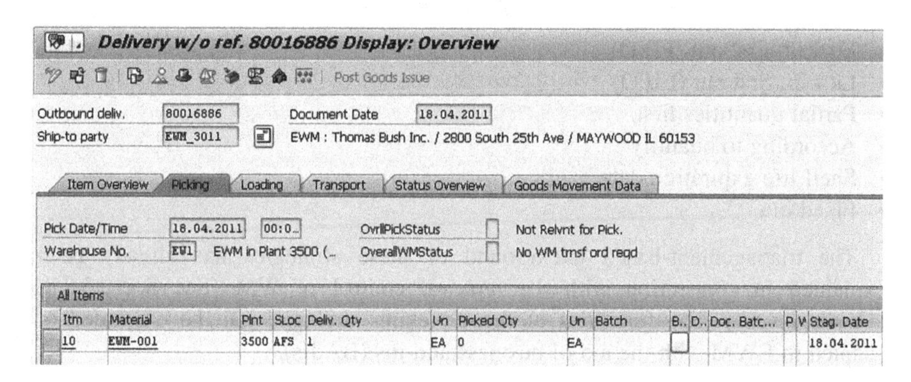

Fig. 3.63 Delivery in ERP

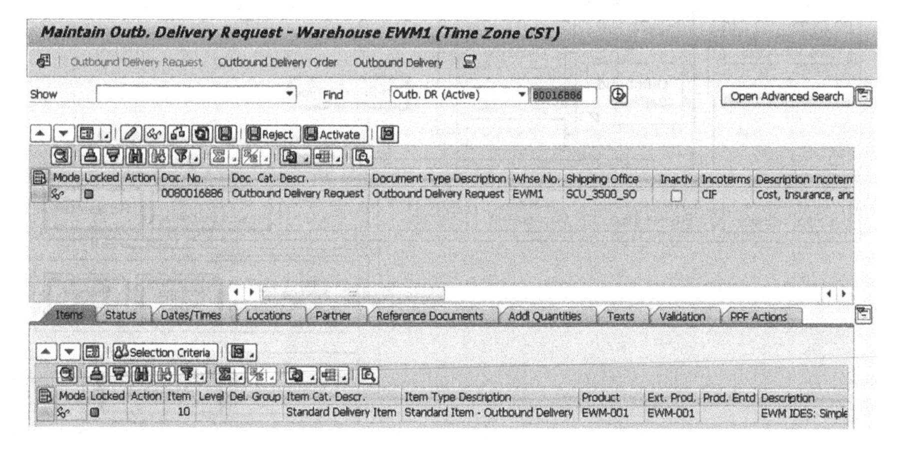

Fig. 3.64 Outbound delivery request in EWM

To determine the source storage bin from which the material is to be picked, the storage type is first determined for picking. Storage type determination is based on a storage type search sequence and a removal rule. The storage type search sequence indicates which storage types are to be searched for the stock to be picked, and is primarily determined on the basis of product characteristics and item data from the outbound delivery order. As a result of storage type determination, the storage type is ascertained for picking (see Fig. 3.65). The determination of the storage bins from which stock is to be picked is the task of the outbound delivery strategy.

For goods issue, the picking strategy aids in finding the optimal picking storage bin. The picking strategy is mapped with a removal rule that is allocated to a specific storage type to be searched via the determined storage type search sequence (see Fig. 3.65). This removal rule represents a type of sorting rule, and contains the criteria with which the determined storage bins are to be sorted. Such sorting criteria include the goods receipt date of the last stock addition or the stock quantity available in the storage bin. Based on this sorting, EWM uses the picking strategies stored in the system. Possible picking strategies include:

- First in, first out (FIFO)
- Last in, first out (LIFO)
- Partial quantities first
- According to quantity
- Shelf life expiration date
- Fixed bin

The management-based background of these strategies has already been explained in connection with the removal control of WM (See Sect. 3.3.4, “Goods Issue”). The functions of these picking strategies can be expanded or adapted in EWM with the aid of business add-ins (BAdIs).

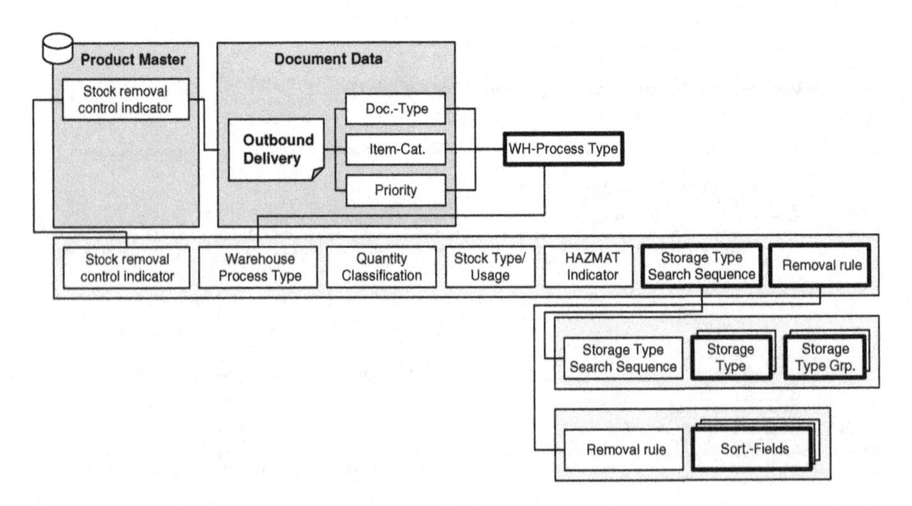

Fig. 3.65 Technical process of removal control

Multi-client capability of SAP EWM

In this regard, we would like to mention the multi-client capability of SAP EWM. By precisely entering the owner and authorized parties as a stock attribute, stock can be logically separated. The system ensures that the permissible stock is always determined in the picking process. This is one of the functions that make EWM suitable for use by logistics service providers whose warehouse management includes maintenance of the stock of various customers. The system not only finds the desired picking storage bin, but also checks the ownership circumstances of the quants stored there and verifies whether the required quantity can be removed at all.

Outbound delivery order (warehouse request)

Figure 3.66 shows Outbound Delivery 113, which was generated from the activated Outbound Delivery Request 80016886. The system has already determined the warehouse process type (2010) for each item and the destination storage bin (GI-ZONE) for staging. In addition, the item data includes information on the type of stock from which picking is to take place, as well as other stock attributes such as authorized personnel and the owner of the materials to be removed.

The path taken for physical outbound delivery to a goods recipient is defined through routes. *Routes* represent a sequence of legs connected by load transfer locations, and they can be divided into various tours. The significance and characteristics of routes and transportation zones is described in Chap. 2, "Transport Logistics".

From the viewpoint of EWM, SCM route determination – known as the *routing guide* – is used to find routes that are best suited to send materials to the goods recipient. The Routing Guide performs the following to determine a route:

- Static route determination
- Scheduling
- Calculation of transportation costs

Actual route determination is based on a hierarchical relationship of means of transport and the profiles stored in the system for the respective carriers and general transportation costs.

The *carrier profile* can be used to match carriers to geographical data, such as locations and transportation zones serviced by them, that are taken into consideration during the selection of a carrier and the route determination process. In addition to the general transportation cost profile, the carrier profile enables the

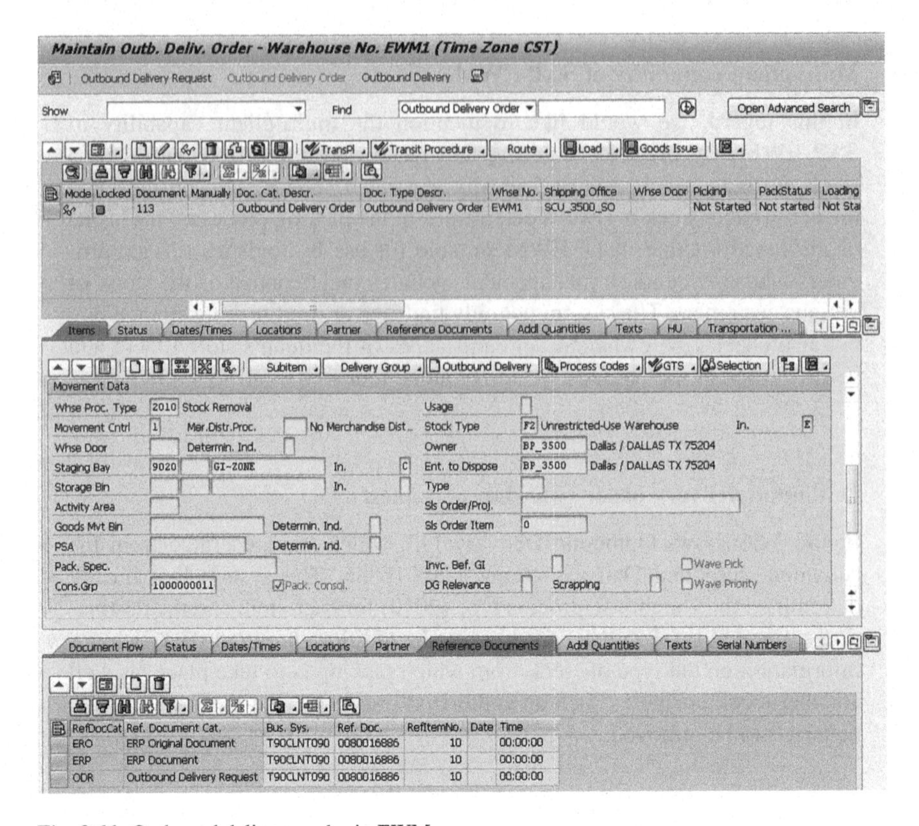

Fig. 3.66 Outbound delivery order in EWM

maintenance of dimension-based transportation costs that take into account time, distance and transport weights.

The general *transportation cost profile* contains cost information relevant to carrier, location, transportation and dimensions. For the location-relevant costs, penalty costs can be defined, depending on the priority of a specific location. These penalty costs would then be valid for inbound deliveries that arrive too early or too late. Transportation-relevant costs are one-time, fixed transportation costs for each means of transport. General and carrier-based profiles are used along with the routes and transportation zones stored in the system for route determination.

Static route determination puts together tours based on the routes and actual transport requirements stored in the system. To do so, it first determines the possible geographical route. It then identifies the individual locations for the route, which is divided into legs. The resulting sequence of locations represents the final tour.

For the *scheduling* process, the system determines the start and end dates for each tour. When doing so, it takes into account load transfer times as well as the duration of departure, transport and arrival at the destination, and calculates these as forward or backward scheduling. In the default setting, EQM uses forward scheduling, which determines the latest planned goods issue date and changes the planned delivery date accordingly. For backward planning, route determination and scheduling starts from the planned delivery date, which is shortened by the times stored in the system. If the resulting planned delivery date is earlier than the original date or current day, backward scheduling leads to a conflict in dates. The system reacts to this conflict according to the respective EWM settings. If there is no conflict, backward scheduling supplies the latest possible date and time in which a material must leave the warehouse or yard.

The system subsequently conducts a *transportation cost determination* for the resulting tours and times. The transportation costs are added and contain the actual cost of transportation for each leg, as well as any necessary penalty costs of the tours for premature or late delivery. These costs are stored as a cost value in the respective transportation cost and carrier profiles. Such cost values, which can be compared to a score, have no dimensions and thus do not represent any particular currency. Instead, their stored values serve solely to determine the most economical route in EWM, and are not, for instance, used for a freight invoice. More information can be found in the previous chapter, Chap. 2, "Transport Logistics".

An important aspect of warehouse logistics is the staging of sales orders. *Stock removal* involves picking certain partial quantities from a total quantity, based on requirement information. From the viewpoint of EWM, these partial quantities are items to be removed based on a warehouse request, and the total quantity refers to the stock of the required material. Finding this stock and determining the picking storage bin from which this stock is to be taken was the focus of the previous section.

EWM generally allows one- and two-step picking. It also enables you to integrate the packaging process into picking. This procedure, called *Pick & Pack*, enables direct picking in a shipping unit. An expansion of this function is the *Pick, Pack & Pass*, procedure, a one-step picking procedure with decentralized picking and static staging. Pick, Pack & Pass allows the work to be passed from one resource or employee performing the work to another after part of the picking has taken place.

The sequence of warehouse tasks for stock removal can be determined according to various criteria, such as the shortest path. EWM and the corresponding ERP system are closely linked for the picking process. This allows picking specifications such as batches or batch characteristics to be taken into account.

Stock removal is done with *warehouse tasks*, which can be generated in a variety of ways and are subsequently confirmed (see Fig. 3.68). Warehouse tasks can be created manually. You also have the option of having them be generated automatically as soon as incoming and outgoing deliveries are received by EWM. The third option is the creation of warehouse tasks in the course of wave processing. The basic functions of wave management are explained above in connection with warehouse tasks and orders in Sect. 3.4.2, "Warehouse Organization and Stock Movement".

Manual creation of a warehouse task for stock removal

Figure 3.67 shows manual creation of a warehouse task for stock removal for the outbound delivery of stock to be picked in accordance with Warehouse

Request 113 from the previous example (see also Fig. 3.64). The quantity to be removed is to be picked from open stock, Bin 01-01-A in Storage Type 0020 and taken to the goods issue area (GI-ZONE).

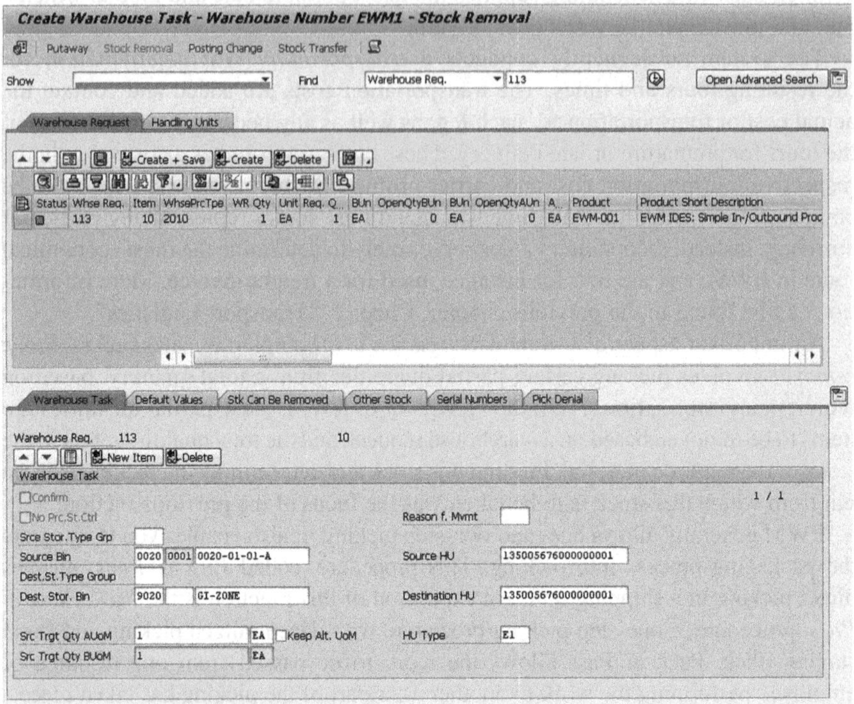

Fig. 3.67 Warehouse task for stock removal

Stock removal with the aid of picking waves enables the bundling of stock removal warehouse tasks into work packages and their collective processing. The advantage of making bundles, or waves, lies in the optimization and rationalization of processes for goods issue. Waves are made by combining warehouse request items based on common activity areas, routes or materials in wave picks. The individual criteria and attributes for wave generation can be stored in the system as specifications and serve as the infrastructure for automatic or manual creation of waves.

In operational practice, physical picking starts with grouping delivery items into waves. In doing so, even the existing conveyor systems are taken into account. We can use a pallet warehouse as an example: If the number of pallet storage locations is known, EWM determines the probable workload. Subsequently, a wave can be released. Certain waves, such as those for full pallets, are generally selected in order to bypass the conveyor system. Instead, they are picked from the reserves and directly staged for outbound delivery, in order to keep the workload as minimal as possible. Other items, for instance, are picked for the conveyor belt, where they can be conveyed to a palleting or package shipping station. Products are typically loaded as soon as they have been packed on a pallet or packaged, and a goods issue is subsequently posted.

Confirming a stock removal

Figure 3.68 shows a manual confirmation of Warehouse Task 3036, generated for the warehouse request in the previous example. After an employee has removed the respective quantity, he confirms the time of removal and thus the warehouse task.

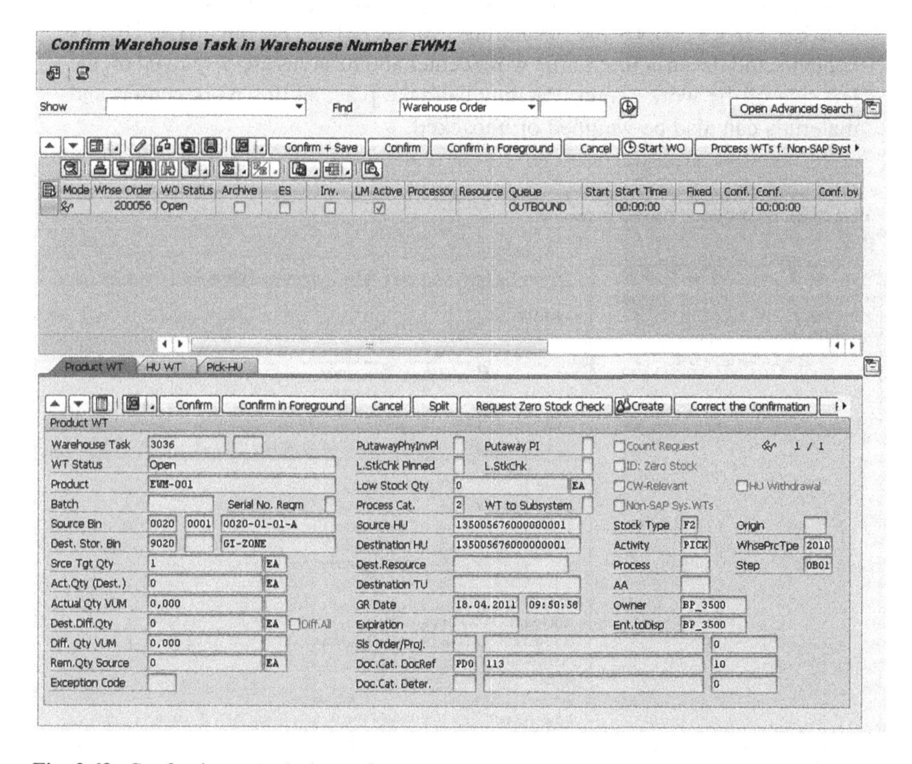

Fig. 3.68 Confirming a stock removal

When a warehouse task for stock removal is confirmed, the stock is removed from its source storage bin and moved to the destination storage bin. If a difference in quantity is detected in the process because the picked quantity deviates from the quantity to be picked, a further warehouse task can be created or the quantity to be delivered is adjusted and diminished accordingly. When a second warehouse task is generated, stock removal is not complete until the second warehouse task has also been confirmed.

3.4.5.1 Packing

In the realm of logistics material flow, in addition to the goods to be transported, focus is increasingly being shifted to shipping units. As a result, EWM offers a variety of options for dissolving, forming, unpacking and labeling shipping units. It can take on information from the ERP system, such as data on how shipments are packed, to process those shipments further.

Packing in a work center

Figure 3.69 shows the manual packing of Material PROD-M02 in a work center. The quantities to be packed are packaged with Packaging Material WIREBASKET, after which Handling Unit JKT01 is generated. The HU, its contents and the structure of the work center and its areas are displayed on the left side of the work center. Packing can take place at this work center, and materials can also be weighed or unpacked.

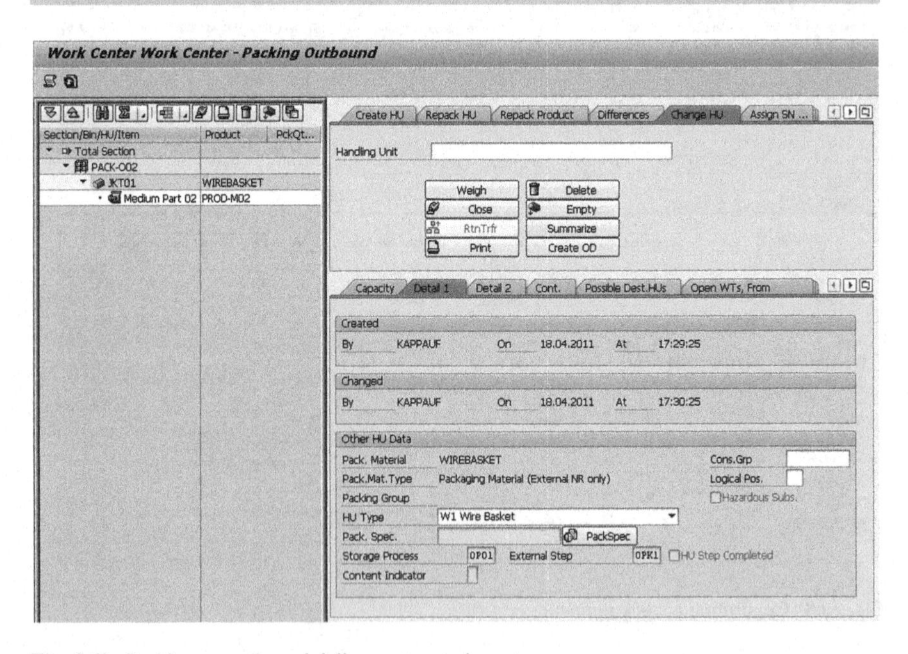

Fig. 3.69 Packing an outbound delivery at a work center

EWM can not only obtain and use the HUs mapped in ERP, but it can also utilize packaging specifications that are stored elsewhere. This master data enables you to define packaging materials and work steps for various products and processes to be accessed and used in subsequent process steps. For example, such packaging specifications can be used when packing products in the shipping department or performing a logistics value-added service. The packaging specs can also be printed and used as work instructions.

3.4.5.1 Loading and Goods Issue

As soon as the materials have been packed, the resulting packages (HUs) are usually loaded, and the goods issue is then posted (see Fig. 3.70). The system can also automatically determine the staging area and door for goods issue in advance, which is based on the route determined when the outbound delivery order was generated.

Like unloading for goods receipt, physical loading can be done in a *simple* or *complex* manner. The goods issue can be posted at various times. Because loading is an optional step, the goods issue can be posted without previous loading, depending on system settings. Alternatively, posting can be done after loading or, if Yard Management is used, at the latest after the transportation unit has left the warehouse grounds. When the goods issue is booked, EWM informs Inventory Management in ERP of the change in stock (see Fig. 3.61).

In certain cases, especially for international shipments, it may be necessary to generate an invoice before the goods issue has been posted. Normally, invoice generation is done as soon as the goods issue has been posted and the delivery status of the outbound delivery allows invoicing. In the case of invoicing before a goods issue, EWM sends an invoice request to the ERP or CRM system, depending on which is used for invoicing. The invoice request is based on one invoice per shipment or on one or more invoices for the complete contents of a vehicle or transportation unit.

The *outbound delivery* is a follow-on document of the outbound delivery order, and represents proof of the physical delivery of the picked materials to the ship-to

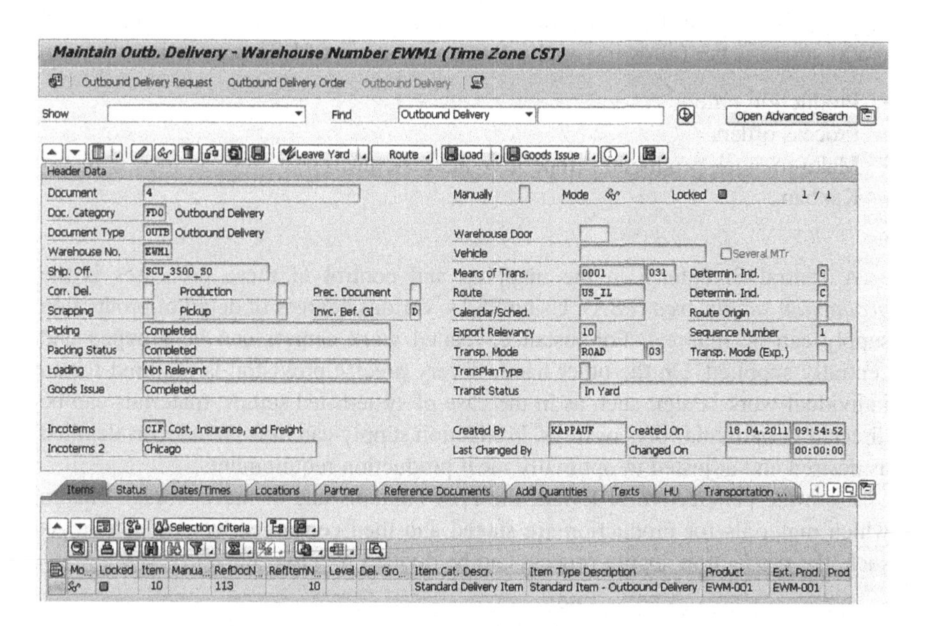

Fig. 3.70 Goods issue and outbound delivery in EWM

party. It is the basis for printing delivery notes or the electronic transmission of a shipping notification. From the goods issue view, it represents the completion of goods issue processing and enables the posting or cancellation of the goods movement (see Fig. 3.70).

3.4.5.2 Production Supply

Physical production in the sense of manufacture is not within the scope of this book. With regard to production logistics, however (see Volume 1, Chap. 5), we explain the requirement-related results of distribution logistics and the integration with the procurement side. In this chapter, we view production supply solely from the perspective of material provision of preliminary products and consumption deliveries in the case of retrograde picking for manufacture.

With its component *Production Supply*, SAP EWM 7.0 fills a gap, enhancing goods issue with an important process. The process of production supply includes activities required for the provision of components to work centers or machines in production (see Fig. 3.71). Depending on the branch of a company, the production processes and materials employed, these processes differ greatly. Production order-related, cross-production order or consumption-controlled provision procedures are common basic strategies that can be tailored to individual needs. Yet within companies, especially in the case of multi-step manufacturing processes or various materials, a variety of procedures is often used for production supply.

In addition to the traditional order-based production supply, EWM allows the mapping of cross-order or consumption-based strategies, such as kanban. SAP EWM supports the following application components of ERP production:

- Production orders
- Process orders
- Make-to-stock production (repetitive manufacturing)
- Kanban

A central SAP term in the mapping and control of these processes is the *production supply area* (PSA). Using PSAs, various degrees of detail of production supply can be mapped. For instance, related work centers can be grouped and centrally supplied. On the other hand, if very precise provision is required for an individual work center, such as in the case of system-led setups, materials can be directed to particular bins or trays. Production supply can thus be made as detailed as desired and adjusted to optimally meet production requirements.

From the perspective of EWM, the PSA contains one or more storage bins in which materials for production are staged and then consumed by production. As such, the production supply process can be divided into two steps in EWM:

1. Staging of products in the PSA
2. Consumption of productions in the PSA through picking for production

For *staging*, products are brought to the production supply area. Depending on whether the PSA is an EWM-controlled storage area, staging is done with the aid of inbound or outbound deliveries or as stock transfer. The result in any case is an increase in stock in the PSA.

First, the materials required for production must be staged with a delivery to the production supply area in ERP. This outbound delivery is replicated in EWM, where it generates an outbound delivery request and, once that request is activated, an outbound delivery order. Warehouse tasks are then generated for the delivery order, and the PSA is supplied with the necessary materials. This increase in stock in the PSA is reported to the ERP system, where it also leads to a goods movement.

Consumption in the case of retrograde stock removal is mapped with the aid of a consumption delivery in the system. After receiving confirmation from the production order, the ERP system generates a consumption delivery that is replicated in the EWM system, where it generates an outbound delivery. The items of this document are not relevant to stock removal, so the goods issue is thus posted

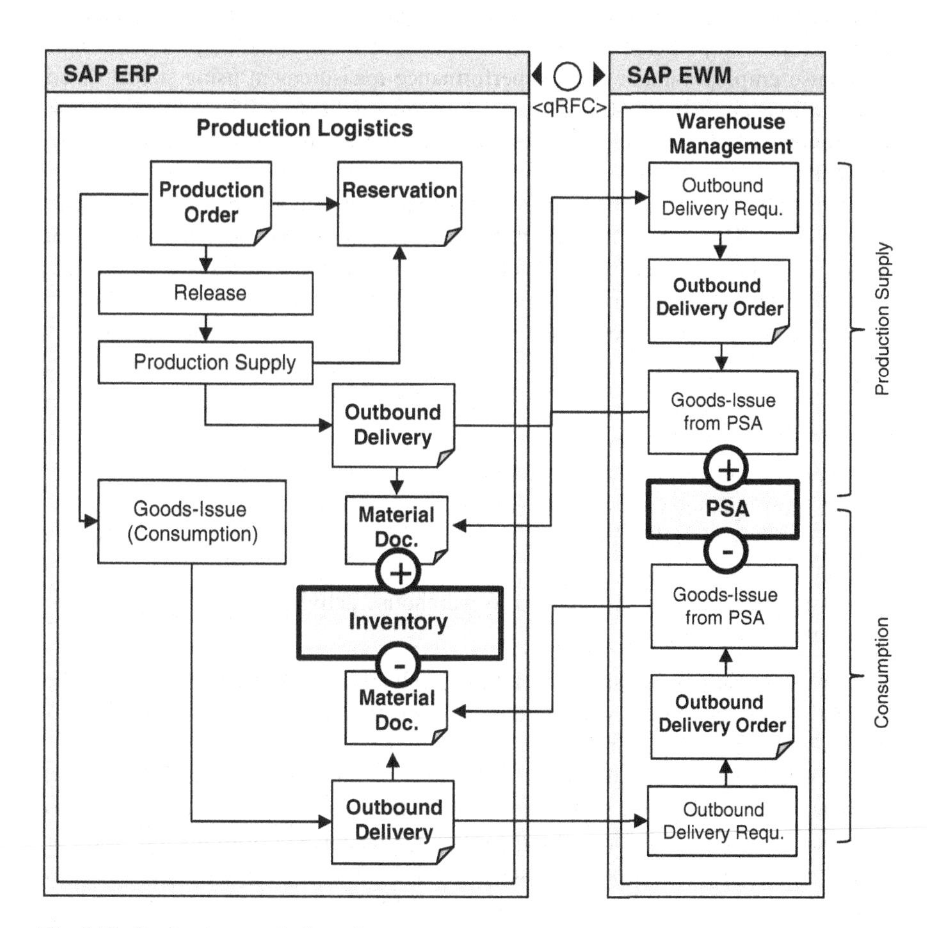

Fig. 3.71 Production supply flow chart

immediately when the outbound delivery order is generated and triggers a corresponding change in stock in ERP.

3.4.6 Cross-Warehouse Functions

SAP EWM offers functions that cannot be allocated to one specific system area, but are available for all warehouse processes. These are cross-process functions in goods receipt and issue – especially including cross-docking and yard management. They also represent various functions for warehouse optimization and control with the aid of modern automation technology.

When the operational size of warehouses increases and logistics processes become more complex, the efficient planning of resources and employees becomes ever more difficult. In order to measure and increase productivity through optimized planning in this realm, EWM has *Labor Management*. Labor Management offers a comprehensive spectrum of functions for the planning and control of warehouse employee labor and for performance measurement using standardized specifications and formula-based key performance indicators. These functions and the preview of the expected work load enable resources to be planned more efficiently.

Labor Management can be activated for a specific warehouse number and certain process steps. Through its activation, EWM requests the start and end time when certain activities are performed, such as when warehouse tasks are confirmed. Using this data, the system can determine whether the times stored in the system for a certain task have been adhered to. It is also possible to communicate this information to the human resources management component HCM to compensate activities in a performance-related manner. Labor Management in EWM supports the following activities:

- Warehouse orders with warehouse tasks
- Tasks for logistics value-added services
- Quality inspection documents
- Inventory documents
- Indirect work not directly related to warehouse activity, such as an employee cleaning a resource

Employees can immediately check their productivity themselves by viewing the transaction results on their wireless data devices or warehouse management monitor, accessible through their desktop (see Fig. 3.72).

Every task in the warehouse is linked to a certain work effort. To plan and evaluate the work load, the system needs the extent of time for an activity, but also the actual duration and information regarding the resource performing the task. For this purpose, a document on the planned work load is generated per activity area and process step. This document allows the planned duration to be compared with the actual duration.

Warehouse Management Monitor for Labor Management

Figure 3.72 shows the Warehouse Management Monitor, with which the operator, warehouse manager or group manager can access information relevant to Labor Management. In addition to receiving precise confirmation times, you can evaluate the efficiency of employees or determine the weight moved within one day in a specific activity area.

3.4.6.1 Resource Management

In addition to Labor Management, *Resource Management* also increases the efficiency of warehouse processes by optimizing the management and distribution of warehouse tasks with the aid of *queues*. A queue is a series of specific warehouse tasks to be performed. The resources are the units represented by a warehouse employee and certain equipment, mentioned at the start of this chapter.

Resources and warehouse tasks are allocated to queues. This optimization primarily lies in the automatic allocation of a required resource to a specific warehouse task and the allocation of resources that are best suited for a particular activity. Various factors are taken into account when this decision is made, such as the latest starting date, the performance priority of a task, the allocated queue, the qualification of a resource and the warehouse order status. A warehouse order already represents an optimally executable work package that a warehouse employee is to carry out within a specified time. The contents of the order are either inventory items or warehouse tasks that are combined into warehouse orders and are ready for processing. The generation of warehouse orders has already been explained; allocation of warehouse orders to warehouse employees is done via Resource Management. Physical allocation can be performed manually or automatically.

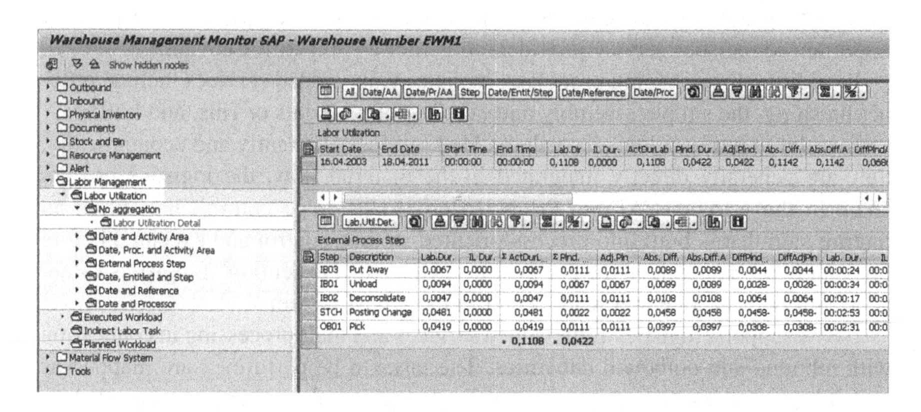

Fig. 3.72 Warehouse management monitor for labor management

3.4.6.2 Catch Weight Management

Especially in the food industry, goods frequently differ with regard to the weight of each individual product, such that a fixed conversion factor to convert a unit such as a "piece" (see below) to a weight unit cannot be determined. Often, an average value is used, which frequently leads to problems. EWM solves these problems with the aid of *Catch Weight Management*. This function enables the stock management of products for which two independent quantity units of equal status are defined.

Catch weight capabilities are completely integrated in the delivery process, such that they do not change anything in the process for inbound and outbound deliveries. For goods movements, the EWM system updates both product quantities. Also, to control internal material flow, you have the option of defining the processes for which a valuation quantity is to be maintained. Catch weight functionalities are also integrated in the inventory and quality-inspection processes.

Material with quantities of equal status

Packaged beef is stored in portion sizes. The logistics quantity, or base quantity, in which the packing unit is maintained is the "piece". The valuation quantity, upon which the valuation price and subsequent sales price are based, is the "gram". In this example, both quantities are independent yet have equal status.

3.4.6.3 Packaging and Logistics Value-Added Services

In the past, packaging served solely to protect goods. Through altered sales and distribution structures in distribution logistics, especially with regard to an increasing internationalization of goods flows, packaging today has storage and transportation functions and facilitates identification and information in the sales process. Packaging processes and logistics value-added services in general are especially important to logistics service providers to garner a competitive advantage.

In addition to packaging processes, logistics value-added services include product finishing, the simple assembly and labeling of materials or Hus, and kitting.

In order to perform logistics value-added services efficiently and economically, they are generally directly integrated in the material flow, the logistics process chain in the warehouse. From the perspective of storage control, it is often an interim step that is built into process-oriented storage control and executed before final putaway or staging in the goods issue area. Actual execution of logistics value-added services is done in a work center in the EWM system.

EWM supports logistics value-added services and their processing in connection with inbound and outbound deliveries. The tasks to be performed are mapped as activities and consolidated in an order for logistics value-added services (VAS). These VAS orders can be created manually as well as automatically. The order consists of an order header, the activities to be performed, the items and any required auxiliary products. As a system document, its basic purpose is to inform warehouse employees of the work steps that need to be carried out. It documents tasks actually performed and any auxiliary products consumed. It can also be used as a calculation base for the invoicing of work done by internal or external service providers.

3.4.6.4 Yard Management

With the increase in size of freight yards, or *yards* for short, the management of vehicles is becoming ever more complex, frequently leading to longer immobilization times. Efficient yard management reduces trucks' immobilization times and increases the number of vehicles that can be processed per hour.

EWM makes it possible to manage your yard – your own parking lot structures, warehouse doors and the individual transportation units – in a clearly arranged and efficient manner. In addition, stock still on vehicles can be taken into account in the running processes of a warehouse. Vehicle parking spaces are mapped by the software as standard storage bins that can be combined into yard areas. The registration of vehicles that arrive in or leave the yard is done at checkpoints. From there, trucks and trailers are led toward a parking space or door for immediate loading and unloading. In order to optimize the use of the facilities, an incoming trailer can be unloaded and then loaded right away as an outgoing trailer. Movement in the freight yard can be achieved via wireless data transmission or desktop transactions.

The functional requirements of a warehouse management system are growing constantly, following the general trend of increased automation and mechanization of work processes. In distribution logistics, when customers are supplied, punctual delivery in perfect quality is the goal. In procurement logistics, suppliers are changing their product range at shorter and shorter intervals. Added to this are legal requirements on the product side regarding traceability and batch management requirement. The resulting high demands to logistics can often only be met by automated processes. *Warehouse automation* has thus been gaining in significance. It not only allows demands to be met, but also triggers a high potential for optimization.

We will thus give you a short overview of the basic functions of SAP EWM-supported automation technologies.

Radio frequency (RF) devices have been an indispensible part of logistics for quite some time now. The decisive advantage is that communication with a warehouse management system is not only possible from special work centers, but can be carried out from virtually every location in a warehouse. In addition, the devices replace all documents and allow a rapid and fault-free readout of coded information, such as barcodes or 2D codes. The direct interaction offered by RF devices enables correct data validation and ensures a high-quality standard in the warehouse by eliminating data-entry errors. SAP software is device-independent, and possesses tools that can convert messages and information from a device.

SAP EWM enables the use of radio frequency in more areas than previous SAP solutions. For instance, EWM lets you perform several activities with RF support, including confirming warehouse tasks, packaging, deconsolidation, loading and unloading, and taking inventory. The RF environment also allows a tailored design of the menu structure and the connection of third-party suppliers that have specialized in visualization on mobile devices.

In addition to pure data entry with the aid of radio frequency devices and the accompanying flexibility and validation, the RF environment makes it possible to optimize tasks and their execution with a variety of functions. Using the *double cycle with RF*, the function known as *interleaving*, you can optimize resource utilization through a reduction in travel times in the warehouse. The system avoids deadheading (empty trips) or travel times by automatically assigning a resource to a warehouse task after another warehouse task has been completed that is geographically located to the new resource item. In addition, you can use an *execution restraint* to control the maximum number of permitted resources in a specific storage area. The restraint enables resources to perform warehouse tasks in a specific area as unhindered as possible and not hinder one another in the process.

For example, for semi-system-controlled processing, the system instructs the warehouse employee to go to a specific storage bin and take out any HU. In this case, the execution restraint determines the probable times in which the resources in a certain area will be leaving that area. The calculation of the optimal route and the times in the warehouse is based on a distance calculation. Depending on the calculated time and the number of permitted resources, access can be allowed or denied.

In addition to radio frequency connection, *radio frequency identification (RFID)* is being used more and more in logistics. In combination with the *auto-ID infrastructure* (SAP AII), EWM supports the use of RFID for all prevalent processes, such as goods receipt and issue (see Chap. 4, "Trade Formalities: Governance, Risk, Compliance"). EWM uses SAP AII to communicate with RFID hardware such as readers. It does not matter whether that reader is a mobile or static device or a tunnel reader.

An increasing trend in logistics is the use of automatic warehouse and conveyor technology. With the aid of the Material Flow System (MTS) integrated in EWM, you can directly connect automated systems or warehouse and conveyor technology to SAP and control them in real time. The connection of *stored program control (SPC)* systems via communication channels and the mapping of checkpoints, rack feeders, conveyor segments and the like can be configured using the system settings. Based on the layout-oriented storage control, MFS lets you control the entire material flow via all automatic components of a warehouse. This close integration enables all strategies to be transferred directly to the execution of storage and conveyor technology.

3.4.6.5 Cross-Docking

The direct transfer of materials or handling units from goods receipt to goods issue without prior storage is called *cross-docking*. Forgoing storage and directly staging goods in the goods issue area reduces warehouse movements and stockholding costs. This process also enables an increase in warehouse efficiency. Lead times can be considerably reduced, because the warehouse only serves as a load transfer site and goods are only in the warehouse as long as is needed for transferring the load.

EWM supports various options for performing cross-docking processes (see Table 3.3), that is, for transporting products directly from the goods receipt area to the staging zones. This can be planned in advance, so that it is determined before goods arrive that they are to be loaded in a cross-docking process. Alternatively, the decision to use cross-docking can also be made after the goods have arrived in the warehouse.

For *planned cross-docking*, when goods are received, it has already been determined that they will not be stored but will leave the warehouse via goods issue. Planned cross-docking is done either as transportation cross-docking or merchandise distribution. The latter is a branch-specific scenario that SAP requires for a retail system and will not be explained in more detail.

Transportation cross-docking (TCD) is planned cross-docking for the optimization of your transportation costs. It supports the transportation of HUs along a variety of distribution centers or stock transfer locations up to the final destination. The advantages of this scenario lie in the fact that several deliveries can be consolidated into new shipments, that the means of transport can be changed if necessary, and that export activities can be processed centrally. The decision as to whether or not cross-docking should be executed does not occur in EWM, but rather ERP.

The process begins with a stock transfer order in ERP, which can be generated automatically using APO or manually. Manual creation of a stock transfer order is described in Volume 1, Chap. 4, "Procurement Logistics". After the goods issue is posted in the issuing warehouse, the EWM system that is to perform the cross-docking receives the necessary inbound and outbound deliveries and information regarding which delivery items are relevant to cross-docking. The reference is the ERP stock transfer order number. As soon as the goods receipt has been posted,

Table 3.3 System requirements for cross-docking scenarios

Scenario	ERP	CRM	APO
Planned cross-docking			
Transportation cross-docking	X		(X)
Merchandise distribution	Retail		
Unplanned cross-docking			
Push deployment	X		X
Pick from goods receipt	X	X	X
EWM-controlled cross-docking	X		

EWM generates the warehouse tasks serving to move those goods to the goods issue area. Confirming these warehouse tasks triggers physical movement of the goods, after which a goods issue is posted to complete the process.

In the case of unplanned or opportunistic cross-docking, the decision to either store goods or move them directly to the goods issue area is based on the individual situation. Performance is either in the form of a *pick from goods receipt* or a so-called *push deployment*.

When goods are picked from goods receipt, cross-docking in the goods receipt area begins with the standard inbound deliveries. EWM attempts to generate the necessary warehouse tasks for putaway, and while doing so, it checks to see if existing warehouse tasks for picking can be replaced. First, EWM determines whether the delivered goods are relevant for a putaway delay. This is determined based on the warehouse process type and the stock type of the goods, and leads to a delay in the generation of warehouse tasks for putaway. The delay gives APO a chance to verify whether any of the inbound delivery items are eligible for cross-docking. If any backorders are present, an outbound delivery document is automatically generated, and the goods are transferred to the goods issue area and brought directly to a customer or another warehouse.

In the case of *push deployment*, APO determines whether the goods are to be moved from one storage location to another. This decision is also triggered upon goods receipt and is based on a turnover forecast stored in APO.

In the case of unplanned, opportunistic cross-docking, the respective decision is not made in EWM but in APO. In exceptional cases, cross-docking can also be decided in EWM.

Opportunistic cross-docking can also be controlled completely by EWM. Whether or not a delivery item is eligible for cross-docking depends on the settings stored in EWM and is not transmitted via ERP or APO. Cross-docking can be performed for inbound as well as outbound deliveries. This type of cross-docking is based on current data and requires the use of a radio frequency environment. In order to avoid inconsistencies and because of the required timeliness, only those warehouse tasks that are allocated to an RF environment are used.

When an inbound delivery is received, the system searches for suitable outbound delivery items as soon as the goods receipt has been booked and the warehouse tasks for putaway have been generated. If no outbound delivery items can be found, the generated warehouse tasks are executed and the goods are placed into storage. If, however, EWM finds open picking warehouse tasks, they are canceled. Subsequently, new warehouse tasks for picking are generated and allocated to the stock to be put away.

The same functions are also available for goods issue. As soon as an outbound delivery and its respective picking warehouse tasks have been generated, the system searches for suitable inbound delivery items. The retrieved warehouse tasks are canceled and the system generates new warehouse tasks for picking and allocates them to the stock to be put away.

3.5 Summary

In this chapter, you have been introduced to the operational basis of warehouse logistics, as well as the major differences between inventory management and warehouse management with SAP. Inventory management serves to physically manage stock in storage bins and manage stock movements in connection with the processes of procurement, production and distribution logistics.

Today, companies expect an efficient warehouse management system to not only offer operative management of materials and their storage bins, but also a control and monitoring system based on continual optimization of material flow, equipment and human resources, beginning with goods receipt and extending through all storage and process levels leading to goods issue.

With Warehouse Management (WM) and the SCM-based SAP Extended Warehouse Management (SAP EWM), SAP Business Suite offers two different systems at once with which stock in a warehouse can be managed and monitored. The respective area of usage ultimately depends on the business needs and economic demands of an enterprise.

Functionally speaking, SAP EWM represents a further development of functions already existing in WM, and also meets demands of third-party logistics (3PL) that previously could only be covered on a limited basis. Therefore, EWM is especially suited to logistics service providers, retail or companies dealing with replacement-parts logistics. Such firms operate several warehouses with a high degree of turnover and automation, as well as complex optimization strategies, and demand uniform, comprehensive control. In addition to process- and layout-oriented storage control, EWM offers multi-level, internal transportation using warehouse tasks, as well as built-in options for warehouse automation and control.

Chapter 4
Trade Formalities: Governance, Risk, Compliance

Trade formalities such as customs laws, trade agreements, environmental regulations, or trade and product safety standards are increasingly influencing logistics processes. This pertains to goods transport as well as internal logistics processes and collaboration with business partners.

4.1 The Basics

In our global economy, the growing commodity flow is accompanied by growing challenges:

- An increasing number of laws and regulations are leading to augmented risks in processing, a higher degree of complexity and higher costs for the execution of processes.
- The increase in globalization automatically causes an increase in the quantity of goods involved in cross-border traffic.
- New types of threats (such as terrorism and environmental problems) are leading to the passing of more and more approval processes and test codes that must be followed in domestic and international logistics.
- With the progression of automatic processes used by the public authorities of individual states, new electronic communications procedures are being provided for data transfer to make processing more efficient. However, this does not mean that international standards exist for information transmission. Every nation or union of nations is developing its own forms of communications. For example:
  - ATLAS for electronic customs registration and administration files (such as import fee notifications) in Germany
  - NCTS (New Computerized Transit System) for transit processing within the European Union

J. Kappauf et al., *Logistic Core Operations with SAP*,
DOI 10.1007/978-3-642-18202-0_4,

 - AES (Automated Export System) for the registration of exports to the United States
 - e-Customs for customs registration in Japan

- Because processing times for logistics processes are becoming increasingly shorter, there is less and less time to fulfill the heightened level of regulations and tests, which in turn raises risk.

Depending on the type of goods, departure and destination locations, transit countries and involved business partners, a cross-border shipment can require the correct generation and punctual submission of up to 35 documents. What is more, up to 25 business partners and institutions have to communicate with one another, including customs agents, freight carriers, logistics service providers, banks and security administrations. Legal processing may require the consideration of as many as 600 laws or 500 trade agreements, all of which can change continually. Figure 4.1 shows a few examples of trade formalities that must be taken into consideration for logistics processing between trade partners in various countries or regions.

4.2 Process Risks

These complex, ever-changing rules mean a high risk that conformity problems will arise in the execution of logistics processes. Such problems can stem from sources including:

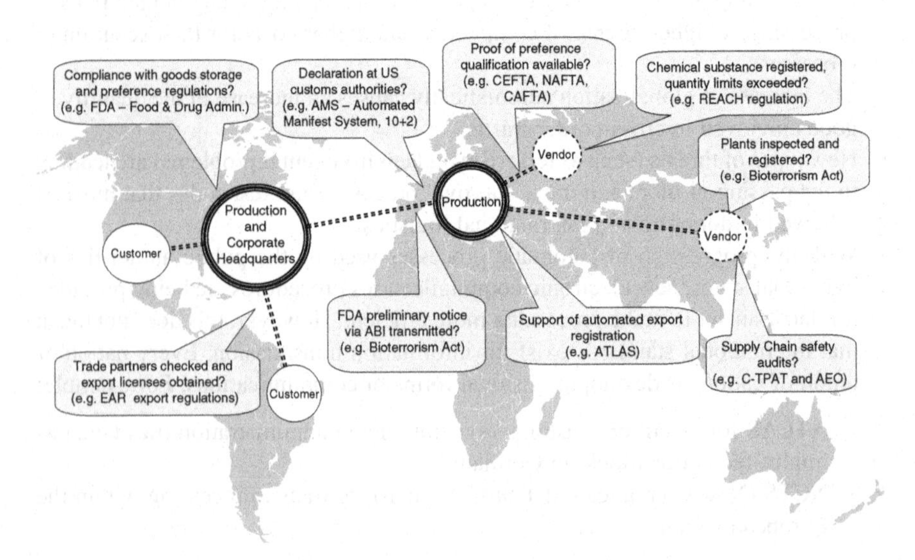

Fig. 4.1 Examples of trade formalities in a global economy

- **Faulty or imprecise data provided by suppliers**
  Imprecise data can cause delays in customs, import or export processing that generally hinder the overall process and thus lead to higher costs or problems with business partners.
- **An absence or lack of collaboration with partners and authorities**
  Collaboration problems also lead to delays or make repeated data entry necessary in various systems that may in turn lead to faulty data.
- **Strict organizational or geographical separation of processing procedures**
  Due to the strict separation (silo principle) often also mirrored in the computer systems, costs for adherence to regulations skyrocket.
- **No possibility of budgeting trade costs**
  If trade costs cannot be budgeted, not all savings opportunities can be achieved when paying customs fees.

The risks described above can be reduced if suitable data processing systems with high-performance integration and connectivity to business partners and authorities are employed for documentation and process execution:

- Through centralized systems, the data and processes required for processing can be standardized, which ultimately leads to a reduction in costs for fulfillment of formalities.
- The connectivity and integration serve to generate an active and rapid distribution of information, such that all involved partners receive the necessary information in a timely manner and in the desired quality.
- An integrated function for risk management enables you to visualize and evaluate company-wide risks.
- Integration with business partners, such as logistics service providers, and the opportunity to submit necessary documents to authorities yourself in electronic form offer further potential for savings in process execution.
- Integration with suppliers and the active monitoring of regulation fulfillment ensure that business transactions with suppliers are processed properly.

SAP offers suitable software solutions for all of the above uses.

4.3 SAP Applications for Logistics Processing According to Trade Formalities

In the SAP logistics portfolio, you will find two applications that will support you in the proper processing of foreign trade and customs procedures: the Foreign Trade component of SAP ERP module SD (Sales and Distribution), and SAP BusinessObjects Global Trade Services (part of SAP BusinessObjects governance, risk, and compliance (GRC) solutions).

The ERP Foreign Trade component offers you the following core functions:

- Efficient integration into the ERP logistics document chain with the operation of all import and export processes
- Automatic verification of export licenses based on current regulations
- Simplified reporting through automatic procedures for the generation, print and submission of reports
- Identification of products that are to be considered for preference processing
- Integration with partners and authorities with the aid of EDI communication (EDI: Electronic Data Interchange)
- Accessibility of all foreign trade data in all relevant SAP components

The Foreign Trade component is integrated in the sales and distribution (SD) as well as the materials management (MM) processes. The link to sales and distribution enables a precise processing of requests in foreign trade matters with regard to goods issues and exports. In materials management, you can process goods receipts and imports, and in the realm of documentary payments, there is integration of foreign trade data with Financing. The Foreign Trade component thus assists you in ensuring that your import and export transactions occur in a lawful manner, and that fees to be paid to customs authorities can be calculated immediately and accurately.

SAP BusinessObjects Global Trade Services (GTS) is a component that acts as an add-on to an SAP NetWeaver system. You can thus use the components in an SAP system environment as well as with non-SAP systems. SAP BusinessObjects Global Trade Services provides you with the following advantages in process execution:

- You can automate most of your international trade processes.
- You can manage a large number of business partners and documents.
- It is easier to comply with continually changing international laws.
- You can cooperate with the modernized systems and electronic communication means of government and customs offices.

Using GTS, you can avoid financial risks and pricey delays in export and import processes, and thus be able to react more quickly to international business opportunities. The component offers the following core functions:

- **Compliance Management**
  - Sanctioned party list check and positive lists (with a simulation function)
  - Proof of official audits
- **Export and import control**
  - Export legal control
  - Import legal control
  - Product classification under import/export aspects
  - Management of import and export licenses
  - Embargo checking

- **Customs Management**
  - Electronic customs declaration procedures, shipping procedures
  - Product tariff classification and customs value determination
  - Customs processing
  - Printing of trade documents
- **Risk Management**
  - Preference processing
  - Letter of credit processing
  - Restitution for EU market-regulated goods
- **Electronic Compliance Reporting**
  - INTRASTAT (intra-European Union trade statistics) reporting

In the following section, we will explain the functions of foreign trade processing with SAP ERP.

4.4 Foreign Trade Processing with SAP ERP

Foreign trade processing in SAP ERP is integrated in all system components of the ERP system and thus in the processes of procurement and sales and distribution logistics. With such close integration, the system fulfills all foreign-trade-related requirements to enable goods receipts and imports as well as goods issue and cross-border sales and distribution in compliance with legal regulations.

In this regard, the system helps you to observe the stipulations that exist between your country and other countries set forth in various trade agreements. Specifically, it supports trade within a customs union, trade based on free trade agreements, and trade with a sovereign third country.

Therefore, the purpose of foreign trade processing with SAP ERP is not only to provide the necessary functions, but also to ensure that import and export transactions are compliant with the law; that foreign trade documents are generated accordingly; and that any fees to be paid to customs offices are calculated immediately and precisely. In accordance with these tasks, ERP foreign trade processes are divided into general foreign trade processing and the following areas, whose core functions will be explained later in this section:

- Legal control
- Preference handling
- Documentary payments
- Communication and printing
- Periodic declarations

Firstly, because of their central significance to general foreign trade processing, we will address the necessary master data, relevant documents and their integration.

4.4.1 Master Data and Integration

The master data used in the procurement and sales and distribution logistics processes, including vendor, customer, material and purchasing information records, contains foreign trade data. The respective process integration is depicted in Fig. 4.2. This data is especially required for control of legal checks and preference processing, and is directly brought over in documents as recommended values or indirectly during the automatic determination of export-related information. The import side includes all processes of cross-border procurement logistics. The export side deals with cross-border goods traffic in sales and distribution.

Complete master data records are a prerequisite for automated foreign trade control and guarantee efficient import and export processing. The customer and vendor master data, information regarding countries and regions, and precise addresses determine the geographical location and thus the customs process and legal controls that apply.

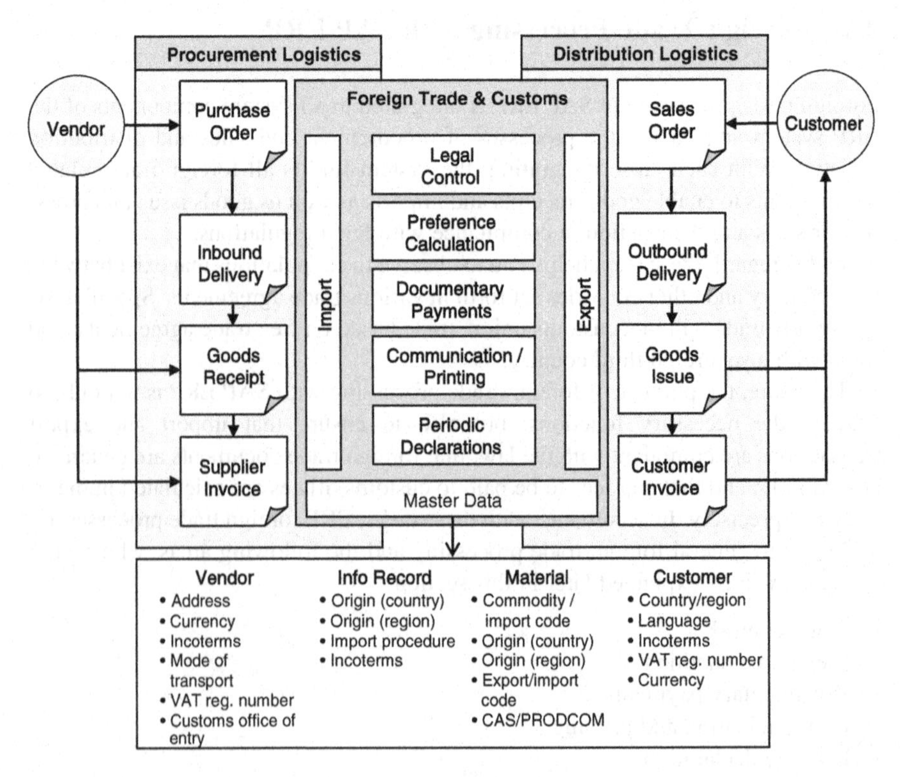

Fig. 4.2 Overview of foreign trade processing with SAP ERP

One of these controls is the *sanctioned party list* check, which the system compares to embargo and boycott lists to prevent a business transaction with certain customers or those from certain regions. The *vendor master* contains further information such as the import customs office of the vendor through which goods generally enter the country, Incoterms and the mode of transport via which the imported goods will cross the border (such as postal, sea, air or road transport). If required, ERP can maintain permissible combinations of modes of transport and customs offices. The customs office refers to the geographical location in which goods are to be registered for import or export before they enter the free commerce of the economic region or leave the sovereign territory of a country.

Among the general data contained in the customer master are the region and country with which export business can be controlled in the sales and distribution documents. The region of a customer is required for reports to authorities. Within the context of risk management and in order to prevent a loss of accounts receivable, you can set in the customer master whether a letter of credit is to be automatically used as a safeguard when generating a sales document for a particular customer (see Fig. 4.4). The documentary payments, or letter of credit processing, will be detailed in the following section.

The *material master record* has its own foreign trade view in which the necessary information can be maintained and tariff classification – the allocation of a commodity code – takes place. In addition to tariff classification, the material master includes information on the region and country of origin. This information is required later for such tasks as preference determination.

Certain goods are to be reported to the respective authorities in stipulated units of measure. The material master thus contains the particular unit of measure in addition to the statistical commodity code.

A purchasing info record represents the link between vendor and material. Purchasing info records facilitate the ordering procedure by saving supplier- and material-related data. Data on foreign trade processing saved in the purchasing info record overrides the vendor master record and enables a differentiated customs-related control of import processes. In particular, data relevant to customs preferences, such as preference zone and the status of a vendor declaration, are automatically determined using preference processing.

Goods received or shipped within the context of procurement or distribution logistics from the member state of a customs union are classified as receipt or issue. If the goods are shipped from or to a country that is not a member of the same customs union or territory, the receipt is classified as an import, and the issue as an export.

The documents required for logistics processing in Purchasing and Sales and Distribution contain in their header and item levels all information necessary for foreign trade processing in SAP ERP. In Purchasing, this data is recommended for purchase orders and inbound deliveries from vendor and material master records as well as purchasing info records. In Sales and Distribution, export-related information is found in the sales order, delivery and billing documents from the customer and material master records. This foreign trade data forms the basis for the correct and legal processing of import and export business transactions, which are explained in more detail in the following sections.

4.4.2 Legal Control

Import and export business transactions must be based on current laws and regulations. In order to comply with legal requirements, the legal control function offers a process-specific check based on the respective transaction and the customer and country of destination of the export, in addition to product-specific checks of the goods to be exported.

Sales processes are initially checked to verify that the shipment is not destined for a country limited by an embargo or boycott or that the customer is not on a *sanctioned party list*.

Subsequently, a check is performed to determine whether the material to be exported requires an *export license* and if such a license has been properly maintained in the system. Whether a material is subject to legal control depends on its classification and the foreign trade data maintained in the material master. Foreign trade processing in SAP ERP offers the possibility of maintaining export licenses and monitoring sales orders and deliveries with regard to their foreign trade relevance, and blocking them if the necessary licenses do not exist.

4.4.3 Preference Determination

Preference determination establishes the ratio of origin and thus the degree of added value of in-house-produced products based on value percentages, source and originating status of the individual components. The originating status is measured according to the *list rules*, which determine how high the value-based third-country proportion of a product may be so that the finished product can still be declared as an originating product eligible for preferential treatment.

To put it simply, the originating status of a finished product is based on the maximum allowed percentage value of third-country precursor materials in comparison to the net value of the finished product in order for the finished product to still qualify for preference.

If a *preference agreement* exists between the country of origin and the country of destination, products eligible for preference can be exported to the country of destination on a duty-free or low-customs-duty basis.

From a customs standpoint, such a preference agreement between nations of the European Union and other countries or groups of countries thus represents preferential treatment for goods from particular countries. Under certain circumstances, it can offer exporters of products with originating status a competitive advantage, since the finished product is not burdened with many, if any, customs fees.

Preferential character of an in-house produced air conditioner

A company with headquarters in Germany manufactures air conditioners. The components for the finished product are procured internationally from various suppliers. The air conditioners have an ex-factory price of €1,000. According to HS (Harmonized System) Item 8,415, air conditioners made with motorized fans are only permitted to have precursor materials procured from foreign countries with a maximum value of 40%, so that the finished product can still be declared as preferential originating goods. In our example, the value of these materials amounts to a total of €400, meaning that German originating status applies to the device.

Customs preferences are determined in the foreign trade component of SAP ERP as follows:

1. First, a vendor declaration is requested.
2. The data from the vendor declaration is aggregated in the material master records.
3. A preference calculation is performed to determine the origin of a material and its eligibility for preference.
4. The preference results are used in the sales and distribution documents.

The determination of preferential treatment eligibility in SAP ERP is initially performed on the basis of *vendor declarations*. The vendor declarations serve as a document of information and certification regarding the preference-based origin of materials supplied by the vendor. Vendors are obligated to submit such declarations if they are directly or indirectly involved in preferential goods trade between the EU and other states.

Requesting and managing vendor declarations and the associated generation of reminders are supported by the system. Vendor declarations can, for instance, be requested on the basis of existing purchasing info records. Vendor declarations are aggregated in the material master, where they set a preference indicator.

Preference determination uses preference indicators in the material master to determine a preference condition and updates the preference indicator of the materials. Per material and preference zone, the system generates a condition record containing the determined preference price. Actual preference calculation is done on the basis of a BOM explosion, in which the components and their origins are determined, and the value of the precursor materials flows into the calculation. Figure 4.3 shows a view of the customs and preference data in the material master.

The results of preference determination are used in the sales documents because the preference condition records are taken into account during the pricing procedure. The system compares the sales price of a finished product with the preference price. In order for a finished product to be accepted at customs as preferential, the sales price of the product to be exported must be greater than its preference price. The business transaction and sales price of a certain product thus individually determine the preference validity of a sales product. As proof of validity of preference in accordance with customs regulations, movement certificates and evidence or certificates of origin are printed.

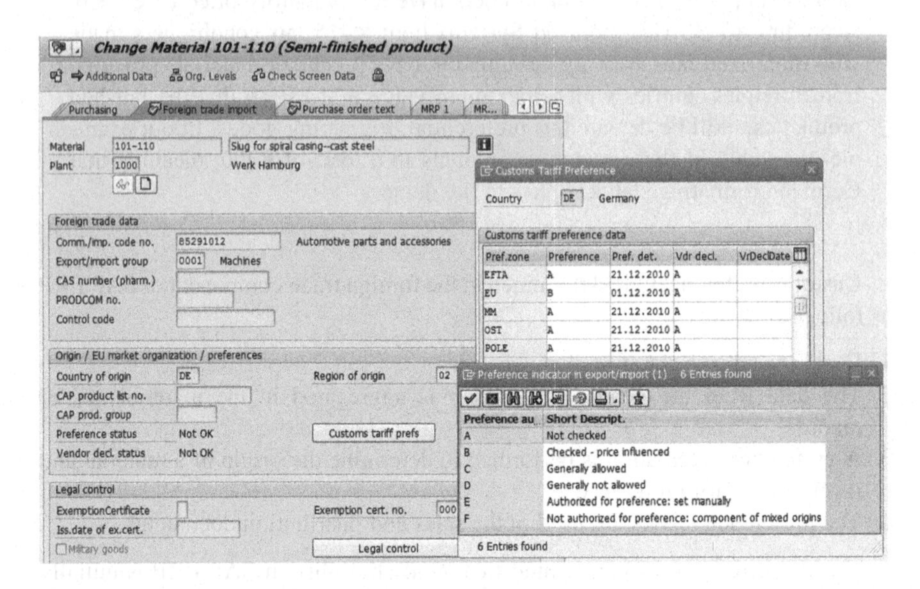

Fig. 4.3 Foreign trade data and customs preferences in the material master

4.4.4 Documentary Payments

A *letter of credit* is a financial document that is used when collateral is required for a particular business transaction with a certain business partner. Generally, this involves the exporting of goods in cases where the business partner's headquarters are in a foreign country. Documentary payments in foreign trade processing use letters of credit for credit management, export credit insurance and as a collateral instrument to minimize the risk involved in export business. Documentary payment in SAP ERP is seamlessly integrated in the sales and distribution component SD. This type of processing serves as confirmation to the exporter of punctual and complete payment, and also confirms to the importer that the goods for which he has paid are in fact the ones he has ordered. In addition, using a letter of credit supplies confirmation that the goods have been shipped.

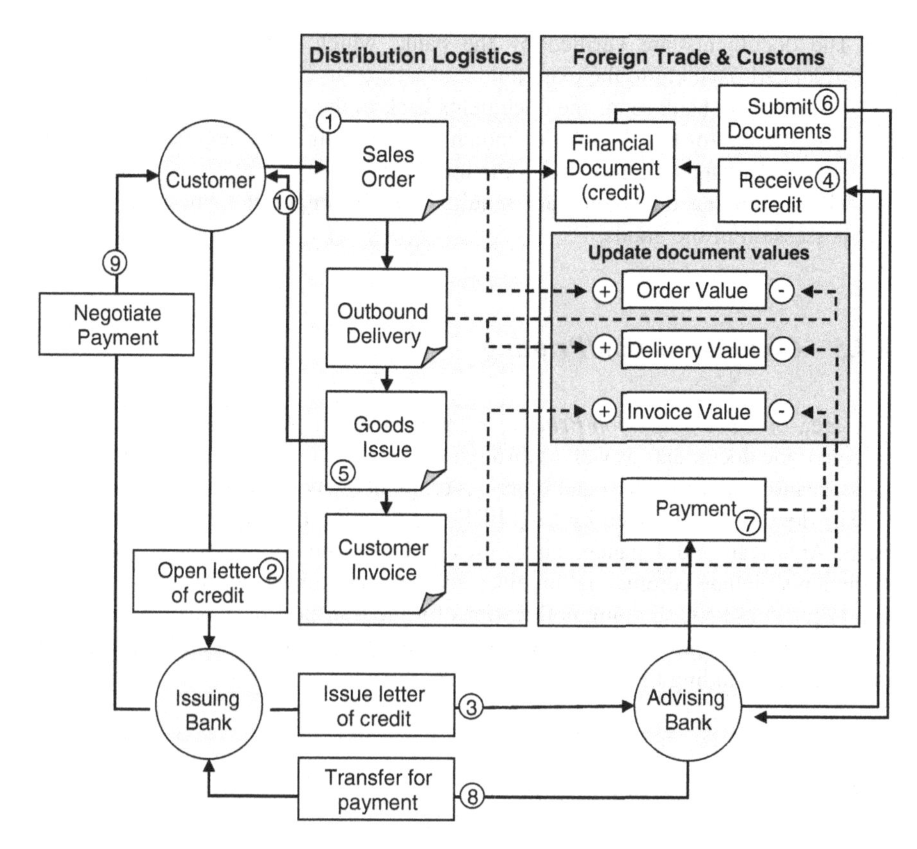

Fig. 4.4 Overview of letter of credit processing

Figure 4.4 shows an example of letter of credit processing, which can basically be done as follows:

1. The customer, the importer of the goods, places an order at a company, in this case the exporter. The sales agreement is established, and importer and exporter exchange the sales order and order confirmation. This confirmation contains the conditions to which this transaction is subject. In our case, the export sale is secured by a letter of credit.
2. According to conditions, the customer, in his role as an importer, opens a letter of credit at his firm's bank.
3. This bank, also called the *issuing bank*, approves the letter of credit request and opens a letter of credit to the benefit of the exporter at a bank in the country of the exporter and on behalf of the importer. The exporter's bank in this case is known as the *advising bank*.
4. The advising bank informs the exporter and forwards the letter of credit to him.
5. Through the opening of the letter of credit, the exporter acknowledges the importer's ability to pay, and ships the goods in observance of the conditions stipulated in the letter of credit.
6. The documents related to shipping the goods are forwarded to the advising bank along with the letter of credit.

7. The documents are verified by the bank, which subsequently makes the stipulated payment to the exporter.
8. The advising bank sends the documents back to the issuing bank for payment.
9. In exchange for the shipping documents, the issuing bank negotiates payment at the stipulated conditions with the importer.
10. These shipping documents are required for the importer to be able to take possession of the goods.

4.4.5 Communication/Printing

The area of *Communication/Printing* supports foreign trade by printing official foreign trade documents as well as through active interfaces for the dispatch of EDI messages to customs offices and other government agencies.

The documents provided by SAP ERP comply with the requirements of European, American and Japanese authorities and need no further adaptation. Such documents include commercial invoices, pro forma invoices, certificates of origin, export packing lists, shipping notifications for the transport of goods between two locations within a customs territory of the European Union and all foreign trade documents, including those required by U.S. customs.

4.4.6 Periodic Declarations

Import and export transactions must be centrally declared to facilitate the recording of goods traffic between the member states of the European Union. These so-called INTRASTAT and EXTRASTAT declarations flow into central statistics and document intracommunity trade.

Corresponding requirements also exist for Japan and states of the North American Free Trade Agreement (NAFTA), as well as for the states of the European Free Trade Association (EFTA). The ERP system supports employees in processing foreign trade by generating these periodic payments, and offers the opportunity to transmit the data required in the declarations as printed documents, discs or via EDI or email.

4.5 Governance, Risk and Compliance with SAP BusinessObjects Global Trade Services

SAP BusinessObjects Global Trade Services, a part of SAP BusinessObjects Governance, Risk, and Compliance (GRC) solutions, is an independent software component with which you can monitor and control customs, foreign trade, conformity verification and risk management processes in cooperation with SAP ERP logistics processes.

SAP BusinessObjects Global Trade Services only requires the SAP NetWeaver component for installation. Thus, it can be used independently on a separate SAP system, as an integrated system component of an ERP system (in the same or another client) as well as a subcomponent on any other SAP NetWeaver-based SAP system. Technical integration with logistics processes running on one or more ERP systems is achieved using interfaces that supply GTS with master and transactional data.

4.5.1 General Process and Integration in SAP ERP

To begin process integration between SAP ERP logistics and SAP BusinessObjects Global Trade Services, the required ERP master data must first be transferred to the GTS master database. This is done via an asynchronous *master data transfer* using ALK (application link enabling) technology. The transferred master data includes the following types:

- ERP customer master data that is used in GTS as business partner master data. Such data also includes organizational master data, which in GTS becomes foreign trade organizations with user plants.
- ERP vendor master data, which in GTS becomes business partner master data.
- ERP material master data, which in GTS becomes product master data.

After the master data has been provided to SAP BusinessObjects Global Trade Services, you can enhance these master data records with the corresponding foreign trade, conformity and customs-related additional master data (see Fig. 4.5).

Process integration is generally achieved by transferring logistics document data from the ERP system to GTS. Purchasing or sales documents are transferred from ERP, undergo various tests in GTS and trigger the generation of a customs document. This customs document can also be subjected to various tests. Its status can be monitored and queried, and corresponding release or block information flows back to ERP logistics processes.

The ERP logistics documents and respective billing documents can be used to generate a customs shipment document, which is used for customs declaration purposes. Figure 4.5 shows the basic process of a conformity check and customs declaration process.

The GTS user interface differs in the appearance of its main menu from the other SAP logistics applications. Instead of the typical menu tree, all application areas are accessible via buttons on the main menu screen. In Fig. 4.6, you can see the GTS main menu with its functional areas Compliance Management, Customs Management, Risk Management, Electronic Compliance Reporting and System administration.

4.5.2 Compliance Management: Sanctioned Party List Check

Due to various events (such as the terrorist attacks of Sept. 11, 2001) in recent years, measures for preventing commercial and financial transactions that could enable terrorist attacks (such as weapons trade and technology transfer) have become considerably more strict. For instance, automatic checks against *blacklists* for trade partners and embargo and boycott lists for countries have been established.

GTS supports implementation of these legal requirements in payment transactions by enabling an automatic check of your master data against regularly published blacklists (such as by the Office of Foreign Asset Control in the United States, the German Federal Ministry of Economics and Technology, and the EU). Test results can then be used to prevent transactions with undesired trade partners, countries or restricted products.

Figure 4.7 depicts the process of a sanctioned party list check. The business partners indicated in the SAP ERP materials management and sales documents are screened by the GTS component against existing sanctioned party lists (see **1** in Fig. 4.7). You can obtain sanctioned party lists **2** from external sources (such as

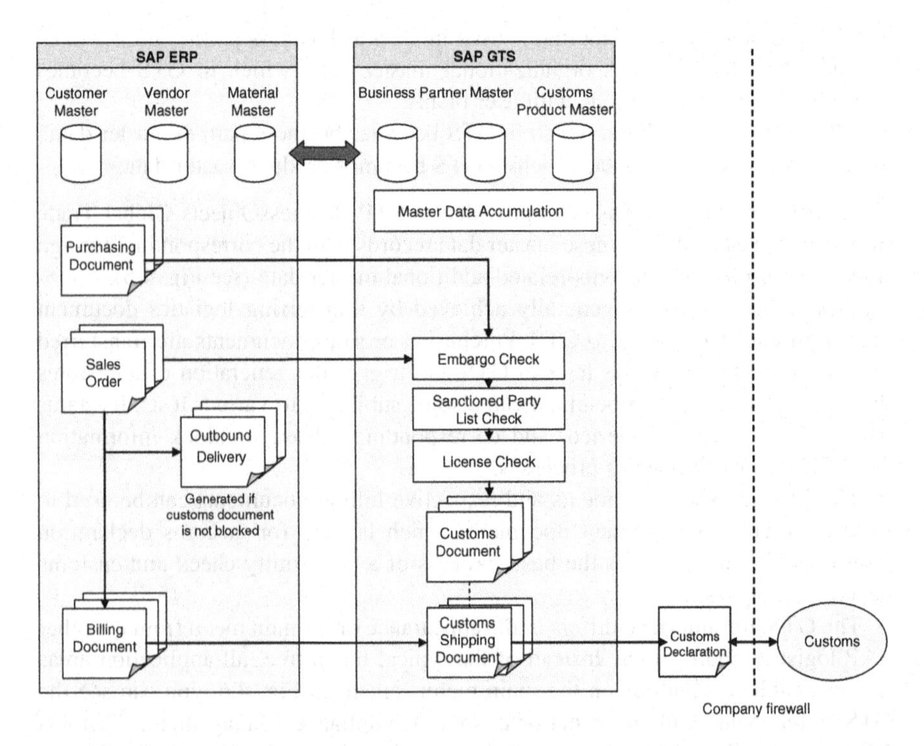

Fig. 4.5 Communications and processing between SAP ERP und SAP BusinessObjects Global Trade Services (example: conformity check/customs declaration)

authorized data suppliers) and load them into the GTS component via the XML interface. You can also create your own lists and apply them in the checks.

Based on those lists, you can perform a periodic screening **3** of your business partner database and, if necessary, identify a partner as *blocked*. The business partners stored in GTS have an initial status of not screened when the data is transferred. The sanctioned party list check is done through a comparison of the addresses. Depending on the results, a business partner receives the status of *blocked* or *released* for each list.

On the other hand, you can compare materials management documents (purchasing, material and inbound delivery documents) and sales and distribution documents (sales, outbound delivery, billing documents as well as the request for quotation and the quotation itself) to the sanctioned party lists via integration of the logistics documents with GTS **4**. Periodic monitoring also allows you to subsequently release a blocked partner if a change has made this possible **5**. Figure 4.8 shows the test result of a sanctioned party list check against the business partner master.

Fig. 4.6 Main menu of SAP BusinessObjects Global Trade Services

When checking financial transactions (incoming and outgoing payment documents), SAP BusinessObjects Global Trade Services supports the checks cited below. If the check results in a business partner being blocked, the entire transaction is prevented; this is done in the following steps:

1. Comparison of the business partner master data, including the account holder, with blacklists
2. Evaluation of the check results in the payment program
3. Check in the payment program to verify whether the country of the payment recipient or payer is subject to an embargo
4. Check of the application of funds by GTS in the payment program

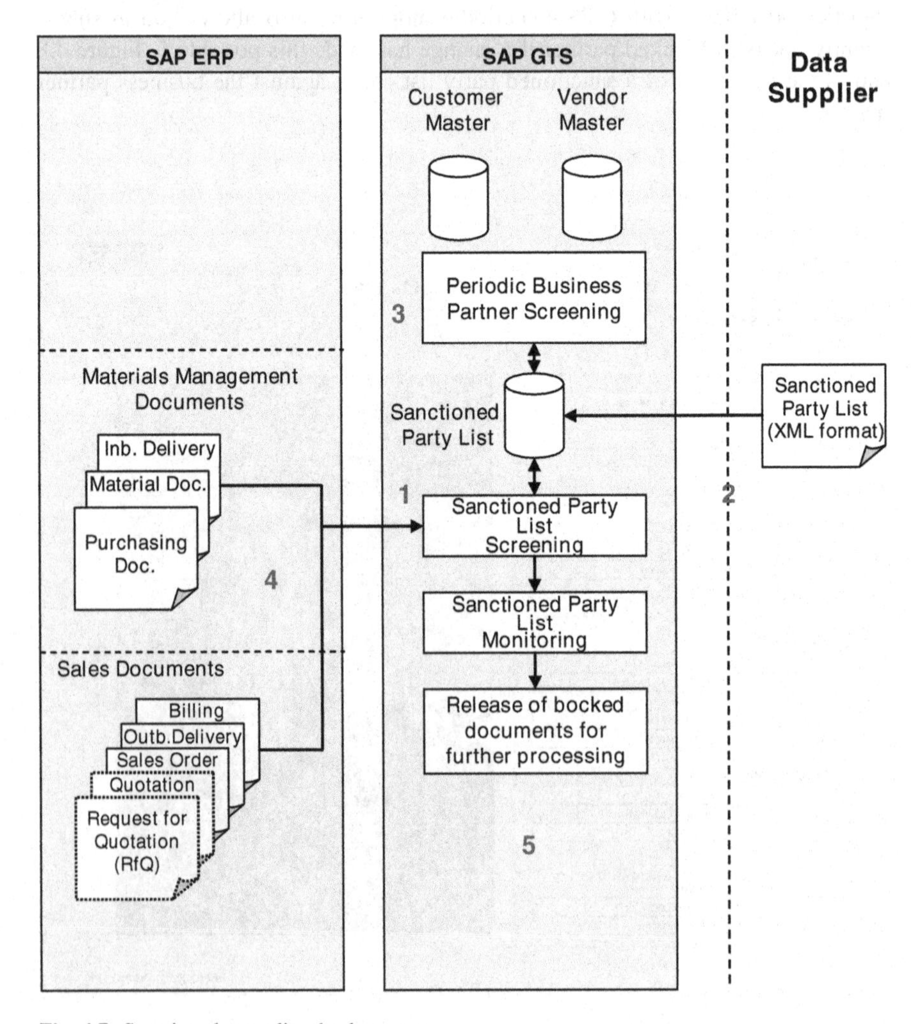

Fig. 4.7 Sanctioned party list check process

The following section provides an overview of the functions of import/export control.

4.5.3 Compliance Management: Import/Export Control

Foreign trade laws strictly regulate the import and export of goods. Such laws include arms control and drug laws, embargo and other regulations. Frequently, licenses are required for import or export to ensure smooth processing. For instance, in the United States, the Department of Commerce, Department of State, the Food and Drug Administration and the Drug Enforcement Administration enforce extensive license requirements. GTS can support you in this area with its Import/Export Control, by providing clear information on existing licenses and embargo regulations.

To enable efficient screening, the goods to be moved must first be classified. For this, you can assign all goods and products commodity codes, export list numbers or other country-specific codes. You can also do a key word or phonetic search for a corresponding number. Import and export lists can also be imported as files that you can obtain from data suppliers.

To determine the relevant *licenses*, you can maintain a search strategy in GTS. You can then run a check of your import or export deliveries in order to ensure that the import or export conforms to the laws of the respective country. For exports, the regulations pertaining to the importing country apply, and for imports, those of the exporting country. The actual check is generally based on a combination of the parameters export list number, country of destination, country group, individual products, customer data, goods quantities and values.

Figure 4.9 shows an overview of the function menu provided by GTS for the execution of various export control areas.

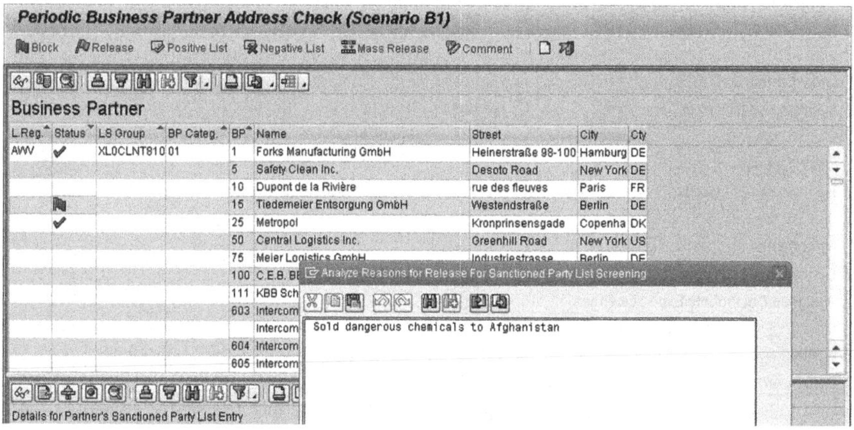

Fig. 4.8 Result of a sanctioned party list check of business partners in GTS

Import/Export Control offers a number of functions:

- **Monitoring functions**
  - Monitoring of blocked documents and payments
  - Management of technically incomplete documents and payments
  - Tracking of control-relevant products
- **Embargo checks**
  - Monitoring of blocked documents and payments
  - Management of technically incomplete documents and payments
  - Analysis of the release situation or business partners in an embargo situation

Fig. 4.9 Functions in Compliance Management for legal export controls

- **Licenses**
  - Maintenance of import licenses
  - Maintenance of export licenses
  - Value and quantity depreciation of existing licenses
  - Remaining quantity, remaining value and validity check of existing licenses
- **Other supported functions**
  - Dissolution and maintenance of bills of material (BOMs)
  - Maintenance of control data, search and check rules
  - Simulation of import and export transactions

Figure 4.10 shows the process of an import and export control. The documents generated in logistics for the processing of imported or exported goods can be transmitted to GTS to check for relevant licenses and embargos (see **1** in Fig. 4.10). When a check ends with a positive result, a release of the documents occurs in SAP ERP **2**. The quantities and values for the licensed imports or exports are updated in GTS, enabling you to have an overview of the contingent currently in use **3**.

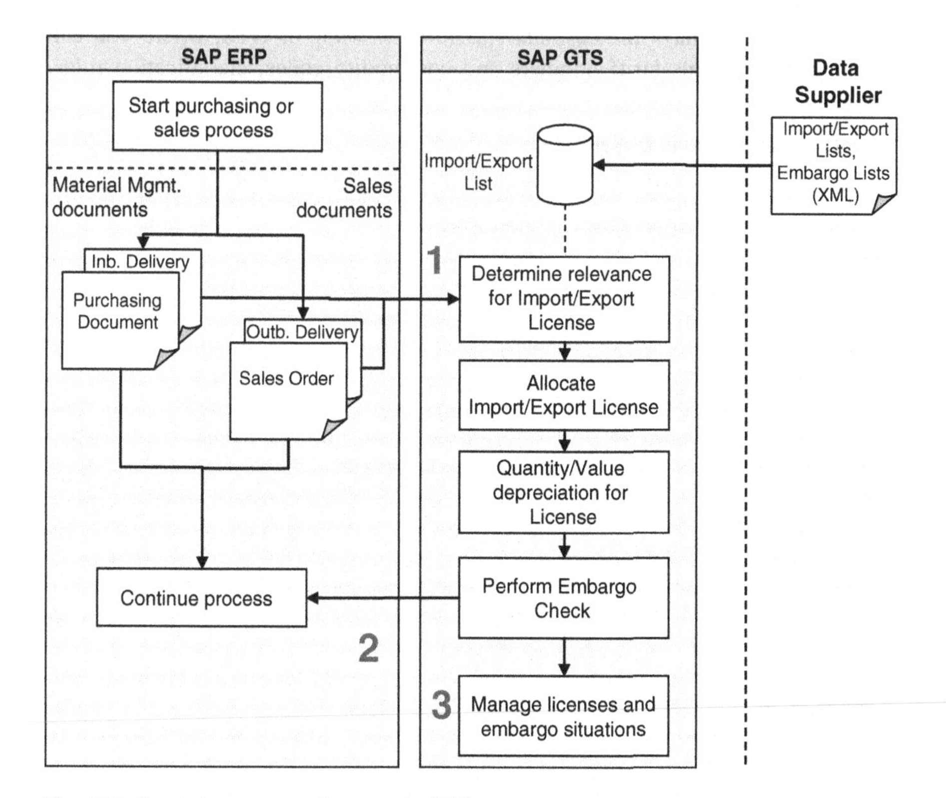

Fig. 4.10 Import/export control process in GTS

4.5.4 Customs Management

The GTS *Customs Management* offers functions that support cross-border goods flow with regard to customs authorities. GTS support of electronic customs declaration and shipping procedures prevents unnecessary delays and thus aids in saving time and costs.

The *Customs Processing Import/Export* cockpit offers a variety of functions, such as:

- The display of import and export activities
- The determination of customs status
- Monitoring of reimport lists and export confirmations
- Manual customs declaration before and after goods receipt
- Entry of supplementary customs declarations
- Monitoring of logistics processes, execution of inventory posting and accessing of stock lists.

The array of functions is based on comprehensive product duty information and duty classification characteristics, which can be maintained as a supplement to the material master.

Figure 4.11 displays the *customs classification view* of GTS, where you can maintain the customs tariff numbers and commodity codes, classification rules,

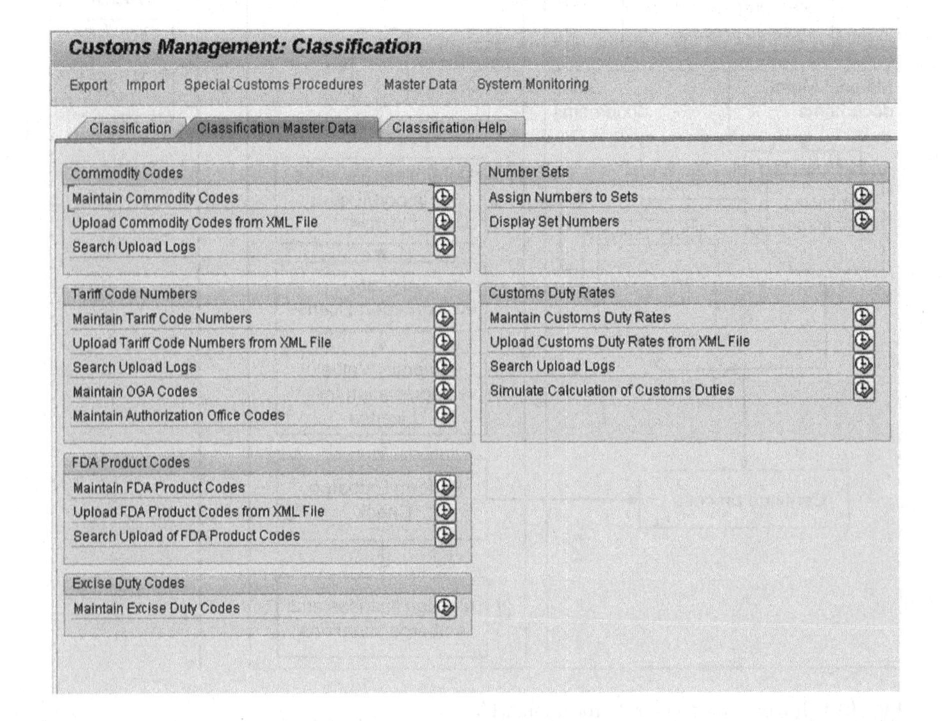

Fig. 4.11 Function menu for customs classification

customs duty rates and other classification characteristics. Of course, you can also import the codes and tariff data to GTS from compatible sources instead of making manual entries.

After updating tariffs, tariff characteristics and rules, you can perform operative customs processes:

- Electronic customs declaration process
- Printing of trade documents for the import and export of goods
- Customs duties calculations

GTS is ready for communication with the customs authorities of several countries. In a few cases, electronic communication is available via direct SAP interfaces, and in other cases you also need an EDI converter and adapter that are linked via *SAP NetWeaver Process Integration* (SAP NetWeaver PI). There are also certified SAP partners who offer such options as supplementary modules for conversion of GTS interfaces to ANSI X.12 601/824 document interfaces of the *Automated Export Systems* (AES), made available by the U.S. Customs and Border Patrol.

4.5.5 Risk Management

The area of *Risk Management* in GTS includes the functions for preference processing, letter of credit processing and for export refunds for EU-market-regulations goods.

With SAP GTS, you can use preference advantages and avoid problems related to violating free trade agreements such as NAFTA or the EU most favored nation agreement. Figure 4.12 shows the primary structure of preference processing.

With *GTS Preference Processing* (see the function menu in Fig. 4.13), you can efficiently obtain the necessary vendor declarations from your vendors and manage them. This can be very complicated without system support, since you have to individually determine the origin of all components used in an end product, especially for extensive bills of material with semi-finished products of various foreign manufacturers.

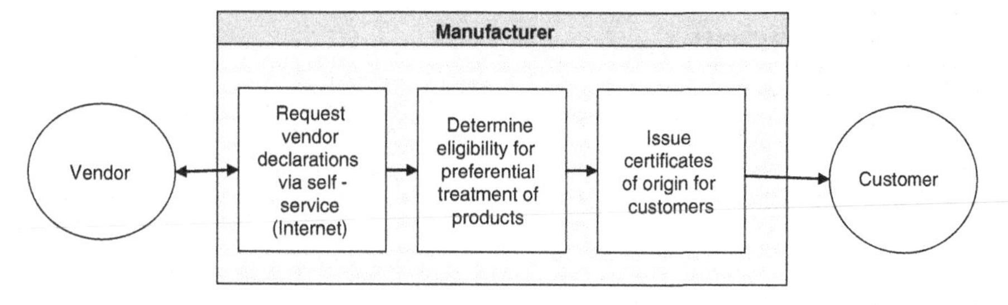

Fig. 4.12 Structure of preference processing

With SAP BusinessObjects Global Trade Services, you can obtain vendor declarations in several ways:

- Forms that can be individually tailored
- Internet-based self-services for vendors
- Fully automatic process using SAP NetWeaver PI

After obtaining the vendor declarations, you can aggregate the data and store it in the material master of the product. Subsequently, you can perform a preference calculation, even for complex products such as automobiles and computer systems, with the automatic use of preference rules and the official origin rules.

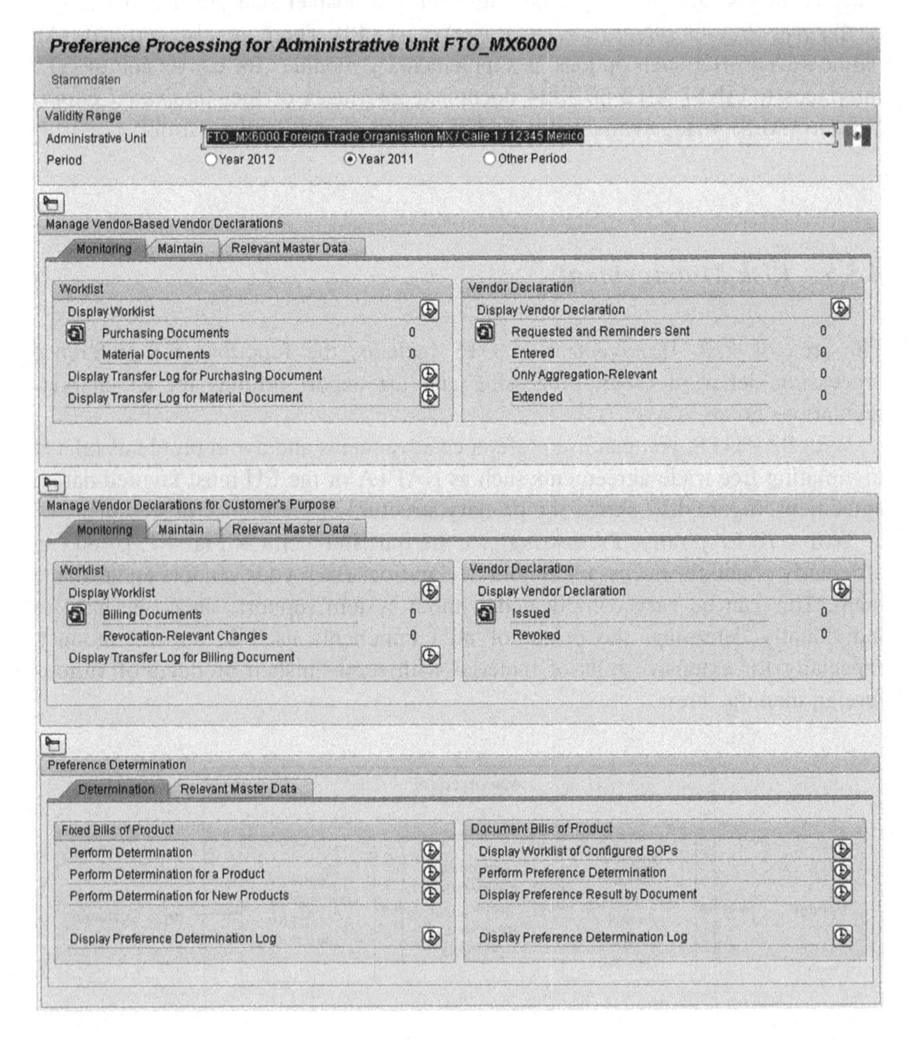

Fig. 4.13 Function menu of GTS Preference Processing

Letter of credit processing supports you in risk minimization when you need to do foreign trade business with financial documents such as letters of credit, collection or guarantees, and wish to manage these documents. Import- and export-related letters of credit are supported and can be allocated to individual transactions. Automatic testing lets you block documents for further processing in risk situations.

4.6 Summary

In the SAP world, you have the opportunity to deal with your trade regulation requirements using SAP ERP components for foreign trade processing in SD as well as SAP BusinessObjects Global Trade Services.

SD Foreign Trade Processing is included in the standard issue of SAP ERP, and is integrated with the ordering and distribution processes as well as finance processing of the ERP system.

The GTS components of SAP BusinessObjects Governance, Risk, and Compliance (GRC) solutions represent the considerably more flexible and efficient solution. It is adapted to current regulations, especially through country-specific interfaces and process support.

Chapter 5
Controlling and Reporting

Controlling logistics processes and the subsequent provision of assessments and reports are significant supplementary functions for the efficient organization and execution of all kinds of supply chain activities.

To conclude this book, we explain SAP Event Management, SAP Auto-ID Infrastructure, SAP Object Event Repository and SAP NetWeaver BW. All of these components enable you to maintain an overview of your processes and keep them under control.

5.1 SAP Event Management (SAP EM)

Universal tool to monitor events: SAP Event Management is a universal, adaptable tool for defining visibility, status tracking and tracing processes, imputing performance data using your own or partner processes, and integrating these processes in your application systems.

SAP Event Management provides flexible support for achieving your process and object status tracking needs. Because there are no "prefabricated", purpose-specific business objects in Event Management (such as sales orders), you can configure suitable object types, process steps and reactions in order to model and construct status processes that are precisely tailored to your requirements.

Event Management has become the standard system for *tracking and tracing* within SAP logistics. Thus, several logistics processes come pre-configured for use with SAP Event Management.

J. Kappauf et al., *Logistic Core Operations with SAP*,

DOI 10.1007/978-3-642-18202-0_5,

5.1.1 Basic Characteristics of SAP EM

5.1.1.1 Event Handlers

The primary objects in Event Management are called *event handlers* (EH). They enable the tracking of individual status processes and the definition of their characteristics. An event handler can represent a material good or object, such as one of the following elements:

- A pallet used as packaging for a goods shipment to be tracked (specifically, the goods are of interest, not the pallet)
- A container that is tracked as an asset during the entire period of possession
- A machine whose function is monitored and recorded
- A shipment (such as an express package) that is traced from pick-up to drop-off

You can also track a process or non-material event, such as:

- A purchase order that proceeds through various processing statuses
- A payment procedure intended to lead to the settlement of an invoice
- An order generated by you, from the requisition to delivery

5.1.1.2 Event Handler Lifecycle

One thing that all event handlers have in common is the presence of a *lifecycle*: Event handlers are created through a certain procedure, after which they exist for a time and are subsequently deactivated by a particular event. The EH lifecycle can be of varying length. Event handlers can process a large number of events. Table 5.1 shows three examples of typical data for the lifecycle of an event handler.

High performance and speed

SAP Event Management is designed to process extensive scenarios and large amounts of data. Several large postal companies use SAP Event Management for package tracking, with several billion event messages per year.

During the lifecycle of an event handler, many different types of events can occur. These are depicted in Fig. 5.1:

Table 5.1 Typical event handler types and their lifecycles

Characteristic	Event handler type		
	Bid invitation	Shipment	Container
Lifecycle	4 h	6 weeks	3–5 years
Number of events	3–5	15–20	>5,000

5.1.1.3 Event Types

Let us take a closer look at the various types of events along the lifecycle:

1. **Regular event**
   This event is known as a process milestone either in the application system or during a process, and is stored in Event Management as an expected event. This event can occur during a defined period of time. The current event occurs within that defined period, as expected.
   Example: A goods shipment is to be delivered between 10 a.m. and 12 a.m., and is delivered at 11:33 a.m.
2. **Premature or overdue event**
   As in the case of a regular event, an expectation definition exists, but the event occurs either too early or too late. The deviation is registered in Event Management but does not necessarily lead to exception reactions.
   Example: An invoice is to be paid by March 30 but is not paid until April 1.
3. **Unexpected event**
   This event is not known to the system as a process milestone, but it happens nevertheless. It can either represent a status or an exception.
   Example: A railway car passes the control center in Boston (status), or a truck breaks down on the highway with engine trouble (exception).
4. **Non-reported event**
   As in the case of a regular or non-delayed event, an expectation definition exists, but the current event does not occur upon expiration of the expected time period. Because it is not known whether the physical event has not occurred or whether it has simply not been reported, exception handling is initiated as soon as the deadline has been reached.

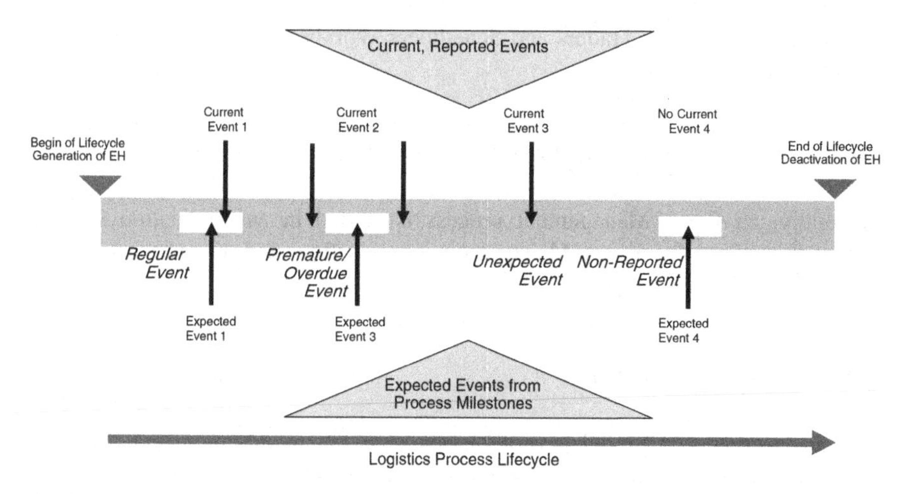

Fig. 5.1 Event types in event management

An example is a message pertaining to a delivery that a service provider must issue within 8 h after delivery. If the service provider does not report the delivery, it does not matter whether he has made the delivery or not, since the measurement criterion is the timely provision of the information (for instance, to facilitate the timely issue of an invoice).

5.1.1.4 Logistics Processes and Event Management

Below, a few logistic processes and process variants are cited that SAP clients from various branches monitor with SAP Event Management. The corresponding solutions for that branch are also indicated:

- Bid invitation and visibility for logistics execution (high-tech and electronics industry; SAP for High Tech)
- Handling unit tracking in output logistics processes (logistics service providers; SAP for Transportation & Logistics)
- Sales and distribution processes in a distributed environment (manufacturing industry; SAP for Industrial Machinery & Components)
- International sea freight, including customs declaration (wholesale and retail; SAP for Retail)
- Integration with a vehicle management system, returns processing (automotive industry; SAP for Automotive)
- Replacement part and tool management (air and space industry; SAP for Aerospace & Defense)
- Tracking of individual packages, including nested loading (postal branch; SAP for Transportation & Logistics)
- Order processing, including production monitoring, delivery and invoicing, railway car management (chemical industry, metal, wood and paper industries; SAP for Chemicals, SAP for Mill Products)
- Integration with the Traders and Schedulers Workbench (oil and gas industry; SAP for Oil & Gas)

5.1.1.5 Event Management Processes

Generally, an Event Management process is started in an application system. Figure 5.2 shows the Event Management process flow and an overview of the functions of Event Management.

5.1.1.6 Process Flow Phases

The process proceeds through several phases. First, the underlying object, such as a shipping document, is generated in the application system. If it is relevant to Event Management, data pertaining to tracking is extracted from the document and sent to

Event Management, where a corresponding event handler is generated. When event messages are received, the corresponding event handler in Event Management is determined, and the events are processed with the aid of the rule type. In doing so, actions such as the sending of messages or actions in the application system can be triggered. Using the Report Manager, you can display the status data and history of event handlers.

5.1.2 Application Interface

5.1.2.1 Integrating SAP Applications

The Application Interface provides a uniform basis for the integration of all SAP applications with Event Management. It is a component of SAP NetWeaver. Based on its configuration options, you can control right from an application how Event Management is to be executed for an object or process.

5.1.2.2 Content-Based Significance of a Business Object

In many cases, the application and business object (such as a shipping document) themselves do not indicate much about the content-based significance of a process.

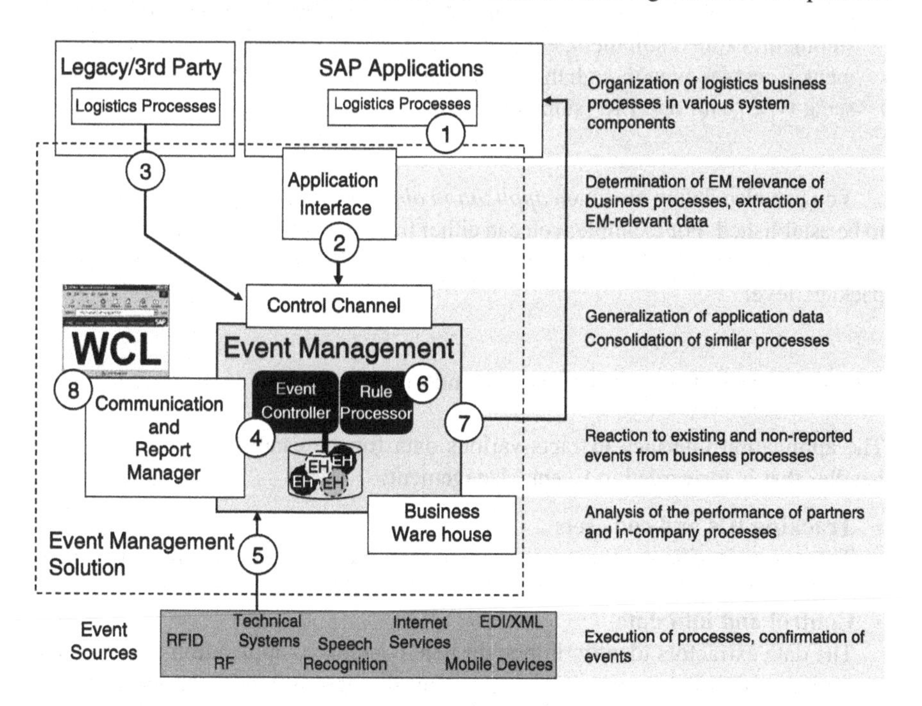

Fig. 5.2 Elements of SAP event management and the integration process flow

In SAP systems, this significance often only becomes apparent through a variety of control parameters, such as document types. A shipment document that represents a consolidated road transport, for instance, is tracked in a different way than a shipment document for a full container designated as sea freight. Only through the transportation type of a shipment document and the existence of a handling unit of the type *Container* does the type of transport and how it is to be tracked become apparent. The event handler is generated accordingly.

There is usually one business process type for every application object. This business process type, used in the application system and in Event Management, is a rough categorization of processes that are supported (such as purchase order, shipment or sales order tracking).

5.1.2.3 Application Objects

The content-based characteristics of a business object are called *application objects* in the Application Interface and in Event Management (examples are container sea freight shipment, dangerous goods road transport and a retail shipment).

Creating application objects

For the business process type *Shipment*, for instance, you can create various application objects (such as container shipment, air freight, dangerous goods shipment, express shipment, etc.), all of which lead to various event management processes even though they were generated from a business object of the same type (with one ERP shipment document).

You can also define on which *application object level* the application objects are to be established. For example, you can either track an entire shipment or generate a separate event handler for individual packages of a shipment and track them on the package level.

5.1.2.4 Data Extraction for Event Handlers

The application interface extracts various data for the establishment of an event handler that is forwarded to Event Management:

- **Tracking IDs and code sets**
  These are identification numbers such as a shipment number, container number or order number.
- **Control and info data**
  The data extractors identify important field contents of application objects, such as the sender and recipient of a shipment, the description of the goods or the name of the driver.

- **Query IDs**
  These are identification numbers that are only used to query information (such as a customer order number).
- **Expected events**
  These are milestones that represent the process flow, such as loading date, dispatch, arrival or unloading date.

5.1.2.5 Creating Event Types

You can create *event types* for a business process type. They define events that take place in the application system within the context of an application business transaction and, from there, are reported as events to Event Management (for instance, you can set the loading status for a shipment within shipment processing). Data is also extracted from the application context in order to report events, and it is added to the event message to be sent.

5.1.3 Event Handlers and Event Messages

The *event handler* is a universally applicable business object that can represent all status and tracking processes in Event Management. It can represent an individual process or object, or a set of objects from different perspectives. An event handler contains a great deal of data that can be individually calibrated for a process with the proper settings. The following data is available for the calibration and control of an event handler:

- **Event handler header and system parameters**
  The header contains control data and references to the application object.
- **Tracking IDs**
  Tracking IDs and their tracking code sets serve to allocate event messages to an event handler. When the tracking ID of an event message matches an event handler, the event is processed by that event handler.
- **Status attribute profiles and values**
  The status can be saved with a single- or multiple-value calibration and an initial value for any purpose. For example, you can define a shipment status that can assume the values "Waiting for dispatch", "In transit" or "Arrived". The statuses and their values can be defined freely.
- **Data container**
  Data containers enable you to save any information on a process or underlying business object.
- **Expected events**
  Events that are defined as process milestones for the event handler can be stored as expected events.

- **Reported events**
  If an event message is received for an event handler, a reported event is generated from it. It can either be expected or refer to an unexpected event.
- **Rules**
  The rule set serves to process events and react to expected and unexpected events.
- **Queries**
  Queries can be used to determine status and event handler data.

5.1.3.1 Mapping Parameters

When an event handler is created, the event handler type is first determined. Then, the transferred parameters of the application system are mapped in the *Event Management parameters*. This allows for a uniform display of data from various applications. This can be necessary if, for instance, various application systems are used for the same process, which is frequently the case in large companies with "expanded" IT system environments.

5.1.3.2 Event Handler Settings

The basic settings for the characteristics of an event handler are defined through the event handler type and associated profiles. The event handler type determines the generation, construction and behavior of an event handler. You can adjust the following settings for an event handler type:

- The *rule set* determines how the event handler reacts to incoming event messages.
- A *status attribute profile* generates one or more status fields in the event handler, each with an initial value and a permissible attribute value list.
- The *extended event profile (EE profile)* determines which expected events are generated and from where the respective dates are derived.
- The *extension table ID* defines a table with which additional event handler parameters can be saved so you can find them quickly and easily with a search query.
- The *BW profile* determines which data from an event handler is to be extracted for transfer to the Business Warehouse.

5.1.3.3 Expected Event Profile

In an event handler, the expected events are generated on the basis of an *expected event profile*. This profile lists all expected events that can occur (such as a dispatch, arrival or inbound delivery confirmation). You can consolidate expected events into groups. For instance, you can create a group in which only one event must occur in order to acknowledge the confirmation for the entire group. Expected events are created from the following data sources:

- **Application object milestones**
  Milestone data can be extracted from an application object (for instance, a delivery date). There are also events that can occur several times in several places (such as an arrival event for every end location of a shipment stage).
- **Events can be determined in Event Management**
  Events are generated in Event Management in relation to one another (for example, an end-of-loading event defined to occur a maximum of 2 h after the start of loading).
- **Events from non-SAP systems**
  Events can be retrieved from non-SAP systems or determined using Web services.

For every event, you can define an earliest and a latest event date (or corresponding time of day) as well as an earliest and latest possible message transmission date (or time). Thus, exception handling can take place for an overdue event as well as for a delayed message if it lies outside the defined time period.

5.1.3.4 Flexible Status Attributes

The *status attribute profile* offers the opportunity to allocate one or more status fields to an event handler. They are filled with an initial value (initial status) when the event handler is generated. You can perform activities during event handling to alter status attributes.

Events change status attributes

A departure event, for instance, can be used to set the shipping status defined in an event handler from *Not started* to *In transit*, and a subsequent arrival event could then set the status to *Arrived.*

For every status attribute, you can save a list of permissible status values.

5.1.4 Event Handling

5.1.4.1 Flow of Event Handling

Event reports can occur at any time, even before the corresponding event handler exists. Events do not get lost, and as soon as an active event handler exists with a suitable tracking ID, the event is processed by it. The following steps are performed:

1. The event is first saved in the database.
2. The event handler is determined and the event is forwarded for further processing.
3. The event is saved as an unexpected event or with reference to an expected event in the event handler.
4. An appropriate rule for event handling is sought. If one is found, all actions defined within it are performed.

5.1.4.2 Rule Sets for Event Handling

Rule processing is based on rule sets, which can contain several rules. Each rule is composed of a condition and actions that are executed depending on the conformity or non-conformity of the conditions. A rule (condition/action), for example, could look something like this: Verify whether the arrival at the destination port has occurred more than 12 h late. If so, send an informational email to the goods recipient. Figure 5.3 shows the principle structure of a rule set.

5.1.4.3 Jump Targets in Rule Sets

For every rule, there is also a definition of where within the rule set the action processing is to continue. For example, if a condition is evaluated to be TRUE and

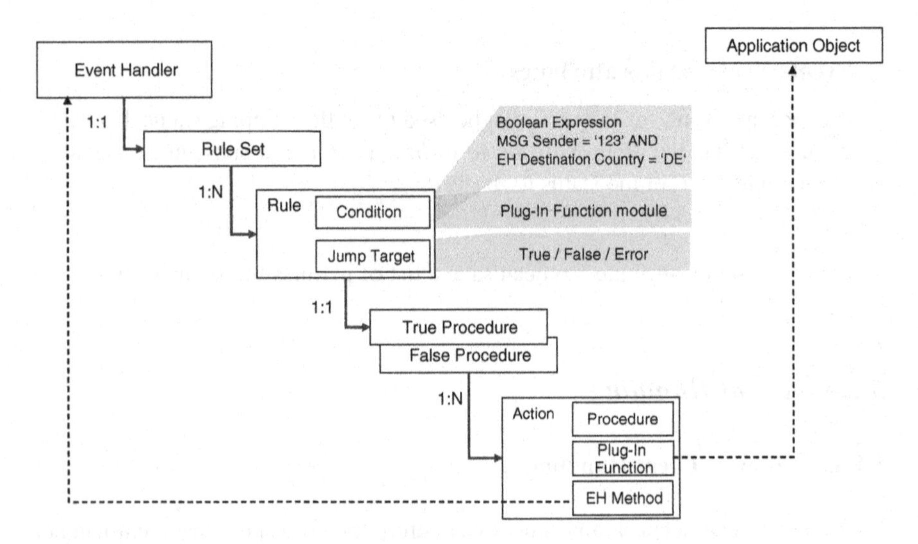

Fig. 5.3 Structure of a rule set in event management

the action has been successfully completed, it is advisable to jump to the end of the rule set to end processing.

5.1.5 Information Input and Output

5.1.5.1 Event Management User Interfaces

SAP Event Management gives you several possibilities for data input and output. There are several functional interfaces and services with which you can create and edit event handlers and send event messages.

In addition, Event Management offers a number of user interfaces to query status information and report events. Figure 5.4 provides an overview of these options.

5.1.5.2 List Display and Event Handler Details

The event handler list allows professional to obtain a detailed look at the processes and conditions of individual event handlers. In addition to overview data and event lists, you can also access all identifications, histories and process logs.

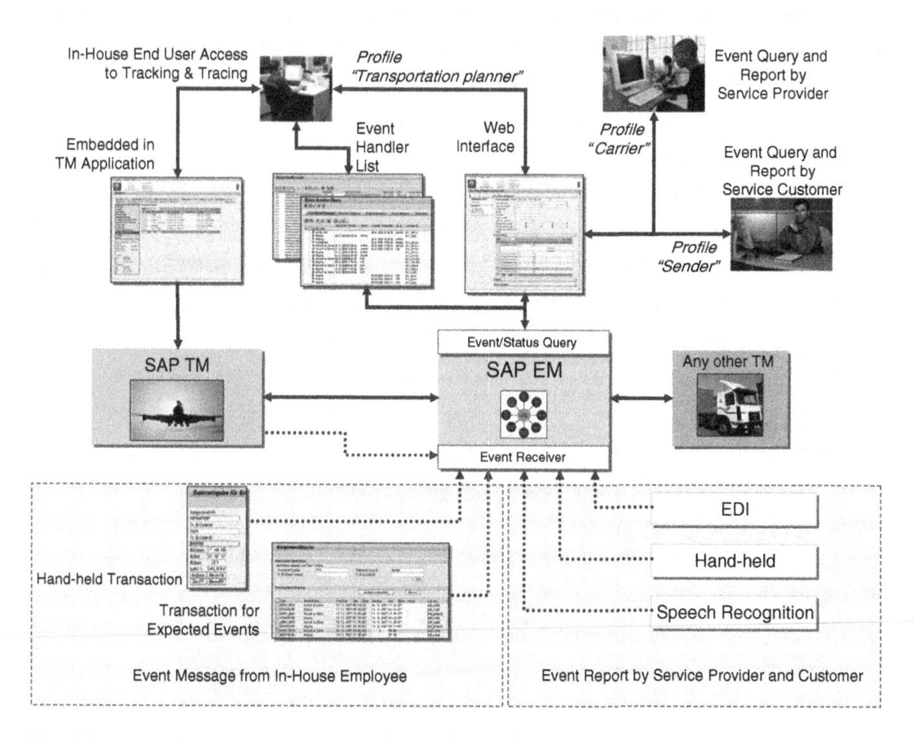

Fig. 5.4 Accessing event management data and reporting events

By double-clicking an entry in the event handler list, you can access the *Event Handler Overview*. Here, you can view consolidated and expected events, event messages, error messages and status details. Status details also indicate any status change history.

When you select the *Detail Overview*, you arrive at the internal view of the event handler, in which you can access all event handler data and the complete history. Figure 5.5 shows the Event Handler Overview with an event list and Detail Overview, including the process history. Here, the individual actions and process steps are listed, including a time stamp of the execution and process status.

5.1.6 *Data Entry for Event Messages*

5.1.6.1 Configurable Hand-Held Scanner User Interface

A configurable user interface is available for the integration of a hand-held scanner. It enables you to scan objects (such as pallets) and send a predefined event to Event Management with each scan (for instance, a goods receipt for pallets including information on the processing employee, location and date and time).

In the event that a user is only authorized to confirm expected events, a transaction for event confirmation can be used. When the tracking ID is entered, all expected events are listed, and you can enter the event date and time, time zone and a reason for any deviance in time.

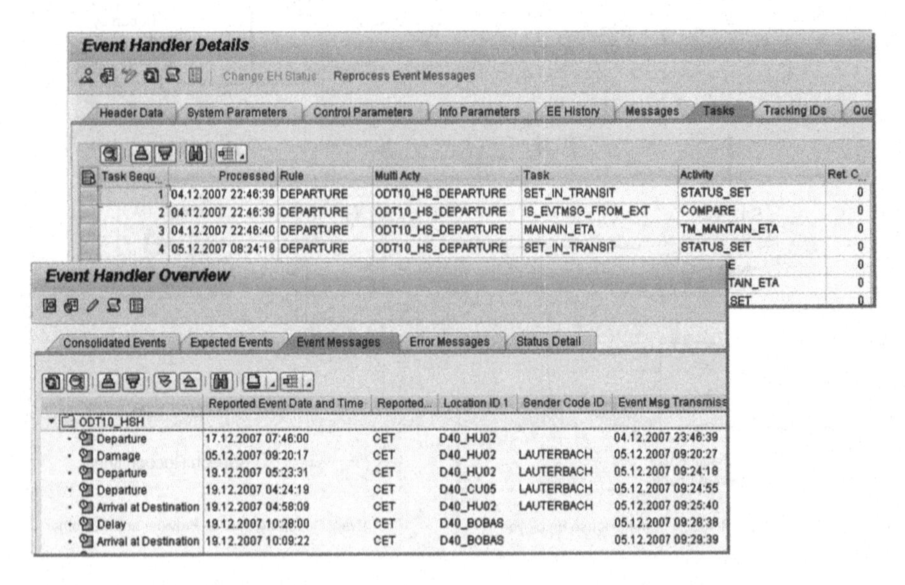

Fig. 5.5 Event handler overview and detail overview

5.1.6.2 Querying Overdue Events

To receive a simple overview of overdue events, you can open a list of overdue expected events. The list appears after entering selection criteria, and by double-clicking an entry, you can link to the corresponding event handler overview.

5.1.7 Web Interface

5.1.7.1 Configurable Web Interface

In Event Management, you can define Web interface transactions and allocate them to users or roles. These transactions can be accessed with any Internet browser. You can define the following profiles for each transaction:

- **User profile**
  The user profile contains the name of the profile and the allocation of selection, display and event message profiles. The user profile is ultimately allocated to a user role.
- **Selection profile**
  This profile contains the configuration of fields that are available as selection fields.
- **Display profile**
  The display profile enables the definition of parameters that are to be displayed in the event handler details and the columns of the event lists.
- **Event message profile**
  This profile is used to define which event messages a user can acknowledge and what additional data he can enter.

Web interface versions

In Event Management – depending on the release version – there are two different versions of the Web interface. Up to and including Release EM 5.0 (SCM 5.0), a Java version is available (Java Server Pages Technology) that is executed on the SAP NetWeaver Java stack. This has been replaced by a Web Dynpro ABAP version in Release EM 5.1.

Figure 5.6 shows the user view of the Web interface when it invokes a transaction with the role of *Shipper*.

Above, you can see selection fields for the event handler. The upper section shows data and parameters of the selected event handler in detail, followed by a list of events.

5.1.8 Standard Processes in Event Management

5.1.8.1 EM Standard Processes for ERP-Based logistics Processing

SAP Event Management has a built-in, comprehensive set of preconfigured monitoring processes with which you can control and visualize logistics processes, such as those in SAP ERP and SAP TM. Figure 5.7 provides an overview of the standard visibility processes that come with SAP Event Management (EM 7.0). In addition

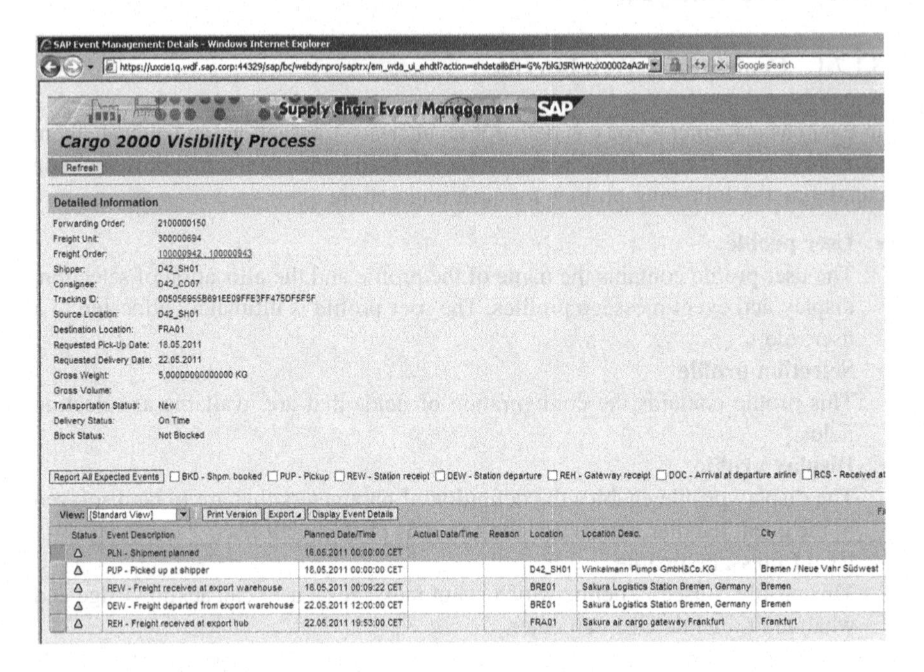

Fig. 5.6 Shipper view of the web interface

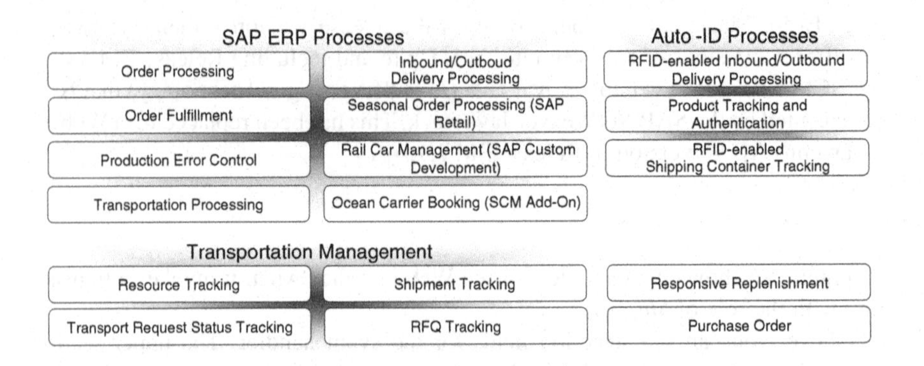

Fig. 5.7 Standard visibility processes in SAP event management (Release SCM EM 7.0)

to ERP and TM processes, you can also view those of automatic identification (*Auto-ID*) and *Supply Network Collaboration* (SNC).

For logistics processes in SAP ERP, there are several visibility processes based on various business objects or process chains.

5.1.8.2 Fulfillment

In the *Fulfillment* area, there is an EM process that begins with the creation of a sales order and then makes delivery processing and subsequent shipment visible. The scheduling of shipment documents containing sales order delivery data triggers the generation of an event handler that includes delivery events (such as picking and packing) as well as transport events and the delivery status.

In the data warehouse SAP NetWeaver Business Warehouse (SAP NetWeaver BW), performance figures are updated regarding execution and reporting quality of the respective service provider and the duration of transportation.

5.1.8.3 Order Processing

In the area of *Order Processing*, you can track the status of an order. If order items exist that are relevant to tracking, the system generates an event handler with which the order confirmation, shipping notification, goods receipt or payment can be controlled. In SAP NetWeaver BW, for instance, key figures regarding cycle time between an order and confirmation, confirmation and shipping notification, or notification and goods receipt are updated to allow you to perform intermediate and in-house performance controls.

5.1.8.4 Production Error Handling

The *production error handling* process is based on a production order that leads to the generation of an event handler. If a production procedure is interrupted by a machine malfunction and a corresponding error notification is created, you can monitor the progress of error elimination. BW integration enables you to access statistics on the frequency of production interruptions and thus to evaluate the reliability of production facilities.

5.1.8.5 RFI-Enabled Transport Packaging Tracking

If your company uses returnable transport packaging and RFID technology (*radio frequency identification*) to label and identify containers, you can employ the *RFID-enabled returnable transport item (RTI) processing*. This process works with SAP Event Management as well as the Auto-ID Infrastructure, which serves to evaluate the label or tag data supplied by the RFID reader (see also Sect. 5.2, "Auto-ID Infrastructure and Object Event Repository"). The process starts with the description of the RFID tags and includes the events loading, unloading at the customer and return from the customer.

5.1.8.6 EM Standard Processes for SAP TM

The standard issue of SAP Transportation Management offers the possibility of having SAP Event Management perform status tracking for important process steps. This support can be activated for individual business objects or process steps, and can be controlled in a higher granularity using document types.

Event Management and the interface between it and SAP TM can be configured and expanded, such that practically any standard status tracking scenario can be mapped. The standard version of SAP TM offers five scenarios and integration points:

5.1.8.7 SAP TM: Order Status Tracking

Shipping order status tracking, the first of these processes, provides information to the sold-to party regarding the current processing status. In order status tracking, no planned events are defined; instead, the current events related to status changes of the shipping inquiry are entered.

5.1.8.8 SAP TM: Shipment Tracking

To track the status of an individual TM shipment, *shipment tracking* – the second important process – is done on the basis of shipment data, and begins as soon as a shipment has received the status *Ready for shipping*. The planned transportation activities of the shipment are then used to generate the expected events for the arrival of the carrier, loading and unloading, etc. The events loading, unloading, departure and arrival can occur more than once, depending on the number of shipment stages and transfer points a shipment has (see Fig. 5.8).

A reported event for the expected events leads to the generation of an executed transport activity in SAP TM. Unexpected events can include late deliveries, customs receipt and issue, load splitting and blocking.

Figure 5.8 shows the interplay of transport request status tracking and shipment tracking. Figure 5.9 illustrates the process integration between the shipping process in TM and shipment tracking in Event Management.

5.1.8.9 SAP TM: Vehicle and Transport Equipment Tracking

The third important process is *vehicle and transport equipment tracking*. The tracking process is triggered based on a TM resource, that is, as soon as a resource is created in the master data that is relevant to tracking and tracing (such as a container), a respective event handler is generated. The event handler for the resource has no expected events, but can register an extremely large number of unexpected events throughout a lifecycle. Events that are traced include departures, arrivals, damage and the allocation and decoupling of tours.

5.1.8.10 SAP TM: Tour Status Tracking

Tour status tracking, the fourth process, aids in tracking tours from the perspective of punctuality and the planned tour steps. Its goal is to inform business partners and service providers of the status of a tour. Based on planned transport activities, all trip and arrival events of a tour are determined and set in the tour event handler as expected events. Unexpected events that can be registered in this regard include any delays reported by the carrier, blocks and any reported omissions of locations (such as port omissions).

5.1.8.11 SAP TM: Automatic RFQ Tracking Process

The last important process is the *RFQ tracking process*, pertaining to requests for bid quotations. It serves to automatically monitor the bidding procedure and track reactions and reply times of service providers for the tendering of shipment requests.

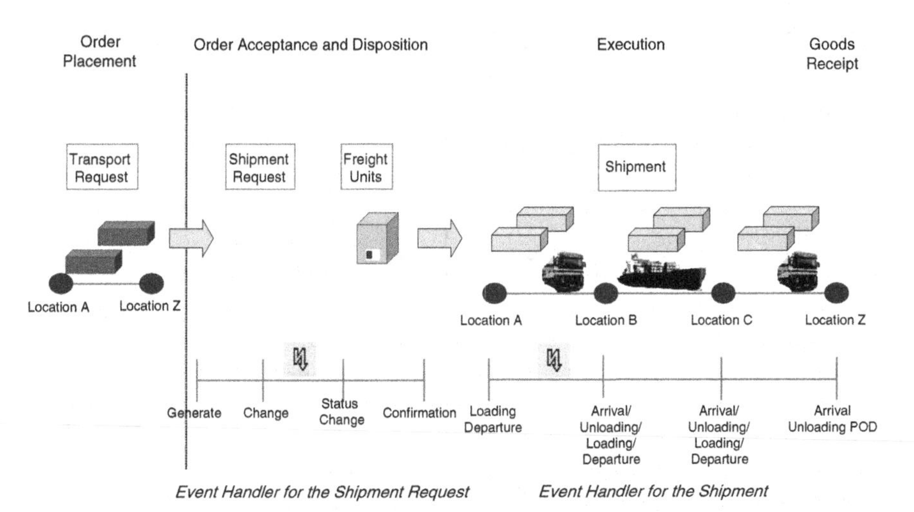

Fig. 5.8 Standard shipment tracking process (SAP transportation management and SAP event management)

The event handler for the tendering process is generated individually from each request for supplier quotations, meaning that every service provider to whom a quotation is allocated is tracked separately. The momentary RFQ status is registered, the bidding process is automatically terminated at the set deadline, and the reaction behavior toward new tenders of the queried service providers is evaluated.

5.1.8.12 Expected Events and Event Groups

The event handler for the tracking of RFQs contains the following expected events:

- Transmission of the request for quotation to the service provider:
  Generation of a document that confirms that the request has been sent.
- Receipt of a reply from a service provider:
  The reply receipt is an event group that can have four event characteristics: *Quotation acceptable*, *Quotation not acceptable*, *Review required* (changes

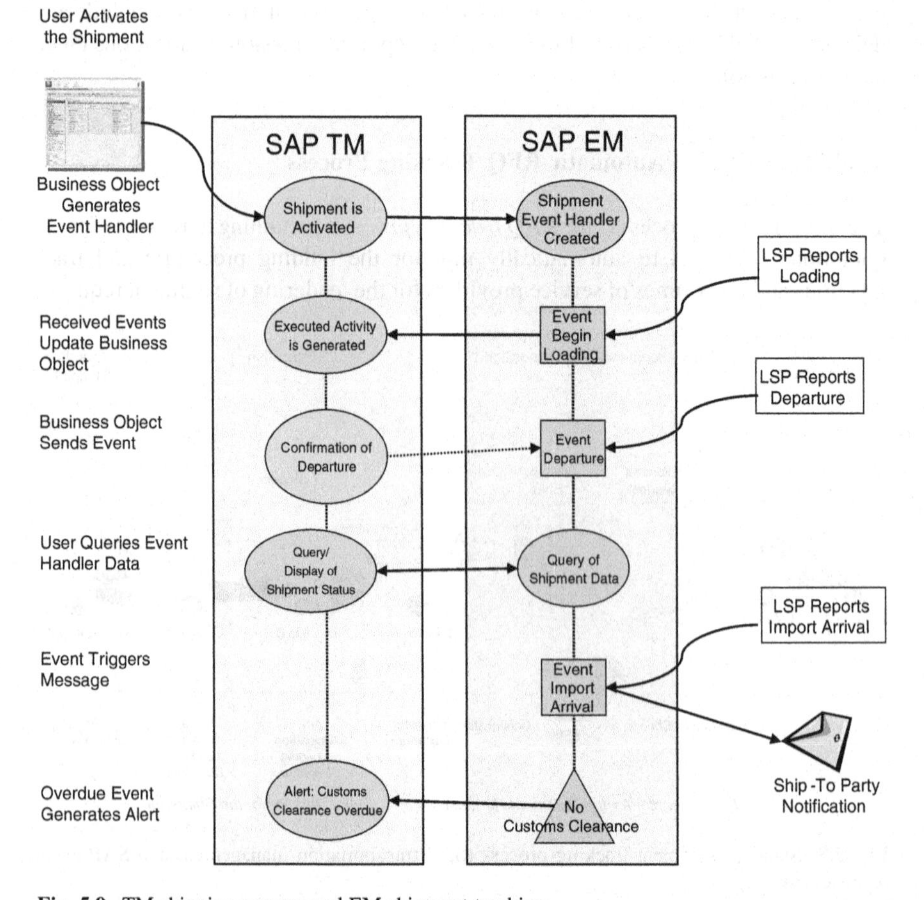

Fig. 5.9 TM shipping process and EM shipment tracking

made by the service provider) and *Rejected* (no answer from the service provider); each of these four events leads to the fulfillment of the expected receipt for that event group.

- End of reply receipt:
  This event is checked by an alert monitor for expected events; if up to that point no reply has been received, the RFQ is automatically with-drawn.
- End of RFQ (occurs automatically after a period of 2 weeks if no service provider has been commissioned based on the RFQ).
- Quotation selection:
  If several quotations are received, the operator can select the most economical option.
- Acceptance of the quotation and order placement with the service:
  The operator confirms the selected quotation with the service provider.

The ability to individually control each quotation enables the bid invitation process to be conducted with several service providers (broadcast) and set varying deadlines for individual service providers. This lets you "reward" reliable or preferred service providers with an extended deadline. If a deadline for a tendering procedure has passed and a quotation has not been accepted or accepted with changes, SAP TM automatically starts the next planned tendering procedure for the shipment or freight order.

Figure 5.10 shows an overview of two RFQ event handlers. For the first EH, a reply from a service provider has not yet been received. That is why the event list

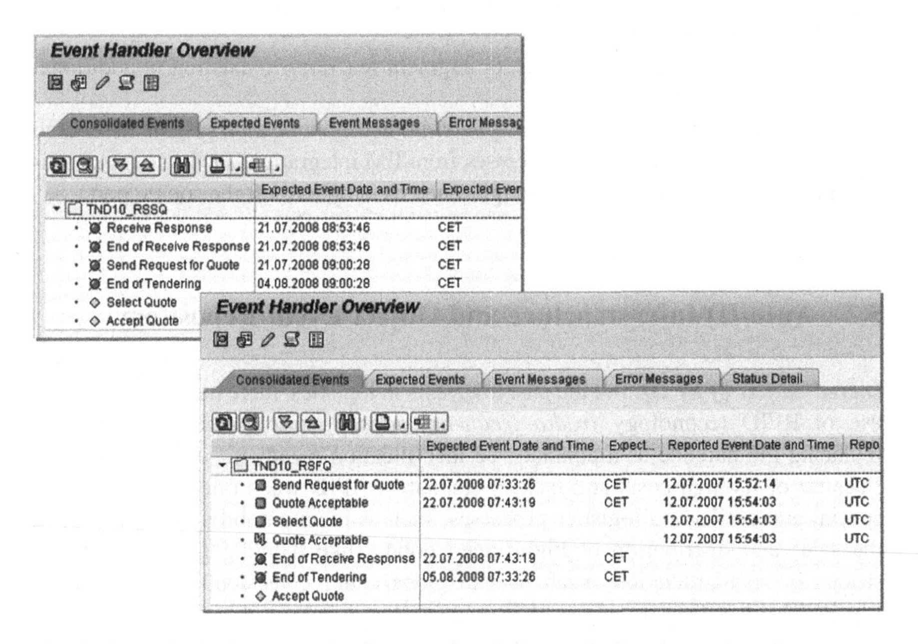

Fig. 5.10 Bid invitation event handler overview before and after the receipt of a response (Accept)

still displays the event group Receive Response. For the second event handler, the service provider has already accepted, so the event group has been replaced by the actually reported event Accept.

5.1.8.13 Constructing Your Own Tracking Scenarios

You can expand the standard tracking processes between the SAP application systems (such as ERP or TM) and SAP Event Management and construct new scenarios with Event Management. In addition, existing scenarios can be expanded by a customer's own process steps to achieve a higher degree of automation.

The following expansion options are available:

- *Expansion of business process types* for the transmission of previously unprocessed application data to the application interface.
- Definition of *additional application object types* for tracking objects of a previously unmonitored type (such as a refrigerated container in a shipment or a packing unit of the type roller container in a shipment order for express goods)
- Definition of *additional extractors and relevance conditions,* so that you can extract other or more extensive data from business project objects of application systems, in order to fill parameters or expected events in Event Management; you can also access the data of other objects.
- *Your own event handler type*, to enable you to define your own complete visibility processes that represent the characteristics of a monitored object or process and save the related data.
- *Expansion of rule sets,* to change or augment reactions to existing or additional events.
- *Expansion of EM-triggered control functions* in the application system, in order to be able to control additional processes from EM integration, such as to automatically generate invoices for sea freight once the ship has left the source port.

5.2 Auto-ID Infrastructure and Object Event Repository

Increased Use of RFID: For the past few years in logistics, there has been increased use of RFID technology (*radio frequency identification*), which is gradually replacing the barcode as a contact-free identification means in several processes. Because of its high costs and lack of standards, RFID was initially only used in special production and logistics processes, such as in automobile production or in the sales and distribution of high-priced goods (like hanging garments). Great strides in standardization (such as EPCglobal) and consistently falling prices for RFID identification chips (RFID tags) have led to this technology also being employed for bulk commodities (for instance, in the retail branch). In the realm of access control, many people have already had direct contact with RFID systems – for example, when using ski lift passes. With its *Auto-ID Infrastructure* and Event-Management-based *Object Event Repository*, SAP offers you two efficient software components to support logistics processes using RFID technology.

5.2.1 Basics of RFID and EPC Technology

5.2.1.1 Advantages of RFID Technology

RFID technology has a few significant advantages in comparison to the widely used barcode system:

- **Tags do not have to be visible**
  RFID enables the identification of logistical units for which, in contrast to a barcode, no direct visibility of the RFID is required. The tag can also be read from any direction.
- **Several tags can be read simultaneously**
  Efficient anti-collision algorithms enable rapid, virtually parallel reading of a large number of RFID tags.
- **Tags are scannable from within packages**
  RFID tags can be used in rough environmental conditions. Unlike a barcode, the tag can also be integrated in packaging to prevent damage.
- **Tags can be inscribed**
  RFID tags can be read as well as inscribed. This enables data enhancement on the tag within the course of the logistics process. The amount of data that can be stored on an RFID tag is relatively large.
- **Reusability**
  RFID tags can be reused.

These characteristics make it profitable to use RFID technology along the entire logistics chain. For example, a product can be supplied with a tag as early as the production phase. Warehousing, transportation to distribution centers, subsequent picking and placing, and sales and distribution can be completely recorded by scanning the RFID tag. This allows you to obtain precise data on the logistics chain.

5.2.1.2 Components of RFID Technology

An RFID system consists of several components that have to work well together. Figure 5.11 shows the basic structure:

- **RFID tags/transponder/chips**
  An RFID tag in its simplest form consists of an antenna coil, a chip and a permanent memory unit. The antenna coil receives radio waves from a reader antenna and thus produces electricity that "wakes up" the chip and makes it upload

the contents of its memory and transmit it in coded form via the tag's antenna. This "answer" from the tag can be received by the reader antenna and relayed.

There are various types of tags. Some of the characteristics named above can be combined in a tag:

- *Passive tags* do not have their own current supply and are fed by the electricity produced by the antenna, as described above.
- *Active tags* have a battery or rechargeable battery and do not need an outside source of current. They are larger than passive tags and are primarily used for the identification of containers and railway cars.
- *Read-only tags* have a programmed identification code in a read-only memory (such as an EPC, or Electronic Product Code), which can be scanned as often as needed.
- *Write-once tags* do not come with an identification code, but have a code programmed into their write-once memories when their identification is defined (for example, when a serial number is issued for a device to which the tag is attached).

Reusable tags can be used repeatedly, that is, deleted and written over with new data. The data can also be enhanced during the process, so that for each process step, the progress can be recorded as data on the tag. The memory used for this purpose is similar to the flash memory of a USB stick.

- **Reader antenna**
  The antenna or antennas of the reader transmit a signal in a frequency synchronized with the tag that stimulates the tag to transmit identification and other data. After the read signal has been sent, the antennas switch to reception mode so they can receive the tag's response. There are several frequency ranges around the world in which RFID antennas operate. Because of dissimilar, country-specific laws, it has not yet been possible to achieve complete synchronization. In the realm of logistics, antennas in the decimeter wave band are used

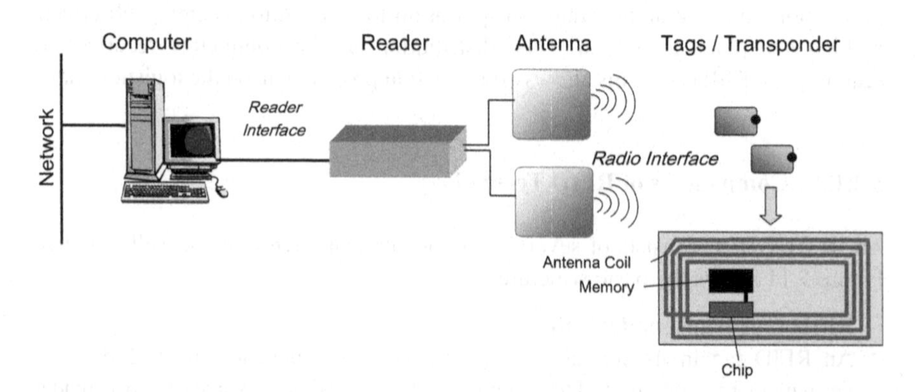

Fig. 5.11 Components of an RFID system

(433 MHz, 869 MHz, 915 MHz or 2.45 GHz). In addition to legal regulations, frequency selection also strongly depends on the operational environment (for example, strongly metallic environments such as storage racks or the presence of shielding objects such as filled chemical canisters).

- **Reader**
  The reader activates the antennas and decodes the data sent back by the tag. It must know the initialization sequence required by the tag to send data back, and it must understand the code in the data transmitted by the tag and convert it to the proper data format required by the underlying logistics system. Reading devices are available as permanently installed standing, wall or gate units. For mobile applications, there are also a number of RFID hand readers that are at least equipped with a display.

5.2.1.3 Specialized Systems

The component characteristics discussed above highlight the fact that despite improvements, the use of RFID technology in logistics is challenging. The individual components, operational environment and material characteristics of the objects to be identified need to be synchronized for successful implementation.

5.2.1.4 EPCglobal

To achieve a logistics standard in identification processes using RFID, the Auto-ID Center was established in the United States (and is now known as EPCglobal, see www.epcglobalus.org), which promotes RFID tag standards and has defined a standardized identification code, the *Electronic Product Code (EPC)*.

5.2.1.5 Electronic Product Code (EPC)

The EPC was developed to create a means of identifying all kinds of physical objects. Of course, it was also designed to include coverage of popular industry norms such as EAN-UCC (*European Article Number/Uniform Code Council*), GTIN (*Global Trade Identification Number*) and SSCC-18 (*Serial Shipping Container Code*). Right from the start, the code was conceived as being expandable, to be able to cover future applications. The Auto-ID standards are based on a minimal amount of data on the chip (only the EPC code) and centralized or distributed storage of the data and information linked to the physical object on server systems (called EPCIS, *Electronic Product Code Information Systems*). This enables you to keep costs for RFID tags as low as possible, making mass application reasonable in terms of price. Figure 5.12 shows an example for a 96-bit EPC.

5.2.1.6 EPC Components

The EPC comprises several components:

- **Code type (header)**
  Classification of the EPC version and information type (such as SGTIN: *Serialized Global Trade Identification Number*), definition of the unit type (such as pallet code) and total code length
- **EPC manager**
  Assigned EPC member number of the number generator, such as a food product manufacturer
- **Object class**
  Object number, such as the item number of a product from the food product manufacturer named above (for example, honey-glazed ham slices)
- **Serial number**
  Identification number of each individual object, such as a single package of honey-glazed ham slices

This structure enables the logistics process to track and identify each item. If, for instance, the honey-glazed ham slices are located in the refrigerated section of a supermarket and scanned with an RFID reader, the EPCIS would allow you to access the entire lifecycle of that package of meat. This facilitates very efficient quality control efforts.

5.2.2 *Auto-ID Infrastructure*

5.2.2.1 Link Between Logistics Applications and RFID Tags

Between the world of process-oriented logistics applications and that of identification- and event-oriented RFID tags, a discrepancy exists with regard to the granularity in which process steps are carried out.

RFID reading processes have the following basic characteristics:

- Large amounts of real-time data with low data content (for instance, only the identification number) are generated and transmitted by readers.

Fig. 5.12 Example of an EPC

- A lack of contextual knowledge about the object or overall process: An object goes through an isolated process step. For the scanning process, no contextual knowledge is available about other connected objects (such as which other cartons belong to a delivery).
- Reading procedures are insufficiently standardized: Several tag and encoding variants can hamper data interpretation.

The following problems must be addressed in logistics applications:

- There is a certain degree of inability to interpret large amounts of very detailed real-time data: A delivery with a pallet of 1,000 yogurt cups may produce performance problems if each individual cup is posted to the goods issue via RFID.
- Many business processes do not require data on the RFID-tag level: For the above delivery, it is only important to know when the pallet was posted to goods issue and how many cups were on it.
- Several business process objects are not designed to process and store detailed information and a history of all logistics objects used.
- There are no direct interfaces to RFID readers available.

Based on these discrepancies, the Auto-ID Infrastructure was developed that builds a bridge between the granular event processing of the RFID world and the bundled view of logistics applications. Figure 5.13 illustrates this issue.

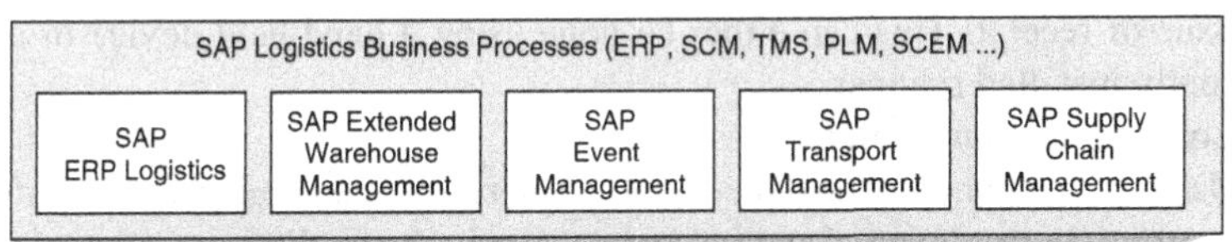

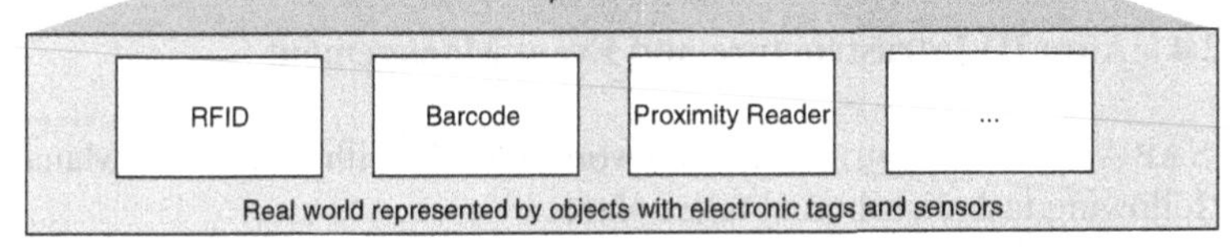

Fig. 5.13 Auto-ID Infrastructure as a bridge between logistics applications and electronic object identification

5.2.2.2 Process Examples of RFID Use

In the realm of logistics, there are a few typical processes in which the use of RFID technology has proved especially effective. These include:

- **Handling unit packing and unpacking**
  In this process, a handling unit is generated with the aid of RFID technology. By scanning the individual item objects and the container, the objects are allocated to a handling unit. A logistics system is thus capable of generating the handling unit based on the RFID information. The Auto-ID Infrastructure collects the information regarding which object is packed into which container and then creates a handling unit with content in the SAP system.The process can also be performed if the individual items are directly supplied with tags on the production line. This occurs frequently in the pharmaceutical industry, where expensive medicine packages are tagged and placed into a container.
- **Content verification of an HU using RFID**
  A known handling unit is checked using an RFID scan to verify its original condition. Without having to be opened, the medicine carton mentioned above can be checked to verify whether all individual packages are still in the box. This process is often used for quality and theft control.
- **Loading and unloading of items and shipping containers**
  The RFID scan can provide you with a complete overview of the loaded or unloaded items and shipping containers and automatically post them to goods issue or receipt. This can either be done using a hand-held device or a permanently installed scanner.
- **Loading sequence check**
  Using a permanently installed gate reader, you can immediately inform the appropriate employee of any inconsistencies in the loading sequence of outgoing pallets and correct the problem.

5.2.2.3 RFID Support in the Warehouse

Figure 5.14 shows an example of a warehouse process with RFID support. In this case, the RFID technology is used on a carton and pallet basis, but not on an item basis. The illustration shows which logistics objects are processed, how RFID support is used, what activities are performed and the tag coding employed.

5.2.2.4 Auto-ID Infrastructure and Event Management

The SAP Auto-ID Infrastructure (AII) works closely with SAP Event Management. The following task distribution takes place:

- SAP Event Management is oriented toward the business process. It undertakes the company-wide collection of visibility and tracking data, the monitoring of

complete and distributed processes, and process monitoring with exception handling.
- SAP Auto-ID Infrastructure is closely oriented toward individual actions with physical objects. It undertakes the local data processing for identification data stemming from RFID readers and other sensors. It also supports data filtering and pre-consolidation, and can issue a direct message to operators if scans lead to unexpected results.

5.2.2.5 Data Processing Using AII

The Auto-ID Infrastructure can be supplied by logistics systems with details on the expected RFID data. The expected data then can be consulted for reference to compare with the scanned data. This enables the determination of whether or not there are deviations from the expected objects. If there are deviations, the system can directly inform the user of the RFID reader.

Figure 5.15 shows a process in which a supplier submits notification of an incoming delivery of a pallet containing 16 cartons with a total of 256 serialized items (1).

All units are equipped with RFID tags. The supplier electronically supplies inbound delivery details with precise RFID tag information (SSCC number for the pallet, SSCC numbers for the cartons, SGTIN for the items. An event handler is created for the pallets via the inbound delivery generated in SAP ERP, and the detailed, expected RFID tag information is transmitted to and saved in the Auto-ID Infrastructure (2).

Once the initial scan of the pallet has been performed, the exact tag information is sent by the reader to AII (3). Here, a comparison of the received information

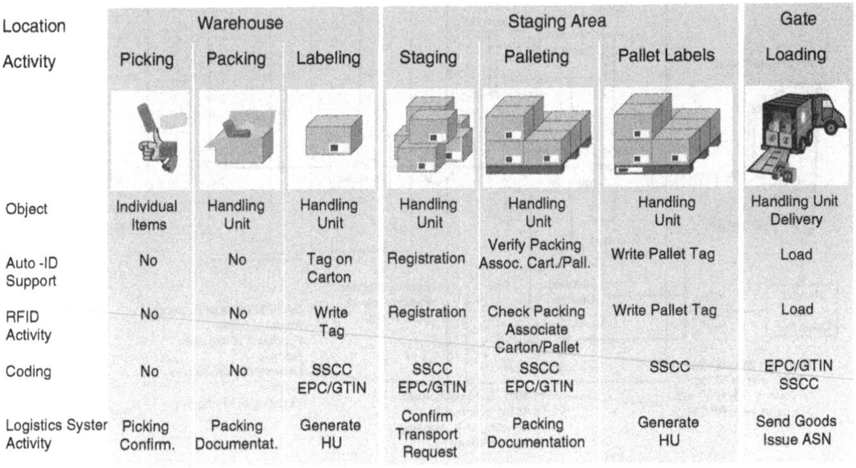

Location	Warehouse			Staging Area			Gate
Activity	Picking	Packing	Labeling	Staging	Palleting	Pallet Labels	Loading
Object	Individual Items	Handling Unit	Handling Unit	Handling Unit	Handling Unit	Handling Unit	Handling Unit Delivery
Auto -ID Support	No	No	Tag on Carton	Registration	Verify Packing Assoc. Cart./Pall.	Write Pallet Tag	Load
RFID Activity	No	No	Write Tag	Registration	Check Packing Associate Carton/Pallet	Write Pallet Tag	Load
Coding	No	No	SSCC EPC/GTIN	SSCC EPC/GTIN	SSCC EPC/GTIN	SSCC	EPC/GTIN SSCC
Logistics Syster Activity	Picking Confirm.	Packing Documentat.	Generate HU	Confirm Transport Request	Packing Documentation	Generate HU	Send Goods Issue ASN

Fig. 5.14 Example of an RFID outgoing delivery process in the warehouse

with the previously saved, expected information can be performed (4). The Auto-ID Infrastructure can inform warehouse workers immediately if the number of scanned objects deviates from the expected amount (such as in a shortfall) or if other items or serial numbers have been delivered than those indicated in the notification.

If the details of the initial scan are correct, the Auto-ID sends a message such as "Pallet received, status OK" to Event Management (5). This establishes the relationship between the 265 individual scans at the gate reader and the business process. The event handler confirms the correct arrival of the pallet (6) and can initiate a goods receipt in ERP (7).

5.2.3 Object Event Repository

5.2.3.1 EPC Information System

Within the context of RFID and EPC technology, we have already demonstrated that with the concept of the central EPCIS (Electronic Product Code Information System), a reduction in costs for the RFID tags can be achieved by storing context and history data regarding the physical object on a central server and making it available there.

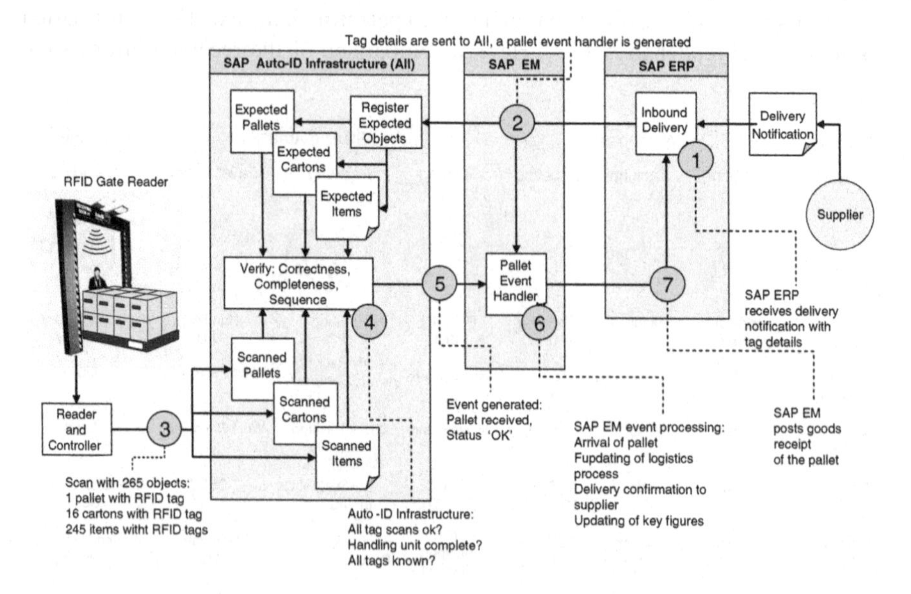

Fig. 5.15 Process for goods receipt of an RFID-assisted inbound delivery

EPCIS is thus a standard to electronically track identifiable objects along the supply chain. The data regarding the what, when, where and why of an object at the individual reading stations is collected centrally and can be queried and distributed via standardized interfaces and services.

5.2.3.2 EPCIS Based on SAP Event Management

The functions and characteristics are a basic component of SAP Event Management. As such, it seemed only logical to expand Event Management in the direction of EPCIS. With its *Object Event Repository* (OER), SAP offers a solution certified by EPCglobal and corresponding to EPCIS Version 1.0 for the implementation of EPCIS based on SAP Event Management. This solution enables you to seamlessly track hierarchically nested objects as well as to authenticate objects in critical processes.

Proof of process for sales and distribution

One example of this is the seamless proof of process for drug production and distribution, which has been or is slated to be implemented in several countries and which, among other points, is designed to support the identification of counterfeit drugs (see Fig. 5.16). High-quality drugs are already supplied with RFID tags on every package during production. This allows the in-warehouse storage as well as transport to wholesale and retail facilities to be tracked and documented. In practical terms, it would also be possible to access the exact history of a medication purchased in a pharmacy and thus be able to detect a counterfeit.

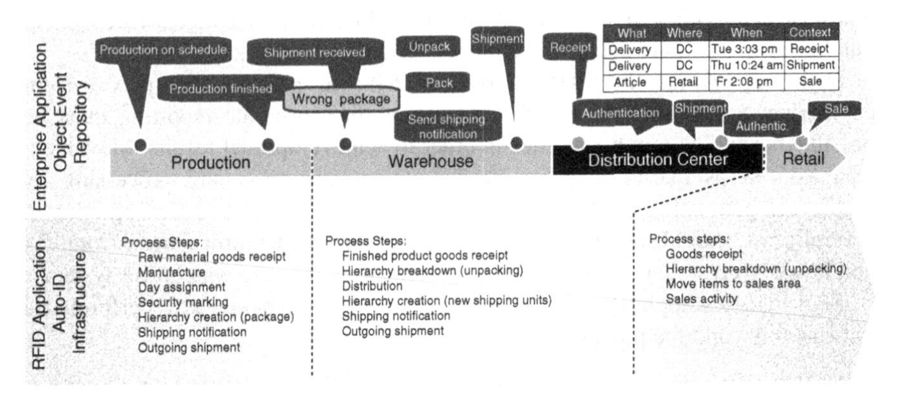

Fig. 5.16 Example of the seamless proof of process in the production and distribution chain with the SAP object event repository

SAP Object Event Repository (OER) offers standard event, collection and query interfaces in correlation with the EPCIS norm in order to collect information on explicitly identified objects and exchange it with other involved partners. In this process, the entire lifecycle of an object can be recorded. If objects are hierarchically packaged, this information can be recorded in a hierarchical display or relational context in OER. The following relationships are possible: simple EPC numbers, tag identification EPC numbers or a pallet-carton-item relationship.

5.3 Reporting and Determining Key Performance Indicators

Economic background of key performance indicators: The components of SAP logistics solutions accumulate a large amount of various data in the business processes in order to meet the demands of diverse parties including one's own company, business partners, freight carriers, official administrations and customs. Much of the data is determined automatically or forwarded via electronic communication. The data volume that is generated in a logistics process can thus be very high. It is not always easy for companies and individual end users to come to conclusions based on the data with regard to the success or failure of a process execution.

Competitive edge: Companies often have a decisive edge over their competitors if they can structure their logistics activities profitably and effectively on the basis of detailed analyses. Data from the past can help you learn lessons for the future and change business structures accordingly. Existing relationships and contracts with customers and suppliers can be newly negotiated based on determined performance and financial data, or processes or business areas can be discontinued.

Reporting in logistics spans details from the structural data analysis in day-to-day business to complex analyses of individual cost components to the determination of *key performance indicators (KPI)*, used for corporate management. Knowledge of this data and access to the data of operative logistics, tracking, financing and purchasing systems are decisive factors for the generation of key indicators.

Data aggregation in SAP NetWeaver BW: In practice, complex system environments can often be found. This means that corporate reporting must be possible beyond system boundaries in order to achieve optimal results. Movement data as well as master data are needed. Data integration and processing by employees generating fly Excel spreadsheets, which is still standard practice in several firms, should soon be a thing of the past; this manual procedure is complex and prone to error, and the task can be performed by the data warehouse system.

SAP logistics applications use two components to generate key performance indicators beyond the processes:

- SAP NetWeaver Business Warehouse (SAP NetWeaver BW), with which you can generate complex analyses and reports.
- The Information System, which offers limited analyses directly in ERP.

In addition, SAP BusinessObjects products are a new component in the SAP Portfolio that can present the data from a data warehouse and other data sources in report or dashboard form.

5.3.1 SAP NetWeaver Business Warehouse

Using SAP NetWeaver Business Warehouse, data from SAP systems, non-SAP systems and structured files (such as Excel spreadsheets) can be processed. Display of the data is either done in a spreadsheet structure or as a conclusive dashboard with signal lights, speedometer graphs or diagrams. For the spreadsheet depiction, you can also select a Microsoft Excel display.

5.3.1.1 Current Data at the Push of a Button

Through SAP NetWeaver BW, the authorized personnel in a company have access to current data at the push of a button to make decisions and control day-to-day business accordingly.

5.3.1.2 New Possibilities

BW consists of the actual data storage with its administration tools for data definition, updating, aggregation and queries as well as the BW content that defines data storage and queries based on content. The BW content is application-dependent and is generally provided as a release for specific application components. In Fig. 5.17, you can see a view of a part of the InfoProvider supplied with SAP from the logistics realm.

The latest technologies like SAP BusinessObjects tools or the SAP NetWeaver Visual Composer use encapsulated Enterprise Services to extract data from various systems and depict them in an easy-to-configure user interface. Reporting is thus no longer programming work, but rather a business process that can be executed by almost any employee.

5.3.2 The Information System in SAP ERP

The information system in SAP ERP can be considered a precursor to SAP NetWeaver BW and serves the statistical processing of logistics data in a pure ERP environment. Because the information system is relatively limited in function and from an architectural perspective is no longer entirely current, we only mention it here briefly. For company-wide reporting, the trend is clearly toward BW/Business Intelligence.

5.3.2.1 Different Characteristics

The information system consists of several individual applications in SAP ERP:

- The Logistics Information System (LIS) generates statistical overviews for warehouse, sales and distribution and transportation processes.
- The Sales Information System (SD-IS) generates statistical overviews for sales.
- The Purchasing Information System (PIS) generates statistical overviews related to purchasing.

5.3.2.2 LIS Standard Analyses

Using LIS as an example, Figure 5.18 shows the basic logistics key figures and overviews that provide the standard analyses.

5.3.2.3 Key Sales and Distribution Indicators

Figure 5.19 shows an example involving the order activities per sales office with a breakdown of weekly key figures for the Hamburg office. In addition, it illustrates the key weight and volume indicators for sales and distribution activities in the shipping points.

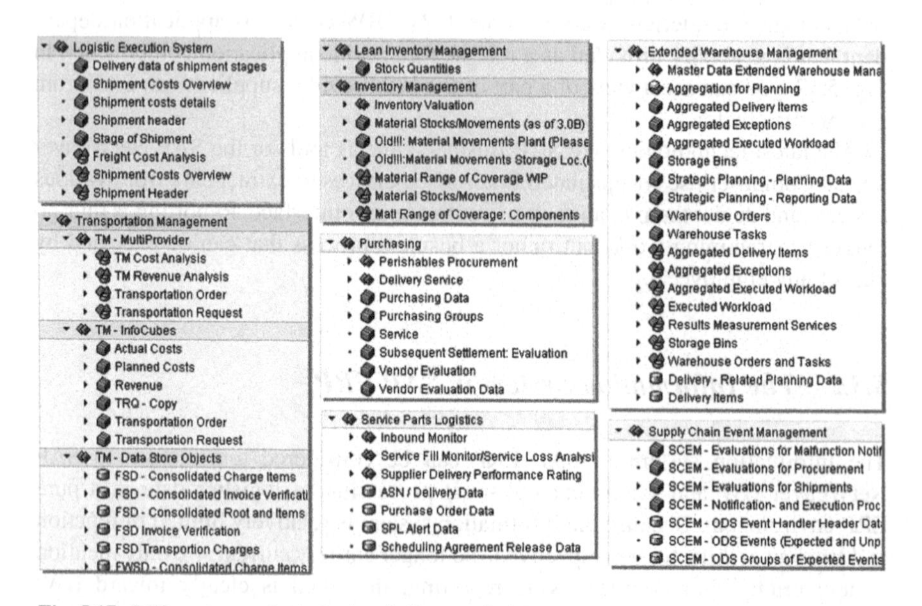

Fig. 5.17 BW content on logistics key indicators (InfoProvider)

5.3.3 The SCOR Data Model

The SCOR model (*Supply Chain Operations Reference*) defined by the Supply Chain Council (see http://www.supply-chain.org), contains a series of significant

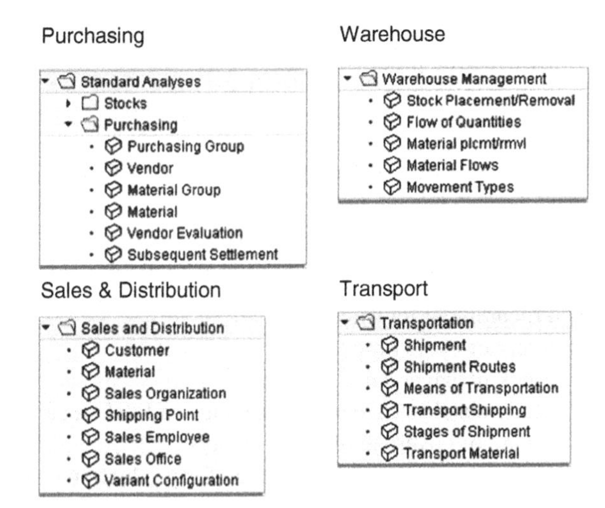

Fig. 5.18 Standard analyses of the logistics information system

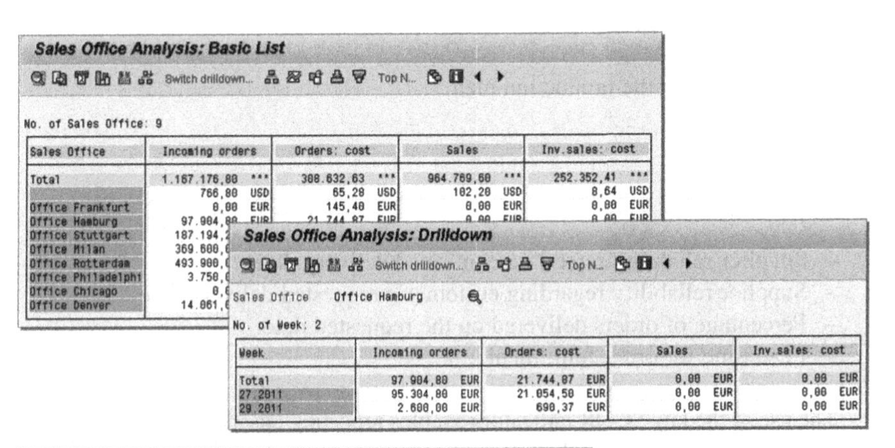

Fig. 5.19 Standard analyses of sales activities per sales office and of the activities of the shipping points

key figures that supply important data from the entire logistics chain. These key performance indicators enable an assessment of logistics efficiency, from suppliers to production to customers, in the following areas:

- Logistics chain planning
- Procurement processes
- Manufacturing processes and stock management
- Distribution processes
- Return processes

5.3.3.1 Key SCOR Indicators in the BW Content

SAP comes installed with standard BW content for the SCOR model. In the areas cited above, the following reports are available:

- Procurement logistics
- Supplier cycle time
- Stock management
- Stock obsolescence as a percentage of total stock
- Range of coverage of stock in days
- Raw material range of coverage for production
- Manufacture
- Capacity utilization
- Production yield
- Compliance with the production plan
- Production cycle
- Costs per unit
- Overhead costs
- Order fulfillment
  - Supplier reliability regarding confirmed delivery dates
  - Supplier reliability regarding customer's requested delivery date
  - Percentage of orders delivered on the requested date
  - Lead time for order fulfillment

The use of the key SCOR indicators enables reporting based on standardized and thus comparable definitions that also have cross-company significance.

5.3.4 Reports and Dashboards: Examples

5.3.4.1 Simple Interpretation of Complex Data

Reporting is a sweeping term, spanning all corporate divisions. In order to arrive at a simple interpretation of frequently complex data, programs have begun using *dashboards*, where "instruments" provide a visualization of the most important key indicators. Parts of the following examples come directly from the Transportation Management or Event Management system, and parts are based on data taken from the BW realm.

5.3.4.2 Example 1: Driver Delays

The first example (Fig. 5.20) shows a typical analysis from the freight forwarding environment. It shows instances of delays per driver and truck so that delivery precision can be monitored.

An analysis of this type requires the seamless tracking of departure and arrival times, which can be achieved with the aid of SAP Event Management and the respective on-board computer and integration of the vehicle's geographic coordinates (geofencing). This data can be provided to a data warehouse and combined with operative data from Transportation Management. A detailed view, as shown in the lower right of the illustration, is a possible result.

5.3.4.3 Example 2: Fuel Consumption

The second example (Fig. 5.21) show similar data with reference to the fuel consumption of a vehicle fleet. In this example, too, not all data comes from a

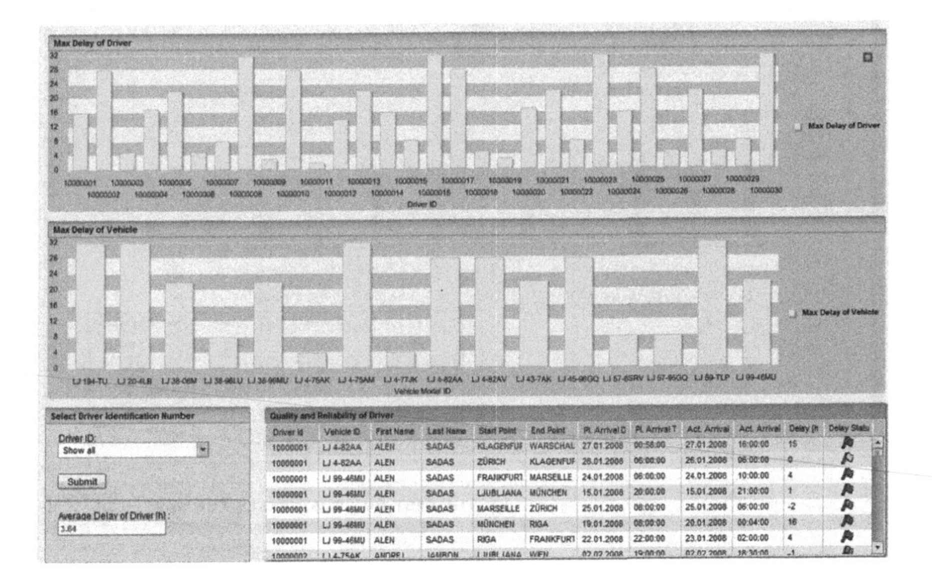

Fig. 5.20 Dashboard for the tardiness of truck drivers

TM system, but also from other systems such as a fuel management system or on-board diagnostic application that sends data from the vehicle to a central server, where it is extracted and forwarded to BW.

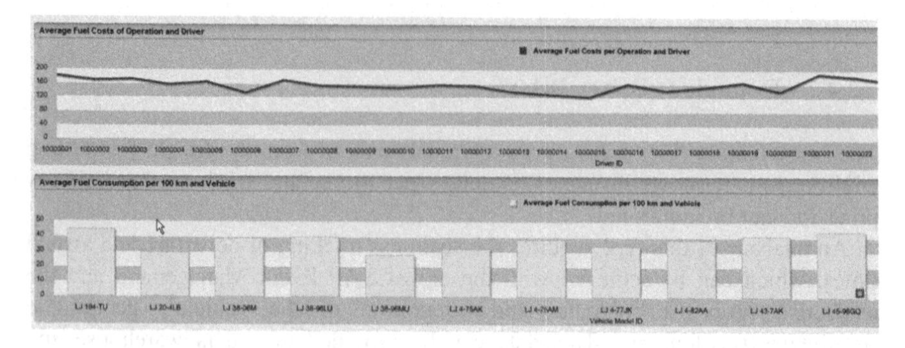

Fig. 5.21 Dashboard for fuel consumption per vehicle

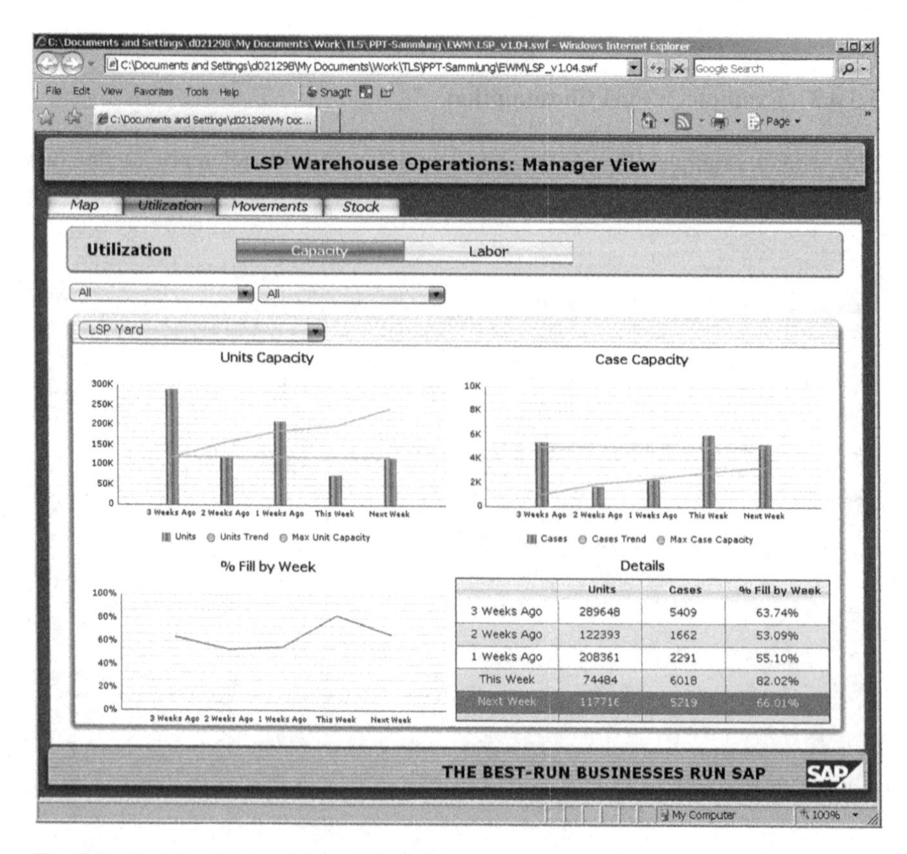

Fig. 5.22 Xcelsius dashboard for contract logistics service providers

5.3.4.4 Example 3: Warehouse Management

The third example shows a dashboard created with SAP BusinessObjects Xcelsius, which offers a manager view of the most important warehouse management data of a contract logistics service provider. The Xcelsius application shown in Fig. 5.22 can be easily integrated in Web applications and sent as a PDF without losing dynamic operability.

5.3.5 Data Extraction from SAP Logistics Applications

5.3.5.1 BW Communication Process

After examining the application examples and economic background of key performance indicators and reporting, we will now conclude with a look at the integration of logistics applications with SAP NetWeaver BW.

Figure 5.23 shows the principle process. When SAP logistics applications run either via the user interface or through planning or batch reports, a data backup generally takes place before the end of the transaction. Once the backup is completed, that is, when the transaction is obviously finished and new data has been updated, an extraction of data relevant to characteristic and key figure determination is performed.

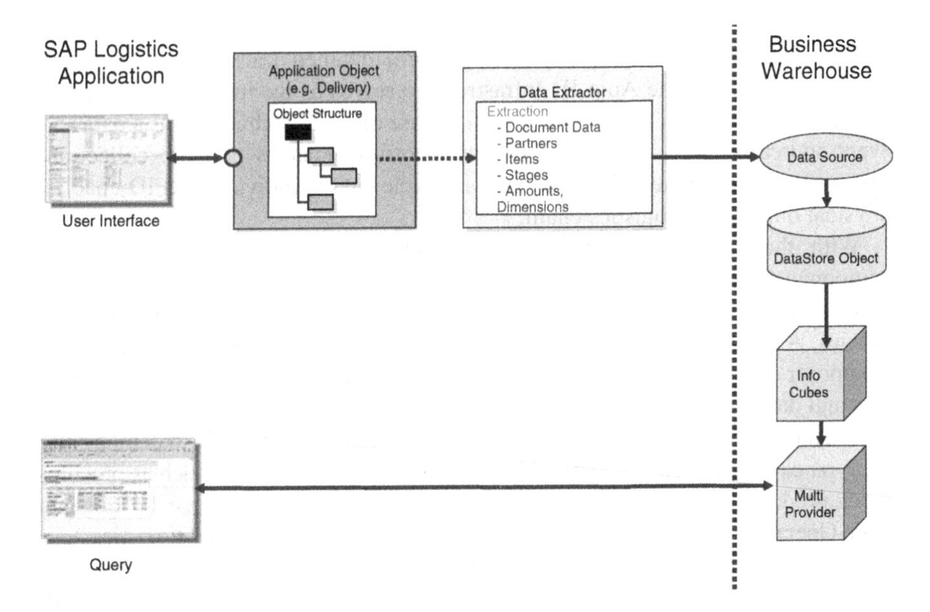

Fig. 5.23 Process of integration and logistics applications and BW

For this, depending on the system components, various technical procedures are used (determined by the individual programs). Data extraction and thus updating of the BW, however, can be switched off for each application in SAP Customizing. The data extraction method is tailored to the application object. The extraction method then puts the data in a DataSource for transfer to BW.

In the Business Warehouse, DataSource information is stored in DataStore objects. From these DataStore objects, the InfoProviders (formerly known as InfoCubes) are provided with information. The InfoProviders save the actual characteristics and key figures. They, in turn, are used by Multi Providers as a data source.

5.3.5.2 Queries

After placement of data in a MultiProvider, you can access information in the Business Warehouse with various queries and combine this information with other available data to perform targeted analyses. These analyses can be displayed in spreadsheets, reports or dashboards.

5.4 Summary

In the SAP world, you can achieve efficient and automated control of logistics processes by utilizing the components Event Management, Auto-ID Infrastructure and Object Event Repository in a form tailored to your processes.

Event Management enables you to perform detailed tracking of status and logistics processes. The Auto-ID Infrastructure serves as the interface of logistics applications and Event Management with the use of RFID technology as a data and event source. Finally, the Object Event Repository enables efficient collection, storage, management and distribution of lifecycle data of individual RFID-enabled physical objects in a logistics chain.

With the aid of the SAP logistics applications and the process control components described above, you can extract a variety of characteristics and key figures that are updated in the BW or information system and made available for queries. A number of visualization methods (from Excel to BW reports, to Visual Composer dashboards to the flexible SAP BusinessObjects tools) facilitate a depiction and data assessment to fit any target group.

We have now reached the end of this book. We hope we were able to demonstrate that you can use the components of SAP Business Suite from areas such as procurement, production, sales and distribution (examined in detail in Volume 1, "Logistic Core Operations with SAP: Procurement, Production and Distribution Logistics"), as well as transportation and warehouse management – the focus of this volume – to seamlessly map a diversity of logistics processes in information systems. SAP Business Suite is thus an ideal basis for system-supported logistics execution.

Glossary

ABAP The programming language of SAP with which most logistics applications are programmed.

Activity area A special feature of SAP EWM representing a logical grouping of storage bin locations with regard to planned warehouse activities.

ALE Application Link Enabling. Technology that establishes and operates shared applications.

APO Advanced Planning & Optimization. SAP APO software contains functions for processing and integrating sales, distribution and production planning, as well as production control and external procurement. SAP APO also offers functions for collaboration with external vendors and their integration in procurement processes (see also VMI).

APO PP/DS Production Planning/Detailed Scheduling. An SAP APO module enabling the planning of production within a factory while simultaneously considering product and capacity constraints, with the goal of increasing throughput and reducing product stock. The result of the planning is a feasible production plan.

ATP Available-to-Promise. Refers to warranted stock, the quantity of a certain material that can be made available at a required date or subsequent time and can thus be used for such purposes as sales orders. The system takes into account the current stock situation and planned inward and outward movements, especially based on purchase orders, production orders and recorded sales orders.

BI Business Intelligence. Business analysis processes and technical instruments in a company used to evaluate companywide data and to provide that data to users.

Business object The copy of a document necessary for a business process in a software system (for example, a sales order in order processing).

CIF Core Interface. Interface for data transfer between an ERP system (SAP R/3 or SAP ERP) and a connected SCM system such as SAP Advanced Planning & Optimization (SAP APO) or SAP Supply Network Collaboration (SAP SNC).

Consolidation The consolidation of goods from several forwarders in a common loading unit (such as a container). Consolidation is executed by a logistics service provider.

J. Kappauf et al., *Logistic Core Operations with SAP*,
DOI 10.1007/978-3-642-18202-0, © Springer-Verlag Berlin Heidelberg 2012

CRM Customer Relationship Management. Supports all customer-related processes within the customer relationship cycle, from market segmentation, lead generation and opportunities to post-sales and customer service. Includes business scenarios such as Field Sales and Service, Customer Interaction Center and Internet Sales and Service.

CTP Capable-to-Promise. Function of the global ATP availability check in which, in contrast to ATP, not only available stock is considered, but also additional sources of requirement coverage, such as production capacities or external suppliers.

Dashboard Display of key logistics data using diagrams and other graphic elements (such as "speedometer"-like displays as on a car dashboard).

EDI Electronic Data Interchange. Cross-company electronic data exchange between business partners (for example, exchanging trade documents).

EHS Environment, Health and Safety. SAP application component for all tasks pertaining to labor, health and environmental safety in a company.

EM Event Management. SAP application component in SAP Supply Chain Management for the monitoring of logistics and other processes.

Embargo list List of people or companies to which it is prohibited to supply certain goods or services.

EPC Electronic Product Code. Code used in RFID chips for product characteristics and identification numbers that is internationally uniform.

ERP Enterprise Resource Planning. The core system with SAP business applications in the areas of logistics, human resources and finances.

Event Handler Generic object in Event Management used to track the status of processes or physical objects (such as shipments tracking).

EWM Extended Warehouse Management. New warehouse management module based on SAP SCM.

FCL Full Container Load. Transport of a full container from a forwarder to a recipient.

Freight invoice Invoice from a logistics service provider to a shipper or recipient of goods that the logistics service provider was commissioned to transport.

gATP global Available-to-Promise. Global availability check with SAP APO on several levels. In contrast to ATP, gATP can check the stock situation of several plants.

GTS Global Trade Services. Component of the SAP system landscape used to process export transactions and consider trade regulations.

HAWB House Air Waybill. Shipment-based house waybill for air freight shipments that is issued by the logistics service provider for the forwarder.

House B/L House bill of lading. Shipment-based house waybill for sea shipments that is issued by the logistics service provider for the forwarder.

HU Handling Unit. Physical unit of packing materials (loading equipment/packaging materials) and the materials kept on or in it. A handling unit has a specific, scannable identification number via which handling unit data can be retrieved.

IMG Implementation Guide. Tool for customer-specific tailoring of SAP systems. The guide has a hierarchical structure based on the application component hierarchy. Central components include the IMG activities, which serve to ensure branching into Customizing and thus the execution of relevant system settings. The following IMG variants exist:
SAP Reference IMGs
Project IMGs
Project view IMGs

Independent requirement A requirement that is created through a direct influence (for example, material requisition for production or a sales order for the material).

KPI Key Performance Indicator. Key performance figure determined from business transaction data.

LCL Less Than Container Load. General cargo shipment in which a forwarder's goods are packed into a container with goods of other forwarders (consolidation).

Lead Pick-ups or movements of goods that are transferred from a local to a long-distance transportation network.

LES Logistics Execution System. SAP application component in SAP ERP with which shipping and transport procedures can be processed.

Letter of credit Documented promise of credit issued by an importer's bank to effect payment to an exporter when the exporter is in possession of the proper documents for an export transaction.

LO General logistics module of SAP ERP.

Master B/L Master Bill of Lading. Consolidation-based waybill that a logistics service provider receives for a consolidated shipment (for example, for several shipments in a single container).

Materials planning Distribution of orders and allocation and provision of resources for the planning of processing.

MAWB Master Air Waybill. Consolidation-based waybill for air freight that an airline issues for the entire cargo of a logistics service provider.

MM Materials Management. Materials management module of SAP ERP.

OER Object Event Repository. Tracking system for RFID-supported logistics processes based on SAP Event Management.

On-carriage Deliveries or shipments of goods that are transshipped from long-distance to local transport.

Operative Planning Planning based on short-term values and goals, such as a transportation plan for the coming day.

Optimization Using targeted methods to find a favorable solution for a complex mathematical or logistics problem. The method generally involves an optimization algorithm (computer program). One example is complex transportation planning.

qRFC queued Remote Function Call. Extension of the transactional remote function call with the option of setting the call sequence.

Requirement The required quantity of a material at a certain time for a specific plant.

RFC Remote Function Call. Calling a function module in a different system (destination) than the one in which the invoked program is running. Connections are possible between various AS ABAP systems or between an AS ABAP and an external system. In external systems, instead of function modules, specially programmed functions are invoked whose interface simulates a function module. There are synchronous, asynchronous and transactional RFCs. Activation of the invoked system is done via the RFC interface.

SAP NetWeaver An open integration and application platform for all SAP solutions and certain solutions from SAP partners. SAP NetWeaver is a Web-based platform that serves as the basis for Enterprise Services Architecture (ESA) and enables the cross-company and technology-independent integration and coordination of employees, information and business processes. Thanks to open standards, information and applications from practically any source can be integrated and can be based on virtually any technology. SAP NetWeaver includes functions for business intelligence, company portals, exchange infrastructure, master data management, mobile infrastructure and a Web application server.

SCM Supply Chain Management contains functions for planning, execution, coordination and collaboration in the supply chain. Among other elements, it is composed of the components and applications APO (Advanced Planning & Optimization), SNC (Supply Network Collaboration) and EM (Event Management). SCM is part of SAP Business Suite.

SCOR Supply Chain Operations Reference Model. Important key data that enables an analytical evaluation of the entire logistics chain.

SD Sales and Distribution. Sales module in SAP ERP.

SNC Supply Network Collaboration. SNC enables the connection of external suppliers to SAP SCM.

SSCC Serial Shipper Container Code. Number of a shipping unit for the identification and labeling of shipping units. A shipping unit under this code is the smallest physical unit of goods and commodities that is not attached to another unit and is or can be treated individually in the transport chain.

Standard software Group of programs that can be used to edit and solve a series of similar or uniform tasks. SAP Business Suite is standard business software.

Strategic planning Planning based on long-term values and goals, such as location planning for production plants.

Tactical planning Planning based on medium-term values and goals, such as production planning for Christmas business.

TM SAP Transportation Management, the transport solution within SAP SCM.

TP/VS Transport Planning/Vehicle Scheduling. Transport optimization in SAP SCM.

Transport request Request of a shipper to a logistics service provider to execute the transportation of goods.

VMI Vendor Managed Inventory. Supplier-controlled inventory for which the supplier has system access to a company's warehouse stock and demand data. The VMI thus enables close cooperation with the supplier and serves to improve the external procurement process.

Warehouse order Generally represents an executable work package to be performed by a warehouse employee within a certain period. Warehouse orders usually consist of the warehouse tasks allocated to them.

Warehouse task A document in SAP EWM containing all necessary information for the movement of a certain material quantity or handling unit in the warehouse.

WM Warehouse Management. A system to define and manage complex warehouse structures within one or more plants. The warehouse management system MM-WM supports warehouse management as well as the execution of all warehouse movement, such as goods putaway, removal and transfer.

Bibliography

3PL Study (2009) The state of logistic outsourcing 2009 third-party logistics. http://www.uk.capgemini.com/services/ceo-agenda/the_state_of_logistics_outsourcing_2009_thirdparty_logistics/
Bradler J (2009) SAP supplier relationship management. SAP Press
Council of Supply Chain Management Professionals. http://cscmp.org/aboutcscmp/definitions.asp. Accessed 9 Dec 2009
Gau O (2010) Praxishandbuch Transport und Versand mit SAP LES, 2nd edn. SAP Press
Glaudig L (2002) Entsorgungslogistik als unternehmensübergreifendes Konzept. GRIN Verlag
Götz T (2010) SAP-Logistikprozesse mit RFID und Barcode, 2nd edn. SAP Press
Gulyássy H, Isermann K (2009) Disposition mit SAP. SAP Press
Hellberg T (2009) Einkauf mit SAP MM, 2nd edn. SAP Press
Hoppe M, Käber A (2009) Warehouse Management mit SAP ERP, 2nd edn. SAP Press
Iyer DR (2007) Effective SAP SD. SAP Press
Kirchler M, Manhart D, Unger J (2008) Service mit SAP CRM. SAP Press
Lauterbach B, Fritz R, Gottlieb J, Mosbrucker B, Dengel T (2009) Transportmanagement mit SAP TM. SAP Press
Liebstückel K (2010) Instandhaltung mit SAP, 2nd edn. SAP Press
Matyas K (2008) Instandhaltungslogistik. Hanser
Melzer-Ridinger R (1995) Materialwirtschaft und Einkauf, vol 2. Oldenbourg
Muir N, Kimbell I (2009) Discover SAP, 2nd edn. SAP Press
Pfohl H-C (2010) Logistiksysteme: Betriebswirtschaftliche Grundlagen, 8th edn. Springer
Rötzel von A (2009) Instandhaltung: Eine betriebliche Herausforderung. VDE-Verlag
Scheibler J, Maurer T (2010) Praxishandbuch Vertrieb mit SAP, 3rd edn. SAP Press
Singh J (2007) Implementing and Configuring SAP Global Trade Services. SAP Press
Wöhe G, Döring U (2008) Einführung in die Allgemeine Betriebswirtschaftslehre, 23rd edn. Vahlen

J. Kappauf et al., *Logistic Core Operations with SAP*,
DOI 10.1007/978-3-642-18202-0, © Springer-Verlag Berlin Heidelberg 2012

Index

J. Kappauf et al., *Logistic Core Operations with SAP*,
DOI 10.1007/978-3-642-18202-0, © Springer-Verlag Berlin Heidelberg 2012